32

Topics in Heterocyclic Chemistry

For further volumes:
http://www.springer.com/series/7081

Aims and Scope

The series Topics in Heterocyclic Chemistry presents critical reviews on present and future trends in the research of heterocyclic compounds. Overall the scope is to cover topics dealing with all areas within heterocyclic chemistry, both experimental and theoretical, of interest to the general heterocyclic chemistry community.

The series consists of topic related volumes edited by renowned editors with contributions of experts in the field.

John P. Wolfe
Editor

Synthesis of Heterocycles via Metal-Catalyzed Reactions that Generate One or More Carbon-Heteroatom Bonds

With contributions by
A. Aponick • N.A. Butt • S.R. Chemler • D.A. Copeland • L.D. Julian • J. Keilitz • J.M. Ketcham • M. Lautens • H.A. Malik • J. Waser • J.P. Wolfe • W. Zhang

Springer

Editor
John P. Wolfe
Department of Chemistry
University of Michigan
Michigan
USA

ISSN 1861-9282 ISSN 1861-9290 (electronic)
ISBN 978-3-642-38879-8 ISBN 978-3-642-38880-4 (eBook)
DOI 10.1007/978-3-642-38880-4
Springer Heidelberg New York Dordrecht London

Library of Congress Control Number: 2013943916

Printed on acid-free paper

Springer is part of Springer Science+Business Media (www.springer.com)

Preface

This volume of *Topics in Heterocyclic Chemistry* is focused on new and innovative metal-catalyzed reactions that effect formation of a carbon–heteroatom bond. The volume is composed of seven chapters, which have been written by a talented group of young scientists who have all made significant contributions to this field.

The first five chapters of this volume are centered on the construction of saturated heterocycles from alkenes bearing appended nitrogen or oxygen nucleophiles. The first chapter, entitled "Synthesis of Saturated Heterocycles via Metal-Catalyzed Alkene Carboamination or Carboalkoxylation Reactions," which I have written, is focused on reactions of these substrates with various carbon electrophiles. These transformations generate both a carbon–carbon bond and a carbon–heteroatom bond and provide stereocontrolled access to a broad range of heterocycles.

In the second chapter, entitled "Synthesis of Saturated Heterocycles via Metal-Catalyzed Alkene Diamination, Aminoalkoxylation, or Dialkoxylation Reactions," Chemler and Copeland outline reactions that generate two carbon–heteroatom bonds.

In the third chapter, entitled "Synthesis of Heterocycles via Palladium-Catalyzed Wacker-Type Oxidative Cyclization Reactions of Hydroxy- and Amino-Alkenes," Zhang and Butt describe the synthesis and highlight the progress that has been made in this field in recent years.

In the fourth chapter, entitled "Synthesis of Saturated Heterocycles via Metal-Catalyzed Hydroamination or Hydroalkoxylation Reactions," Julian provides a highly comprehensive look and includes a considerable amount of useful information about the mechanism of these transformations.

In the fifth chapter, entitled "Synthesis of Saturated Heterocycles via Metal-Catalyzed Allylic Alkylation Reactions," Aponick and Ketcham outline recent progress made and illustrate the utility of these transformations for the construction of complex molecules.

The final two chapters in this volume are also largely centered on the reactivity of alkenes and alkynes in heterocycle-forming processes, but focus on different types of substrates as compared to the first five chapters. In the sixth chapter, entitled "Synthesis of Saturated Heterocycles via Metal-Catalyzed Domino/One-Pot Reactions that Generate a C–N or C–O Bond," Lautens, Keilitz, and Malik provide an update on recent developments in the synthesis.

In the seventh chapter, entitled "Synthesis of Saturated Heterocycles via Metal-Catalyzed Formal Cycloaddition Reactions that Generate a C–N or C–O Bond," Waser rounds out the volume with a new look at the synthesis, which nicely illustrates the utility of strained molecules in heterocycle synthesis.

I would like to thank all of the contributing authors for providing interesting and insightful chapters, and I also appreciate the hard work of the staff at Springer (especially Anette Lindqvist and Tanja Jaeger). Finally, I am particularly grateful to series editor Bert Maes for the opportunity to organize this volume.

Ann Arbor, MI John P. Wolfe

Contents

Top Heterocycl Chem (2013) 32: 1–38
DOI: 10.1007/7081_2012_98

Published online: 28 February 2013

Synthesis of Saturated Heterocycles via Metal-Catalyzed Alkene Carboamination or Carboalkoxylation Reactions

John P. Wolfe

Abstract This review describes recent advances over the past decade in the field of heterocycle synthesis via Pd-catalyzed alkene carboamination or carboalkoxylation reactions. These transformations effect the coupling of a carbon electrophile with an unsaturated alcohol or amine and provide heterocyclic products via difunctionalization of the substrate alkene. These reactions provide stereoselective access to a broad array of oxygen and nitrogen heterocycles, including compounds that contain more than one heteroatom. The current scope and limitations of these transformations are discussed, along with relevant mechanistic details.

Keywords Alkene difunctionalization · Catalysis · Copper · Cross-coupling · Gold · Heterocycles · Palladium · Stereoselective synthesis

Contents

J.P. Wolfe (✉)
Department of Chemistry, University of Michigan, 930 N. University Avenue, Ann Arbor, MI 48109-1055, USA
e-mail: jpwolfe@umich.edu

1 Introduction

Saturated five-, six-, and seven-membered heterocycles are common subunits displayed in a broad array of interesting and useful molecules. These types of compounds are found in biologically active natural products and have historically been of considerable importance in the development of pharmaceuticals and agrochemicals. As such, there has been a longstanding interest in the invention of new strategies and tactics for the synthesis of saturated heterocycles.

Over the past decade considerable efforts have been dedicated towards the development of new approaches to the construction of saturated heterocycles via metal-catalyzed alkene carboalkoxylation or carboamination reactions. These transformations typically involve the coupling of a carbon electrophile (such as an aryl halide) with an alkene bearing a pendant nucleophilic heteroatom functional group (such as an alcohol or amine). The reactions effect difunctionalization of the alkene unit with the formation of one C–C bond and one carbon-heteroatom bond, along with 1–2 stereocenters. In addition, many alkene substrates that contain stereocenters are transformed to products with high diastereoselectivity. These methods are highly convergent and are also generally amenable to the rapid construction of analogs of a particular compound, as a wide array of suitable carbon electrophiles can be readily obtained from commercial sources.

This review covers the most significant developments in this field over the past 10 years and illustrates the broad array of different heterocyclic structures that can be accessed using these methods. The primary focus of this chapter is on aryl, alkenyl, and alkynyl-derived electrophiles. However, reactions that employ CO and related electrophiles are briefly noted.

2 Metal-Catalyzed Alkene Carboalkoxylation Reactions

2.1 Metal-Catalyzed Alkene Alkoxyarylation, Alkoxyalkenylation, and Alkoxyalknylation

2.1.1 Palladium-Catalyzed Reactions of Alkene-Appended Alcohols with Halogenated Carbon Electrophiles

Palladium-catalyzed carboalkoxylation reactions between γ-hydroxy terminal alkenes and aryl bromides provide an efficient and stereoselective means for the generation of substituted tetrahydrofurans. These transformations generate 2,5-*cis*- and 2,3-*trans*-disubstuted products with good to excellent diastereoselectivity (Scheme 1) [1–3]. The reactions are effective with a number of different primary, secondary, and tertiary alcohol substrates. Alkenyl bromides can be used as electrophiles in some instances, although chemical yields are not as high as those

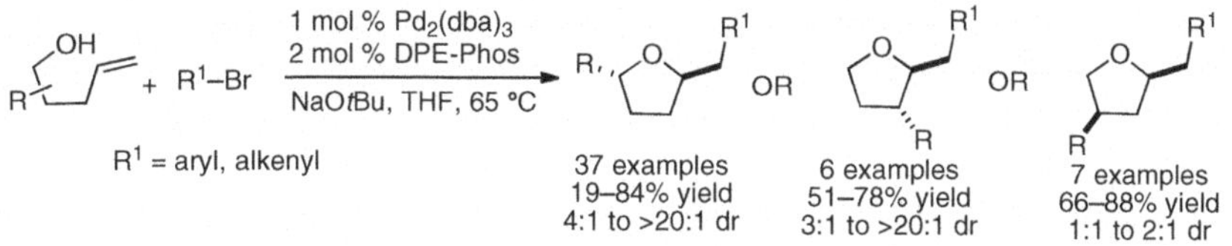

Scheme 1 Pd-catalyzed coupling of γ-hydroxy terminal alkenes with aryl or alkenyl bromides

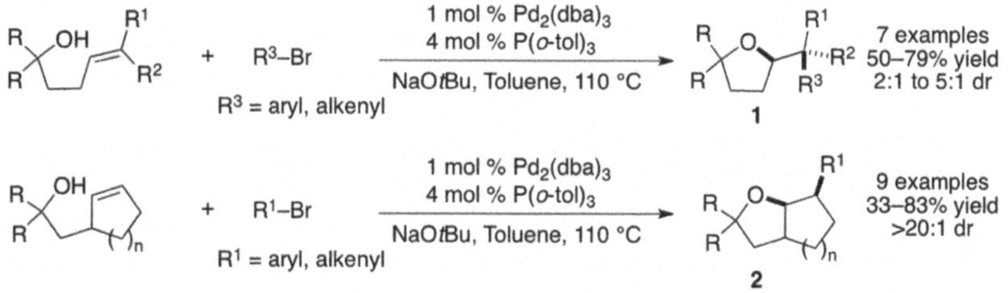

Scheme 2 Pd/P(o-tol)$_3$-catalyzed coupling of γ-hydroxy internal alkenes with aryl or alkenyl bromides

obtained with aryl bromides. Aryl chlorides have occasionally been employed as coupling partners, although the scope of these reactions is currently limited to tertiary alcohols [4].

Tertiary alcohol substrates bearing pendant internal alkenes are converted to tetrahydrofuran products with stereocenters adjacent to the thf ring (Scheme 2) [5]. The major diastereomers in these reactions result from *syn*-addition of the oxygen atom and the aryl or alkenyl group across the double bond. The use of a catalyst composed of $Pd_2(dba)_3$/P(o-tol)$_3$ for reactions of acyclic internal alkene substrates provides products **1** in moderate diastereoselectivity (ca. 2–5:1). However, tertiary alcohols bearing pendant cycloalkenes are converted to bicyclic products **2** with excellent stereocontrol (>20:1 dr).

The mechanism of these transformations (and the other Pd-catalyzed alkene carboalkoxylations described in this section) involves initial oxidative addition of the aryl or alkenyl halide to the Pd(0) complex **3** to provide intermediate **4** (Scheme 3). Deprotonation of the alcohol substrate **5** followed by reaction with **4** affords **6**. Intramolecular *syn*-migratory insertion of the alkene into the Pd–O bond of **6** provides **7**, which undergoes C–C bond-forming reductive elimination to yield the tetrahydrofuran product **8**. The migratory insertion step (**6**–**7**) appears to proceed through a highly organized chairlike transition state in which the substituents on the tether between the alkene and the oxygen atom are oriented to minimize nonbonding interactions. This leads to stereoselective formation of 2,5-*trans*- and 2,3-*trans*-disubstituted products. The relative stereochemistry of the C2 and C1′ stereocenters is controlled by the *syn*-insertion, which ultimately leads to net *syn*-addition of the oxygen atom and the aryl/alkenyl group across the double bond.

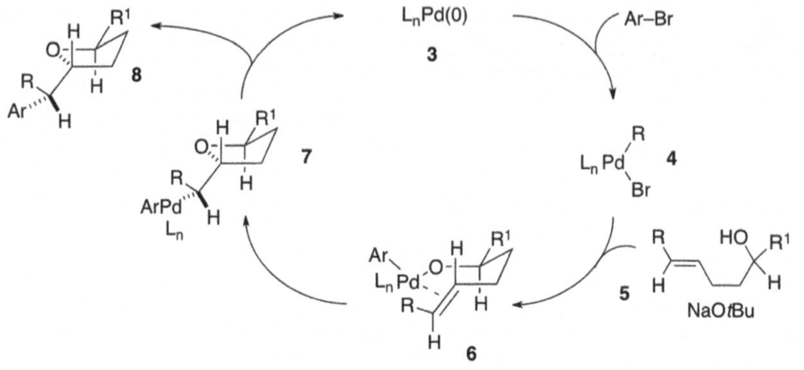

Scheme 3 Mechanism of Pd-catalyzed alkene carboalkoxylation reactions

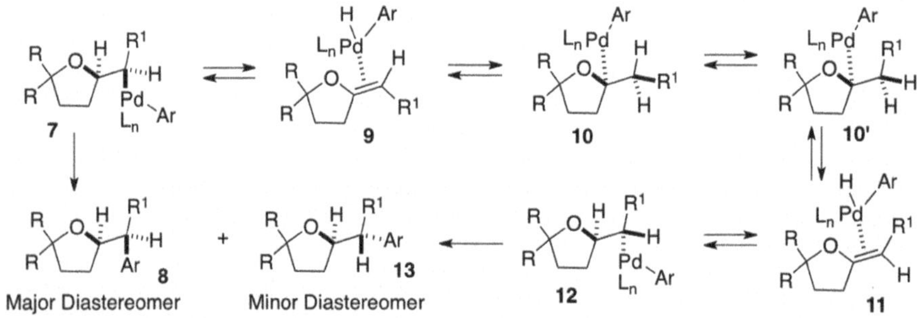

Scheme 4 Mechanistic origin of minor diastereomer

The modest diastereoselectivity obtained in Pd/P(o-tol)$_3$-catalyzed reactions of acyclic internal alkene substrates (Scheme 2) results from competing β-hydride elimination side reactions that occur prior to the reductive elimination step (**7**–**8**) [5]. As shown in Scheme 4, if the reductive elimination from **7** is relatively slow, the metal can migrate from C1′ to C2 via β-hydride elimination to generate **9** and then hydridopalladation to yield **10**. Rotation around the C1′–C2 σ-bond of **10**–**10′** followed by another β-hydride elimination/hydridopalladation sequence provides **12**. Reductive elimination from this latter intermediate then affords the minor diastereomer **13**.

The diastereoselectivity in Pd-catalyzed carboalkoxylation reactions of acyclic internal alkene substrates can be greatly improved by using a catalyst composed of $Pd_2(dba)_3$/S-Phos (Scheme 5) [6]. This catalyst facilitates the reductive elimination step (**7**–**8**), and thereby minimizes competing β-hydride elimination that leads to the minor diastereomer. Diastereoselectivities in reactions of E-alkenes usually exceed 20:1 dr, although reactions of Z-alkene substrates proceed with lower levels of selectivity. Model studies illustrate this transformation may provide a practical approach to the natural product simplakidine A.

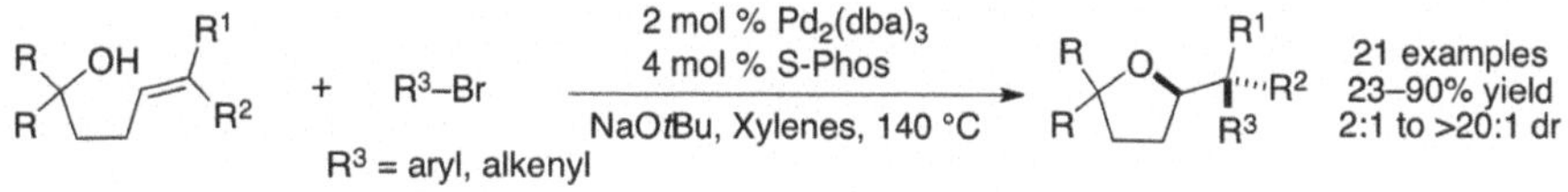

Scheme 5 Pd/S-Phos-catalyzed coupling of γ-hydroxy internal alkenes with aryl or alkenyl bromides

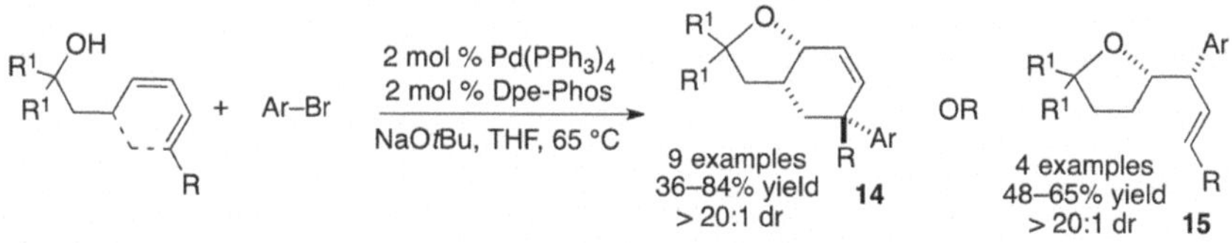

Scheme 6 Pd-catalyzed coupling of 7-hydroxy-1,3-dienes with aryl halides

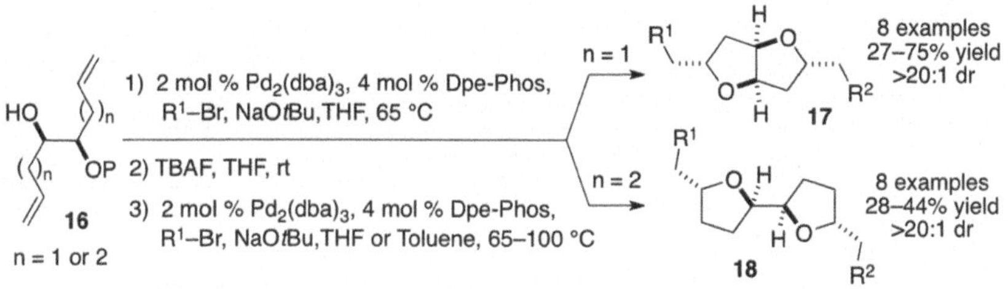

Scheme 7 Synthesis of fused-ring or attached-ring bis-tetrahydrofurans

Alcohols bearing pendant conjugated dienes are also viable substrates in Pd-catalyzed alkene carboalkoxylation reactions (Scheme 6) [7]. Acyclic diene substrates provide the expected products of 1,2-addition to C-4 alkene (**15**), whereas cyclic dienes undergo 1,4-addition by way of intermediate π-allylpalladium complexes to provide products **14**.

The synthesis of fused-ring or attached-ring bis-tetrahydrofurans has been accomplished by a three-step reaction sequence involving initial Pd-catalyzed carboalkoxylation of monoprotected 1,2-diols bearing pendant alkenes **16** (Scheme 7) [8]. Deprotection of the resulting tetrahydrofuran products followed by a second Pd-catalyzed carboalkoxylation affords products such as **17** and **18** in good yield and excellent diastereoselectivity over the three-step sequence.

The intramolecular Pd-catalyzed carboalkoxylation of alkene substrates such as **19** provides access to tetrahydrofurans bearing attached carbocyclic rings (Scheme 8) [9]. The stereochemical outcome of these reactions is dependent on the structure of the phosphine ligand. The use of electron-rich monodentate ligands such as PCy_3 leads to products **20** resulting from *syn*-addition of the alcohol and the aryl group across the double bond. In contrast, the use of chelating bis-phosphine ligands with small bite angles, such as BINAP or 1,2-bis(diphenylphosphino)benzene results in *anti*-addition to the double bond to yield **21**. The change in product stereochemistry appears to

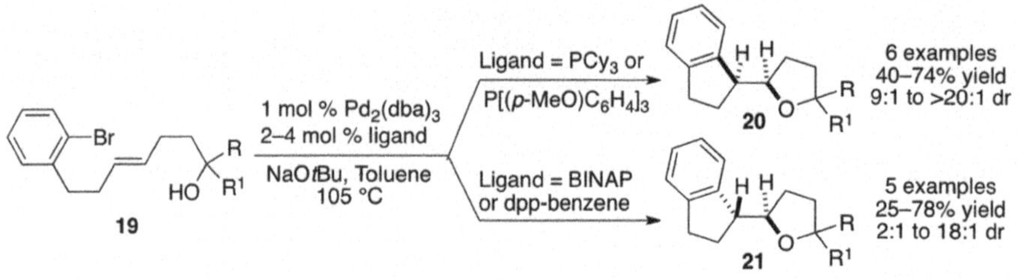

Scheme 8 Intramolecular Pd-catalyzed alkene carboalkoxylation

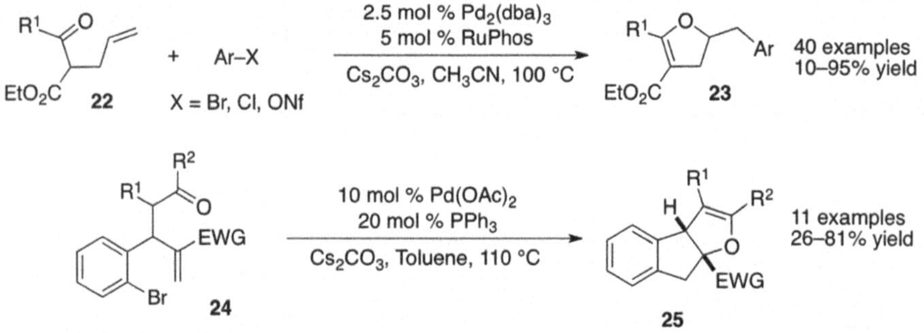

Scheme 9 Pd-catalyzed coupling of β-ketoesters with aryl or alkenyl halides

result from a ligand-dependent change in reaction mechanism. The monodentate ligands favor a *syn*-oxypalladation pathway whereas reactions involving chelating bis-phosphines proceed via *anti*-oxypalladation.

Palladium-catalyzed alkene carboalkoxylation reactions have also been employed for the construction of dihydrofuran derivatives. The coupling of 2-allyl-β-ketoesters **22** with aryl halides or nonaflates affords the 4-(arylmethyl)dihydrofurans **23** in moderate to good yields (Scheme 9) [10]. The transformations proceed via deprotonation of the substrate followed by Pd-catalyzed carboalkoxylation of the alkene with the resulting enolate. Intramolecular variants of these reactions have been used for the conversion of **24** to substituted dihydroindenofuran products **25** [11].

The synthesis of isoxazolidines has been accomplished by Pd-catalyzed alkene carboalkoxylation reactions between aryl bromides and *N*-butenyl hydroxylamine derivatives (Scheme 10) [12]. Stereocontrol is modest in reactions that afford monocyclic heterocycles. However, transformations that generate bicyclic products proceed with good diastereoselectivity and provide access to isoxazolidines that cannot easily be generated via nitrone dipolar cycloaddition reactions. A related strategy has been employed for the construction of 2-substituted isoxazolines from 2-alkenyl oximes (Scheme 11) [13].

Pd-catalyzed carboalkoxylation reactions between unsaturated alcohols and aryl or alkenyl halides have most frequently been employed for the construction of five-membered heterocycles. However, these transformations have also shown some

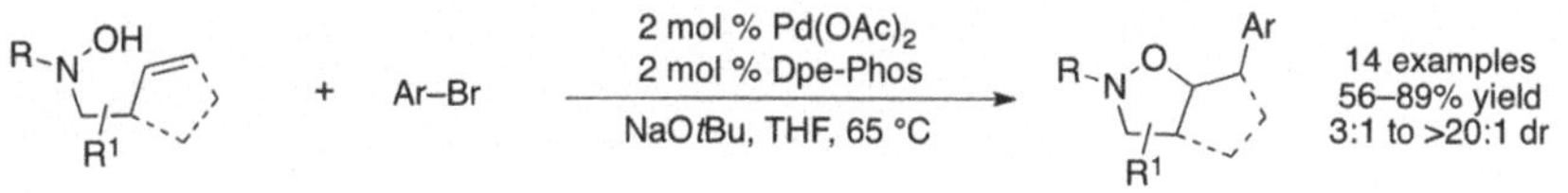

Scheme 10 Pd-catalyzed carboalkoxylation of *N*-butenyl hydroxylamines

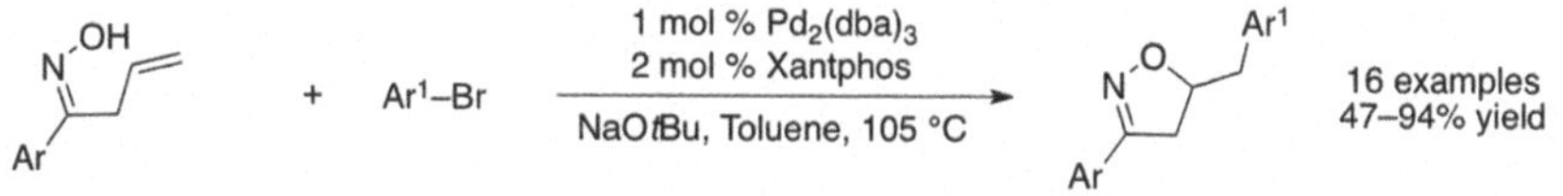

Scheme 11 Pd-catalyzed carboalkoxylation of unsaturated oximes

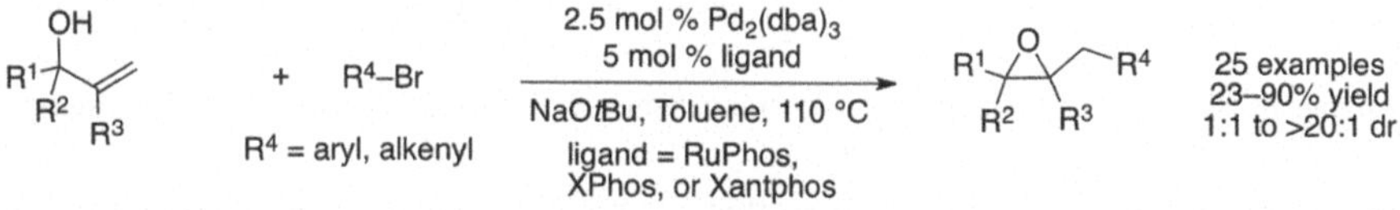

Scheme 12 Pd-catalyzed coupling of tertiary allylic alcohols with aryl or alkenyl bromides

utility for the formation of other ring sizes. Oshima has illustrated that tertiary allylic alcohols can be converted to epoxides via Pd-catalyzed alkene carboalkoxylation (Scheme 12) [14]. Reactions of chiral tertiary allylic alcohols proceed with moderate to good diastereoselectivity and without loss of enantiopurity. Currently the scope of this method appears to be limited to tertiary alcohol substrates, as the use of analogous primary or secondary alcohols has not yet been described. Nonetheless, this represents a fundamentally new approach to epoxide synthesis.

The generation of six-membered oxygen heterocycles via Pd-catalyzed carboalkoxylation of alkenes has proven to be considerably more challenging than the formation of five-membered ring products. However, recent work has illustrated that a catalyst composed of $Pd_2(dba)_3$ and S-Phos is effective for the coupling of 2-(but-3-enyl)phenols with aryl or alkenyl bromides to provide substituted chroman derivatives (Scheme 13) [15]. The transformations proceed with excellent diastereoselectivity in cases where fused tricyclic products are generated. The influence of alkoxide nucleophilicity/basicity on the facility of alkene carboalkoxylation reactions appears to be dependent on ring size, as the preparation of fully saturated tetrahydropyrans from 5-hexen-1-ol derivatives via this method has not yet been achieved, and the Pd/S-Phos catalyzed conversion of 2-allylphenol to a substituted dihydrobenzofuran proceeded in only 37% yield.

Most studies on Pd-catalyzed alkene carboalkoxylation reactions have been focused on the use of aryl or alkenyl halides as the electrophilic component. However, Waser has recently described Pd-catalyzed carboalkoxylation reactions between γ-hydroxyalkenes and the alkynyl halide triisopropylsilyl ethynyl bromide. These transformations generate substituted tetrahydrofuran derivatives with good to excellent levels of diastereoselectivity (Scheme 14) [16]. The stereochemical outcome of

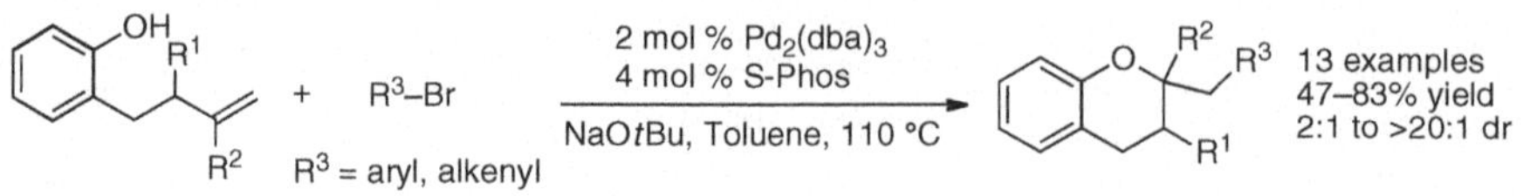

Scheme 13 Pd-catalyzed coupling of 2-(but-3-enyl)phenols with aryl or alkenyl bromides

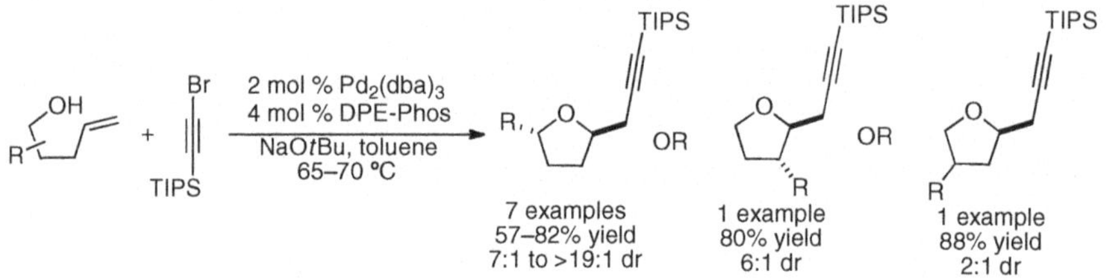

Scheme 14 Pd-catalyzed coupling of γ-hydroxyalkenes with triisopropylsilyl ethynyl bromide

these transformations is analogous to that for reactions of aryl and alkenyl halides described above (Scheme 1). Highest diastereoselectivities are obtained in the formation of *trans*-2,5-disubstituted products (up to >95:5 dr). Currently the scope of this transformation is limited to a single alkynyl bromide substrate. The use of other alkynyl bromides such as phenyl ethynyl bromide leads to formation of complex product mixtures.

Hypervalent iodine reagents have been used as alternative sources of carbon electrophiles in Pd-catalyzed alkene carboalkoxylation reactions. For example, Pd-catalyzed reactions between the hypervalent iodoalkyne reagent **26** and unsaturated phenols or carboxylic acids provide 2-alkynylmethyl benzofurans, benzopyrans, and γ-lactones **27–28** in moderate to good chemical yield (Scheme 15) [17]. The mechanism of reactions that employ reagent **26** are believed to differ from those of the reactions shown above in Schemes 1–14. These latter transformations may proceed via a Pd(II)/Pd(IV) catalytic cycle that is initiated by oxypalladation of the alkene by the $Pd^{II}(hfacac)_2$ complex to generate an alkylpalladium(II) intermediate. Oxidative addition of **26** to this intermediate would generate a Pd(IV) complex that could undergo reductive elimination to yield the observed products.

A related hypervalent iodine reagent **29** has been used as an electrophilic trifluoromethyl group source in Cu-catalyzed carboalkoxylation reactions of unsaturated alcohols and carboxylic acids (Scheme 16) [18]. The transformations are effective for the generation of three-, four-, five-, and six-membered ring products with moderate to excellent diastereoselectivity. The mechanism of these reactions is not yet clear, but may involve either addition of a trifluoromethyl radical to the alkene followed by oxidation and cyclization, or trifluoromethylcupration of the alkene followed by C–O bond forming reductive elimination of the resulting alkylcopper intermediate. These transformations constitute rare examples of metal-catalyzed processes that generate C–CF_3 bonds.

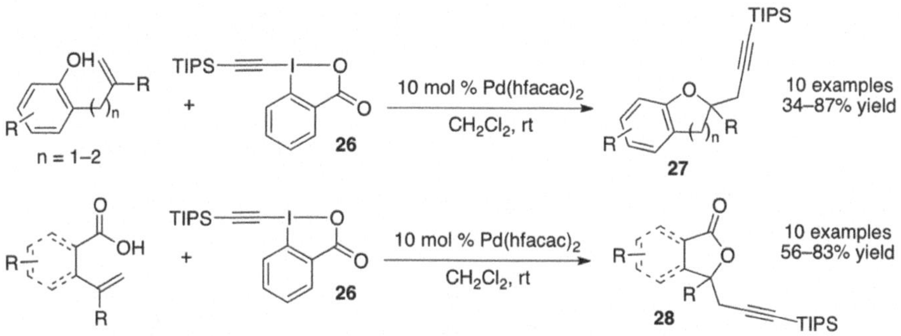

Scheme 15 Pd-catalyzed coupling of unsaturated alcohols and acids with hypervalent iodine reagent **26**

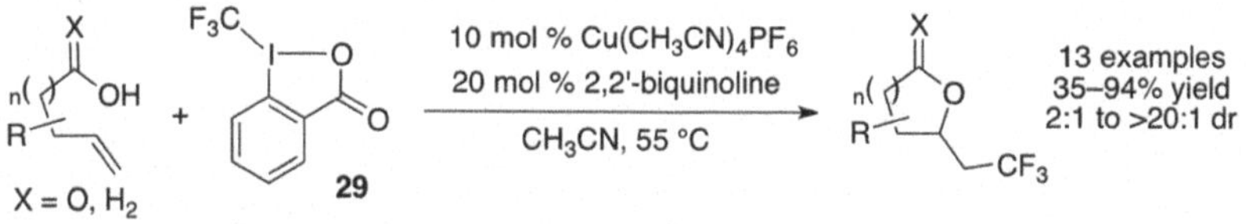

Scheme 16 Pd-catalyzed alkene alkoxytrifluoromethylation

2.1.2 Metal-Catalyzed Reactions of Alkene-Appended Alcohols with Arylboronic Acids and Arylsilanes

Zhang and Lloyd-Jones have independently developed Au-catalyzed alkoxyarylation reactions between phenylboronic acid [19] or phenyltrimethylsilane [20] and γ- or δ-hydroxyalkenes (Scheme 17). The mechanism of these reactions appears to involve initial oxidation of the Au(I) complex **30** to Au(III) complex **31** by selectfluor (Scheme 18) [19, 21]. The Au(III) complex then binds to the alkene to afford **32**, which then undergoes *anti*-oxyauration to yield **33**. Complex **33** is then intercepted by phenylboronic acid or phenyltrimethylsilane to afford the observed tetrahydrofuran product. The Au-catalyzed alkene carboalkoxylation reactions proceed in moderate to good chemical yield, but in contrast to most of the Pd-catalyzed transformations described above (Sect. 2.1.1) the diastereoselectivities obtained in Au-catalyzed reactions are quite low (ca. 1:1 dr). These low stereoselectivities may be due to the outer-sphere alkene *anti*-oxyauration mechanism, which appears to proceed through a less constrained transition state than the corresponding inner-sphere *syn*-oxypalladation pathway for the Pd-catalyzed reactions (Scheme 3, above). As a result, the differences in transition state energies leading to the two diastereomers are relatively small in the Au-catalyzed reactions as compared to the Pd-catalyzed transformations.

Falck has developed a Pd-catalyzed carboalkoxylation reaction between arylboronic acids and β-hydroxyalkenes that generates 2-aryltetrahydrofuran products (Scheme 19) [22]. The diastereoselectivity of these reactions is modest (ca. 1:1) in most cases with the exception of those that generate 2,3-*trans*-disubstituted products

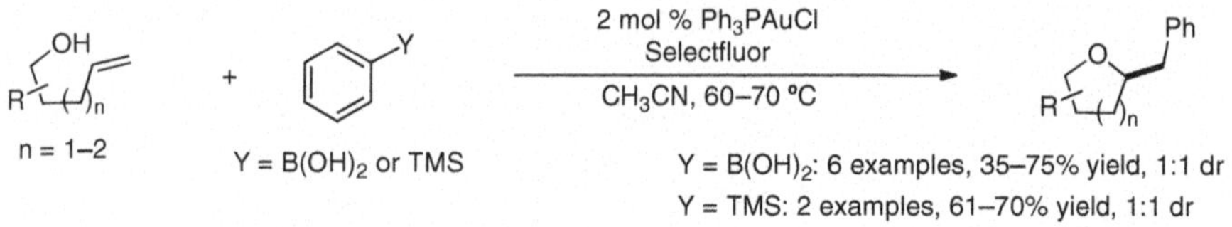

Scheme 17 Au-catalyzed coupling of alkene-appended alcohols and acids with phenylboronic acid or phenyltrimethylsilane

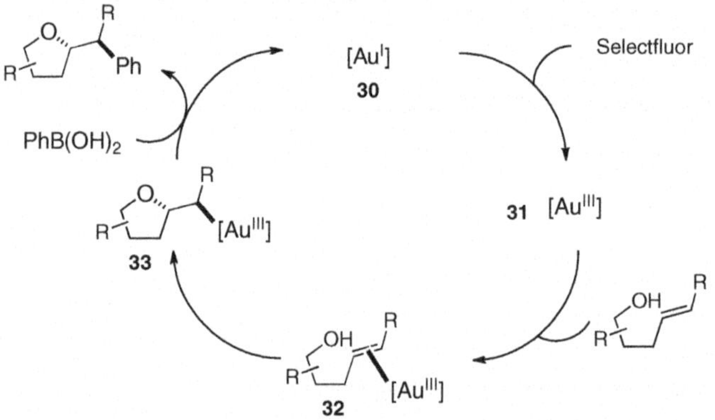

Scheme 18 Mechanism of Au-catalyzed carboalkoxylation reactions

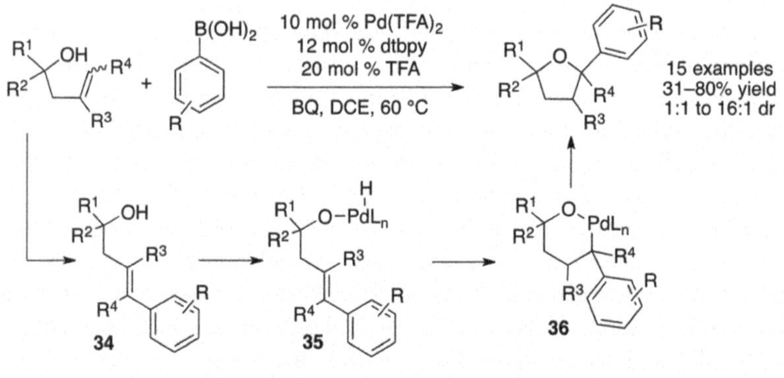

Scheme 19 Synthesis of 2-aryltetrahydrofurans via 1,1-carboalkoxylation of alkenes

(16:1 dr). The mechanism of these reactions is believed to involve initial oxidative heck arylation of the substrate alkene to afford a 4-arylbut-3-en-1-ol intermediate **34**. The conversion of this intermediate to the observed product is proposed to involve oxidative addition of the O–H bond of **34** to provide **35** followed by 6-*endo*-hydridopalladation to give **36** and then reductive elimination to yield the product. This method complements the Pd- and Au-catalyzed carboalkoxylation reactions described above, which all effect 1,2-addition of the O-atom and aryl/alkenyl/alkynyl group to the alkene to afford 2-benzyltetrahydrofuran derivatives.

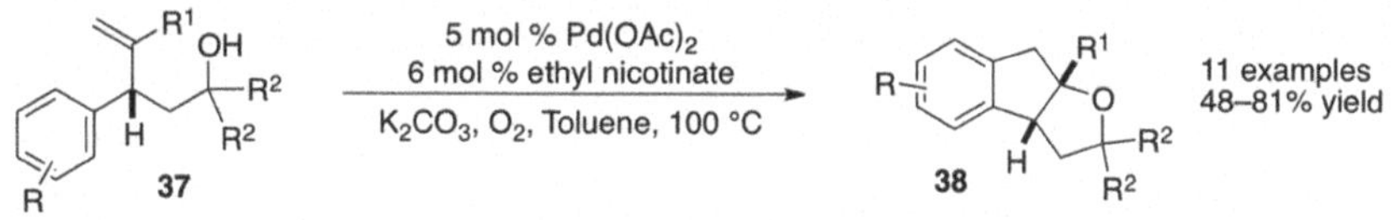

Scheme 20 Buchwald's Pd-catalyzed intramolecular alkene alkoxyarylation

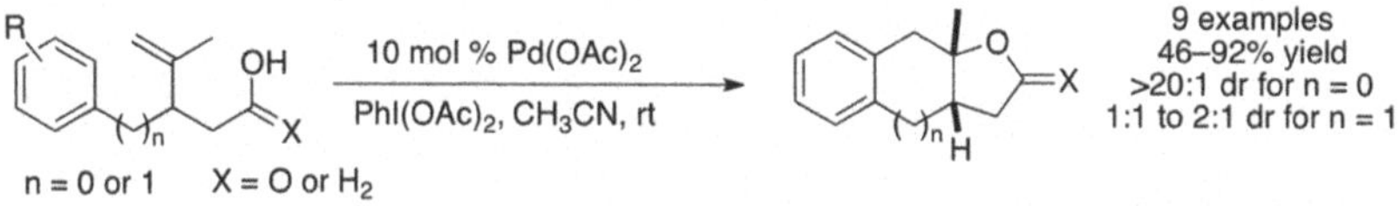

Scheme 21 Stephenson's Pd-catalyzed intramolecular alkene alkoxyarylation

2.1.3 Metal-Catalyzed Reactions of Unsaturated Alcohols or Carboxylic Acids Bearing Pendant Aryl Groups

A number of interesting transformations have been developed that involve arene C–H functionalization as a key component of alkene carboalkoxylation. These transformations alleviate the need for a halogenated arene electrophile, although transformations reported thus far have described only the generation of polycyclic products (as opposed to substituted monocyclic tetrahydrofurans). For example, Buchwald has devised intramolecular alkene carboalkoxylation reactions of 4-pentene-1-ol derivatives **37** bearing aryl groups at C3 (Scheme 20) [23]. These transformations provide tricyclic products **38** in good chemical yield with perfect diastereoselectivity favoring generation of *cis*-fused tricyclic products. The reactions appear to proceed via Pd(II)-mediated *anti*-oxypalladation followed by C–H functionalization of the arene by the resulting alkylpalladium intermediate.

Stephenson has reported a closely related approach to the construction of fused polycyclic tetrahydrofurans (Scheme 21) [24]. A variety of ring sizes can be generated with moderate to excellent diastereoselectivity. Stephenson's reaction conditions are similar to those employed by Buchwald, except that in the Stephenson system $PhI(OAc)_2$ is employed as an oxidant as opposed to the use of O_2 by Buchwald. Interestingly, this change in reaction conditions appears to lead to a change in mechanism, as Pd(IV) complexes are believed to be intermediates along the catalytic cycle in the Stephenson reactions.

Chemler has developed Cu-catalyzed intramolecular alkene alkoxyarylation reactions of substrates **39** and **41** that afford fused- or bridged polycyclic tetrahydrofurans **40** and **42** in good yields with excellent diastereoselectivities (Scheme 22) [25]. The transformations are believed to proceed via initial *syn*-oxycupration of the alkene to give **43**, which undergoes homolytic C–Cu bond cleavage to provide an intermediate alkyl radical **44**. This transient radical is then captured by the pendant arene to generate the observed products. The use of substrates bearing halogen "leaving groups" is not necessary, as the radical capture step leads to the net substitution of an alkyl group for an aromatic H-atom.

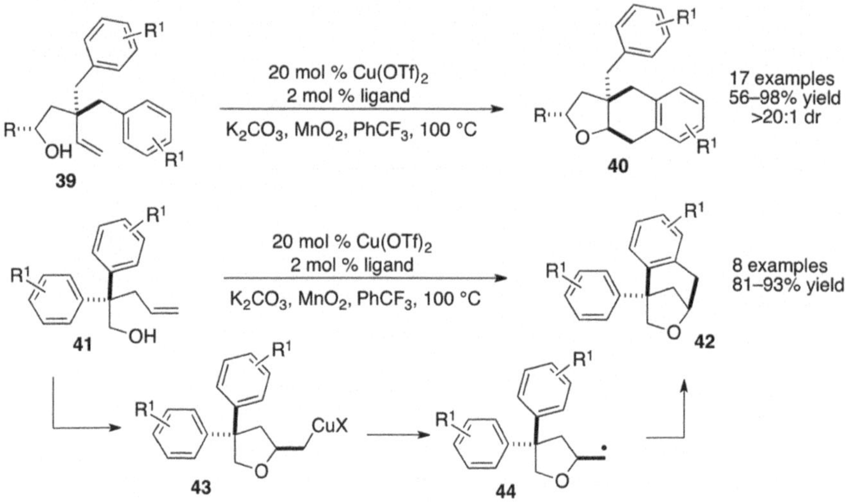

Scheme 22 Cu-catalyzed intramolecular alkene alkoxyarylation

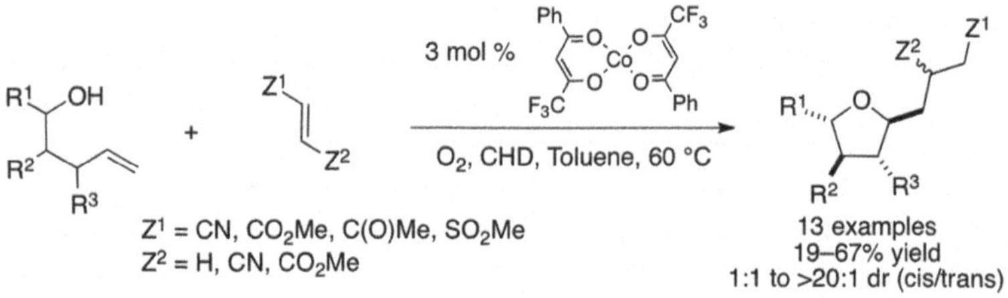

Scheme 23 Co-catalyzed alkene alkoxyalkylation

2.1.4 Cobalt-Catalyzed Reactions Between Alkene-Appended Alcohols and Activated Alkenes or Alkynes

Hartung has illustrated that cobalt-catalyzed reactions between 4-penten-1-ol derivatives and activated alkenes or alkynes provide a concise approach to the construction of mono- and bicyclic substituted tetrahydrofurans (Scheme 23) [26]. These transformations proceed via a mechanism similar to that of the Chemler work noted above (oxymetallation followed by free-radical formation). However, in these cases the intermediate alkyl radical is captured in an intermolecular reaction with the activated alkene or alkyne electrophile. The reactions proceed with good control of stereochemistry around the tetrahydrofuran ring. However, most products are generated as ca. 1:1 mixtures of diastereomers epimeric at the carbon bearing the Z^2 group. The presence of two activating groups on the alkene is required in order to obtain satisfactory yields.

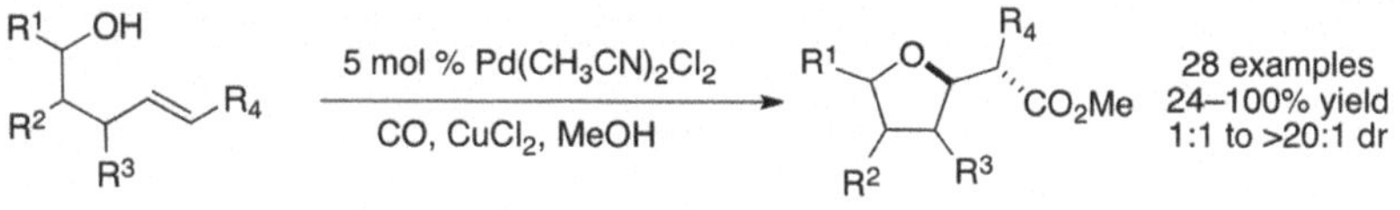

Scheme 24 Pd-catalyzed alkene alkoxycarboalkoxylation

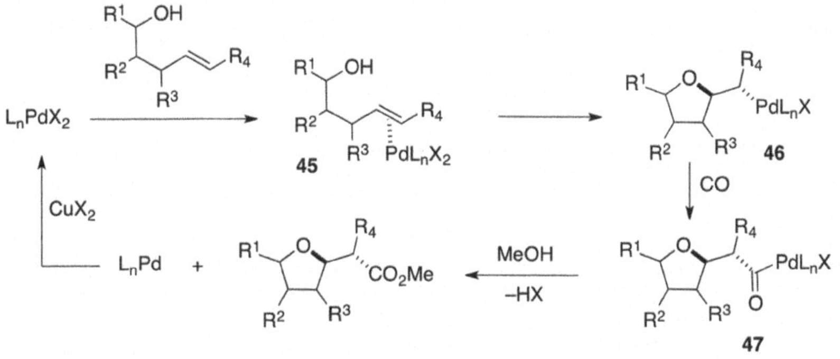

Scheme 25 Mechanism of Pd-catalyzed alkoxycarboalkoxylation reactions

2.2 *Metal-Catalyzed Alkene Alkoxycarboalkoxylation, Alkoxyacylation, and Alkoxycyanation*

The Pd-catalyzed alkoxycarboalkoxylation of 4-penten-1-ol derivatives with carbon monoxide and methanol was initially reported by Semmelhack in 1984 and has been extensively studied by several groups (Scheme 24) [27–32]. These reactions effect the net *anti*-addition of the substrate oxygen atom and a methoxycarbonyl functional group across the double bond to provide tetrahydrofuran products bearing a methyl ester at C1′. High diastereoselectivity is obtained for the product of *anti*-addition. However, modest stereocontrol is occasionally observed in reactions of terminal alkene substrates bearing substituents on the tether between the alcohol and alkene moieties. Nonetheless, many reactions are highly efficient, and this methodology has been applied to the synthesis of several complex molecules including natural products such as tetronomycin and goniothalesidol [33–40]. Further details on these types of reactions are described in a recent review [41].

In contrast to reactions of related substrates with halogenated carbon electrophiles that involve *syn*-oxypalladation of a palladium alkoxide intermediate (Scheme 3), the alkoxycarboalkoxylation reactions proceed via complexation of the Pd(II) catalyst to the alkene to provide **45** followed by *anti*-oxypalladation to give **46** (Scheme 25). Insertion of CO into the C–Pd bond of **46** provides **47**, which is captured by methanol to afford the ester product with concomitant generation of a Pd(0) complex. This complex is then oxidized to Pd(II) by $CuCl_2$ to complete the catalytic cycle.

Douglas has recently described Rh-catalyzed intramolecular alkoxyacylation reactions of 2-allylphenol derived esters that afford dihydrobenzofuran or chroman

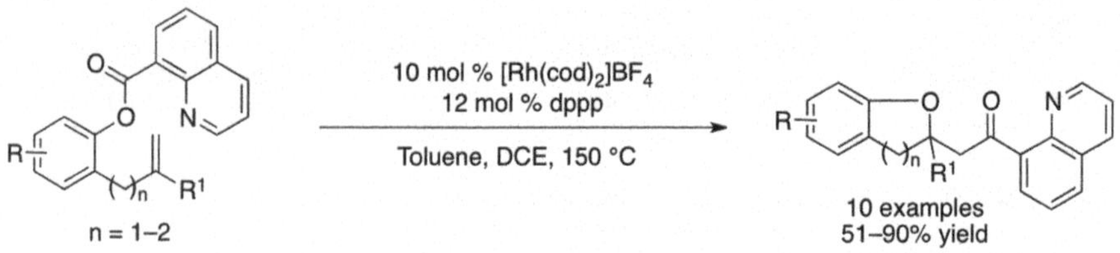

Scheme 26 Rh-catalyzed intramolecular alkoxyacylation of 2-allylphenol-derived esters

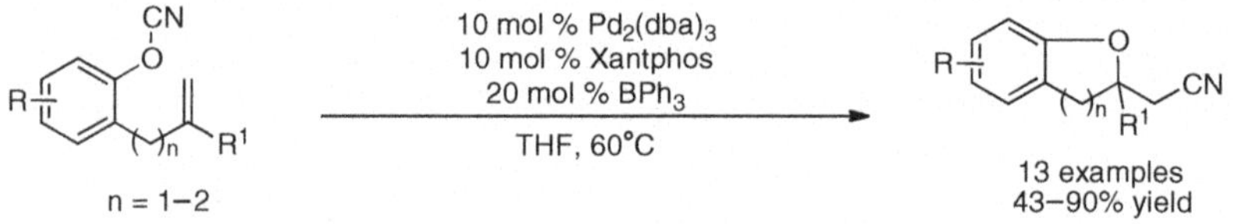

Scheme 27 Pd-catalyzed intramolecular alkoxycyanation of 2-allylphenol-derived cyanates

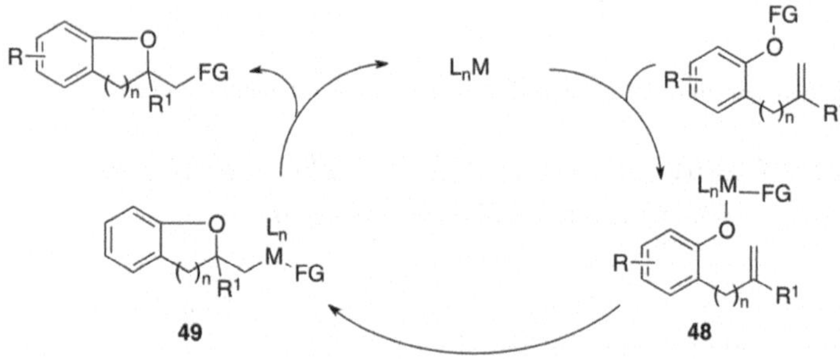

Scheme 28 Mechanism of metal-catalyzed alkoxyacylation and alkoxycyanation

products [42] (Scheme 26). The scope of this method is currently limited to 8-acylquinoline ester substrates, as the quinoline group is required to prevent competing decarbonylation of intermediate rhodium complexes along the catalytic cycle. In addition, the presence of an alkyl group at the internal alkene carbon center ($R^1 \neq H$) is also needed to avoid competing β-hydride elimination side reactions. However, the substrates are readily available and the transformations proceed in good to excellent yield.

The intramolecular alkoxycyanation of 2-allylphenol-derived cyanates was recently reported by Nakao [43] (Scheme 27). A dual catalyst system composed of $Pd_2(dba)_3$/Xantphos and BPh_3 was used to effect these transformations. The BPh_3 acts as a Lewis acid to activate the nitrile towards oxidative addition to the Pd(0) catalyst, which then facilitates the formation of the C–N and C–C bonds. Functional groups such as esters, ethers, and halogens are tolerated under these reaction conditions. The transformation is effective for the generation of either five- or six-membered heterocyclic products.

The mechanisms of the Rh-catalyzed alkoxyacylation and the Pd-catalyzed alkoxycyanation reactions are similar, and both are initiated by oxidative addition of the O–FG bond to the metal catalyst (Scheme 28). The resulting intermediate **48** undergoes *syn*-migratory insertion of the alkene into the M–O bond to yield a new complex **49**. Reductive elimination from **49** then provides the heterocyclic product with concomitant regeneration of the catalyst.

3 Metal-Catalyzed Alkene Carboamination Reactions

3.1 Metal-Catalyzed Alkene Aminoarylation, Aminoalkenylation, and Aminoalknylation

3.1.1 Palladium-Catalyzed Reactions of Alkene-Appended Amines and Related Nucleophiles with Halogenated Carbon Electrophiles

Synthesis of Pyrrolidines and Indolines

Palladium-catalyzed carboamination reactions between *N*-aryl-pent-4-enylamine derivatives provide a concise and convergent approach to the stereoselective construction of substituted pyrrolidines (Scheme 29) [44, 45]. The reactions proceed through a mechanism analogous to that described above for Pd-catalyzed carboalkoxylation reactions between unsaturated alcohols and aryl/alkenyl halides (Scheme 3) [46–50]. These transformations provide access to *cis*-2,5- and *trans*-2,3-disubstituted pyrrolidines with good to excellent diastereoselectivity (usually 10:1 to >20:1 dr). However, stereocontrol is modest in transformations that generate 2,4-disubtituted products (ca. 2:1 dr). The reactions are effective with a range of aryl and alkenyl bromide coupling partners. Aryl chlorides have also been employed as electrophiles, although the scope is not as broad as with the corresponding aryl bromides [4]. Both electron-rich and electron-poor *N*-aryl groups are tolerated on the substrate, although in many instances the use of electron-rich *N*-aryl groups results in the formation of small amounts of regioisomeric side products. Reactions of substrates bearing acyclic internal alkenes provide complex mixtures of regioisomeric products that result from competing β-hydride elimination side reactions. However, related transformations of cyclic internal alkenes such as **50** proceed in good yield, and the selective synthesis of different regioisomeric products such as **51** and **52** from the same substrate can be accomplished by the use of an appropriate phosphine ligand [51].

Intramolecular Pd-catalyzed carboamination reactions of alkenes bearing both a tethered aryl or alkenyl bromide and a pendant amine proceed in good yield with good to excellent diastereoselectivity (Scheme 30) [9, 52]. Transformations of substrates such as **53** provide pyrrolidine derivates **54** that bear attached carbocyclic rings. In contrast, intramolecular reactions of **55** provide arene- or cycloalkene-fused tropane derivatives **56**. The cyclizations occur with no loss of optical activity when enantiomerically enriched substrates are employed. The utility of the tropane-forming reactions was demonstrated through a short synthesis of an NMDA antagonist related to the pharmaceutical lead compound MK-801 [52].

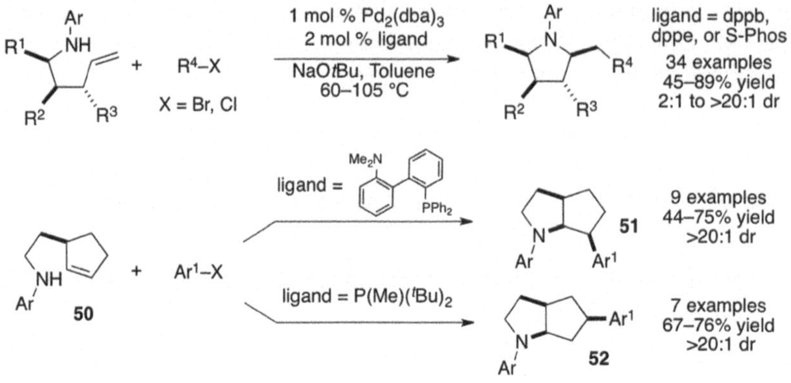

Scheme 29 Synthesis of *N*-aryl pyrrolidines

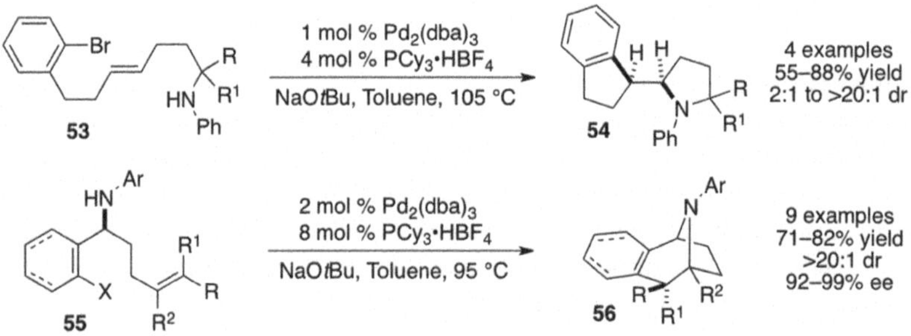

Scheme 30 Intramolecular Pd-catalyzed carboamination of unsaturated *N*-aryl amines

Primary amines and anilines have not yet successfully been employed as substrates in Pd-catalyzed carboamination reactions (to provide NH heterocycles), as the rate of Pd-catalyzed *N*-arylation of these derivatives appears to be considerably faster than the carboamination. However, this reactivity trend has been exploited in one-pot tandem *N*-arylation/carboamination reactions (Scheme 31). The coupling of two different aryl/alkenyl bromides with either 2-allylaniline or pent-4-enylamine derivatives has been used for the generation of *N*-aryl-2-benzylindolines and *N*-aryl-2-benzylpyrrolidines [53, 54]. High chemoselectivity was achieved through an in-situ ligand exchange protocol that allows for modification of catalyst structure and reactivity without the need for isolation of intermediates. Cascade intramolecular *N*-arylation/intermolecular carboamination reactions between substrates such as **57** and aryl chlorides provide a stereoselective route to benzo-fused pyrrolizidine derivatives **58** [55]. The ligand exchange procedure was not needed to obtain high selectivity in reactions of **57** due to the inherent differences in reactivity between aryl bromides and aryl chlorides.

Palladium-catalyzed carboamination reactions between aryl or alkenyl bromides and *N*-boc, *N*-acetyl, or *N*-Cbz-protected pent-4-enylamine derivatives provide

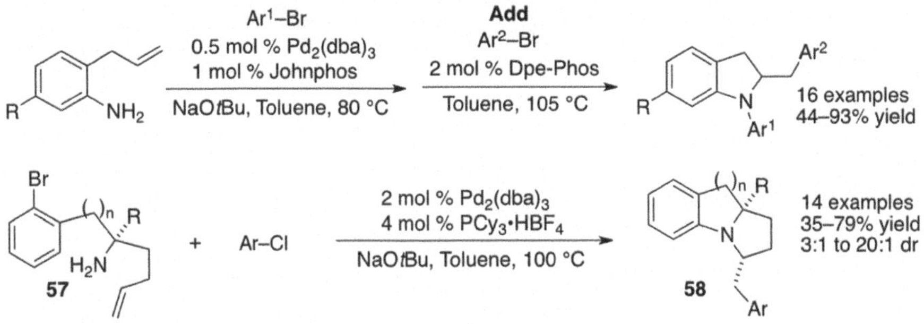

Scheme 31 Tandem *N*-arylation/carboamination reactions

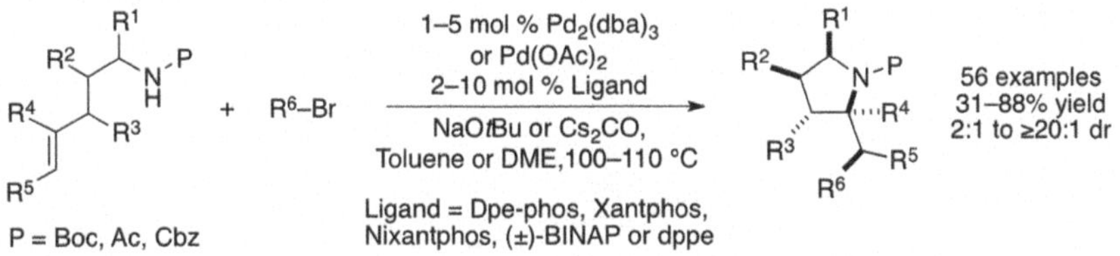

Scheme 32 Synthesis of *N*-protected pyrrolidines

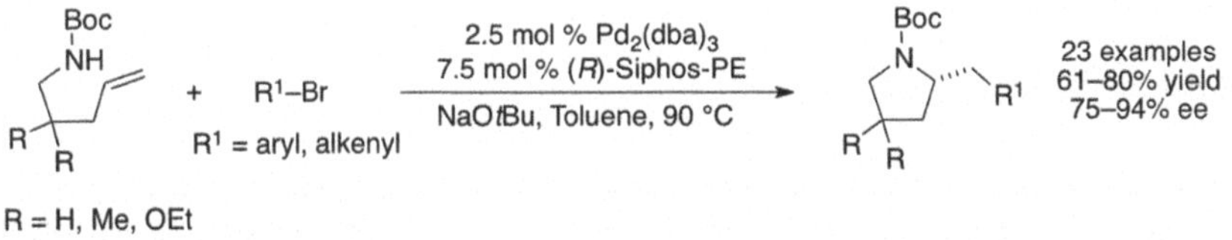

Scheme 33 Asymmetric synthesis of 2-(arylmethyl)pyrrolidines

N-protected pyrrolidine derivatives in good yield with high stereocontrol (Scheme 32) [56–59]. Diastereoselectivity trends in these reactions mirror those described above for related *N*-aryl-pent-4-enylamine substrates; *cis*-2,5-disubstituted products are generated with >20:1 dr. In many instances Cs_2CO_3 can be employed as the base in place of NaO*t*Bu [58, 59]; this modification of reaction conditions leads to dramatically enhanced substrate scope. Under these conditions, a variety of functional groups are tolerated, and reactions of internal alkene substrates can also be effected. These transformations have been applied to the stereoselective synthesis of the natural products (+)-preussin [60] and (±)-tylophorine [61].

A catalyst composed of $Pd_2(dba)_3$ and (*R*)-Siphos-PE has been used to effect enantioselective carboamination reactions of *N*-boc-pent-4-enylamines (Scheme 33) [62]. The reactions proceed in good chemical yield and provide products with up to 94% ee. This catalyst has been used to achieve the key step in asymmetric total syntheses of the natural products (+)-tylophorine and (+)-aphanorphine [63].

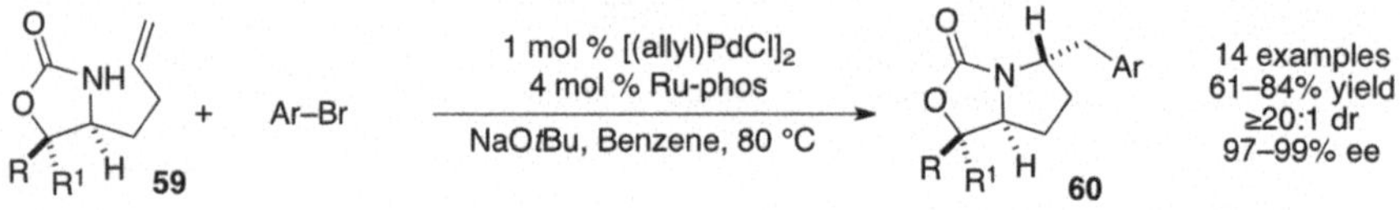

Scheme 34 Synthesis of bicyclic carbamates

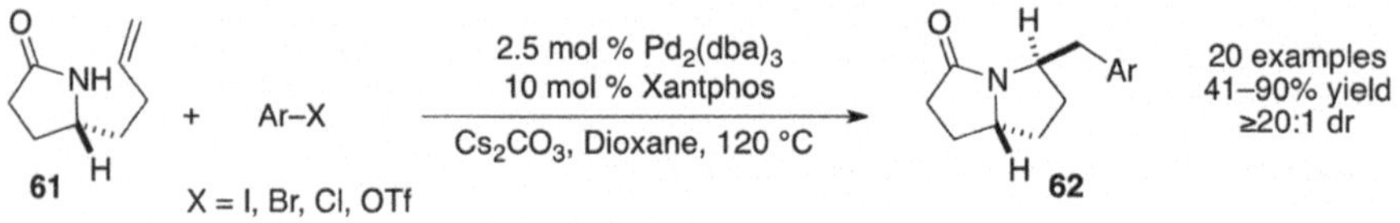

Scheme 35 Synthesis of bicyclic lactams

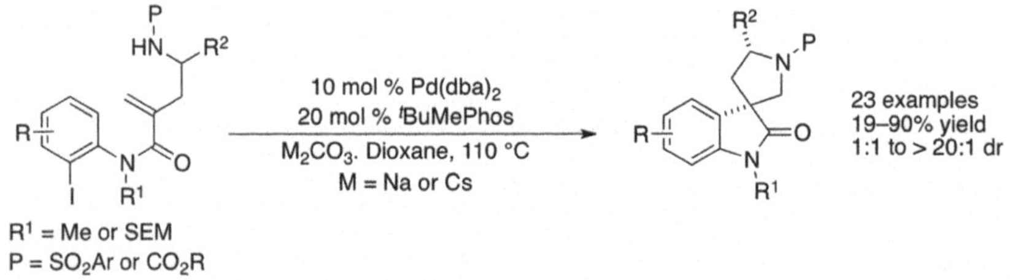

Scheme 36 Synthesis of spirooxindoles

The Pd/Siphos-PE catalyst functions well in transformations of terminal alkene substrates, but efforts to employ these conditions for reactions of internal alkene derivatives have thus far been unsuccessful.

Pd-catalyzed carboamination reactions of oxazolidin-2-ones **59** bearing pendant alkenes afford bicyclic carbamate products **60** that contain a 2,5-trans relationship between the substituents on the pyrrolidine ring moiety (Scheme 34) [64]. These reactions proceed through highly organized transition states in which ring strain is minimized. The carbamate group can be hydrolyzed or reduced under standard reaction conditions to afford 2,5-*trans*-disubstituted pyrrolidine products. The starting materials can be prepared in enantiopure form in a few steps from readily available amino alcohol precursors, and the carboaminations proceed with no loss of optical purity. A related synthesis of bicyclic lactams **62** has also been achieved by Pd-catalyzed carboamination reactions between aryl halides and 4-(but-3-enyl) pyrrolidin-2-ones **61** (Scheme 35) [65]. The stereochemical outcome of these transformations is analogous to that of the related oxazolidin-2-one substrates, and stereoselectivities were uniformly high for all substrate combinations that were examined.

Zhu has developed intramolecular carboamination reactions of unsaturated carbamates and sulfonamides bearing pendant aryl iodides (Scheme 36) [66]. These transformations afford spirocyclic oxindole derivatives in good yield with up to >20:1 dr. The structure of the phosphine ligand has a significant influence on the outcome of these reactions. The use of ^{t}BuMePhos provides satisfactory results. However, many other ligands lead to formation of undesired side products resulting from competing carbopalladation of the alkene. In contrast to most of the reactions described above, which proceed through *syn*-aminopalladation pathways, these transformations appear to result from *anti*-aminopalladation of the alkene.

Synthesis of Five-Membered Nitrogen Heterocycles That Contain Two Heteroatoms

Palladium-catalyzed alkene carboamination reactions have also shown considerable utility for the synthesis of a wide variety of five-membered nitrogen heterocycles that contain two heteroatoms. For example, *O*-vinyl-*N*-boc-1,2-amino alcohols have been coupled with aryl and alkenyl bromides to afford 1,3-oxazolidines in moderate to good yield (Scheme 37) [67]. The products are obtained with good to excellent diastereoselectivity, and the reactions occur with no loss of enantiopurity. These transformations provide a new means of accessing cyclic N,O-acetals in which one C–N bond and one C–C bond are generated (in contrast to classical condensation routes to these molecules in which two carbon-heteroatom bonds are formed).

Palladium-catalyzed carboamination reactions between *N*-allylureas and aryl/alkenyl bromides have been used for the stereoselective construction of substituted imidazolidin-2-ones (Scheme 38) [68, 69]. The substrates are easily prepared in one step from allylic amines and isocyanates. The reactions of internal alkene substrates proceed stereospecifically with net *syn*-addition, and substrates that contain allylic substituents are transformed to *trans*-4,5-disubstituted products with good to excellent diastereoselectivity. A few examples of the synthesis of six-membered cyclic ureas via this method have also been described, and this transformation was used as a key step in the synthesis of the alkaloid natural product (+)-merobatzelladine b [70].

The asymmetric synthesis of imidazolidin-2-ones has been accomplished via enantioselective Pd-catalyzed reactions of *N*-allylureas with aryl or alkenyl bromides (Scheme 39) [71]. The enantioselectivity is dependent on the nature of the aryl group on the cyclizing nitrogen atom, with highest selectivities obtained using substrates bearing *p*-nitrophenyl groups. These transformations are mechanistically related to the enantioselective pyrrolidine-forming reactions described above (Scheme 33). However, the two reactions appear to proceed through different enantiodetermining steps. The alkene aminopalladation step appears to be enantiodetermining in the pyrrolidine-forming reactions, whereas the urea-forming transformations likely involve enantiodetermining C–C bond-forming reductive elimination.

Palladium-catalyzed carboamination reactions have shown considerable utility for the formation of five-membered heterocycles bearing a heteroatom-heteroatom bond. For example, Pd-catalyzed reactions between aryl/alkenyl bromides and *N*-but-3-enyl hydrazine derivatives provide stereocontrolled access to 3,5-disubstituted pyrazolidines (Scheme 40) [72]. The product stereochemistry can be controlled by

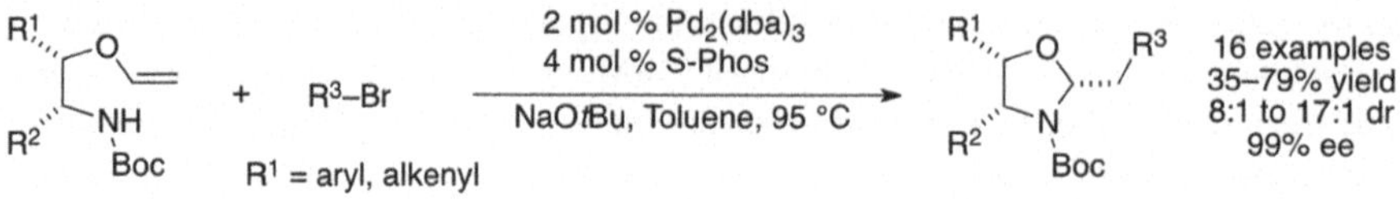

Scheme 37 Asymmetric synthesis of 1,3-oxazolidines

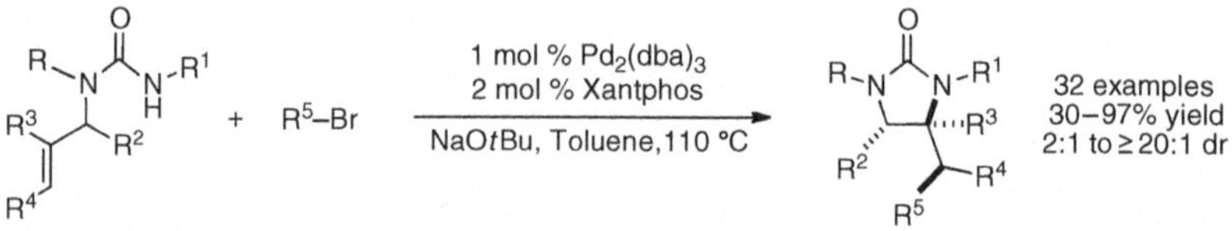

Scheme 38 Synthesis of imidazolidin-2-ones

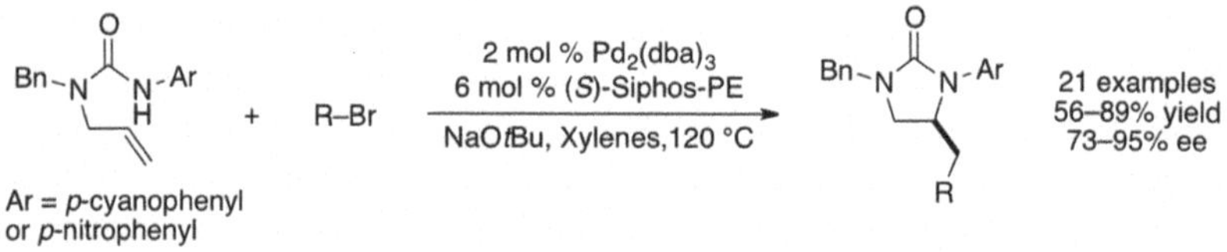

Scheme 39 Asymmetric synthesis of imidazolidin-2-ones

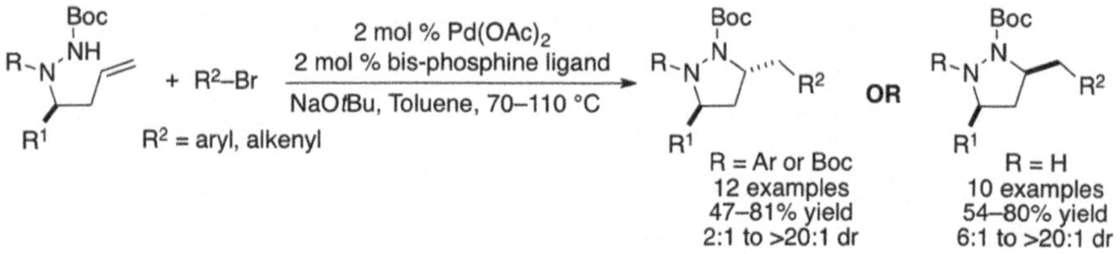

Scheme 40 Synthesis of disubstituted pyrazolidines

either inclusion or omission of a protecting group on the internal (non-cyclizing, N^2) nitrogen atom. Substrates bearing N^2-aryl or –boc groups are transformed to *trans*-3,5-disubstituted products, whereas substrates bearing an unprotected N^2 atom (R = H) are converted to *cis*-3,5-disubstituted pyrazolidines. The former transformations proceed through transition states in which the R-group is pseudoaxial to minimize allylic strain interactions between the R-group and the N^2-substituent. In contrast, the latter processes proceed with pseudoequatorial orientation of the R-group to avoid developing 1,3-diaxial interactions.

A conceptually related synthesis of isoxazolidines has been achieved by Pd-catalyzed carboamination reactions of *O*-but-3-enylhydroxylamines (Scheme 41) [73]. These transformations generate 3,5-*trans*- and 3,4-*trans*-disubstituted isoxazolidines with generally high diastereoselectivities (>20:1 in many cases). The stereoselectivities obtained in these reactions are frequently superior to those obtained

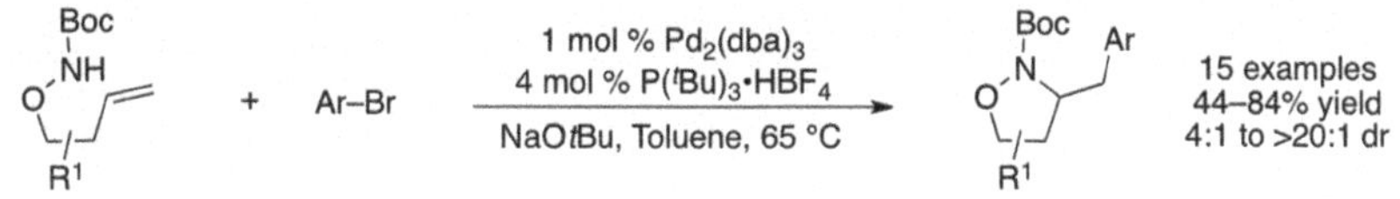

Scheme 41 Synthesis of substituted isoxazolidines

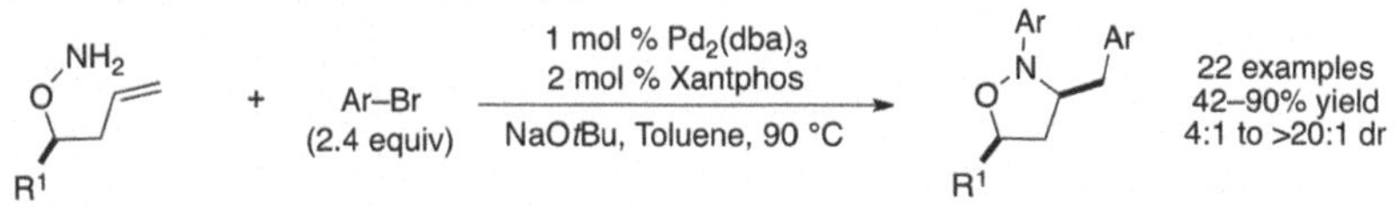

Scheme 42 Synthesis of substituted isoxazolidines via tandem *N*-arylation/carboamination

using dipolar cycloaddition methods for isoxazolidine synthesis. Interestingly, diastereoselectivities in these reactions are considerably higher than for related carboalkoxylation reactions of *N*-but-3-enyl hydroxylamines described above (Scheme 10). This effect appears to be due to the conformation of the *N*-boc-hydroxylamines and the positioning of the boc-group in the transition state [73]. A tandem *N*-arylation/carboamination sequence has been employed for the synthesis of *N*-aryl-3-benzyl isoxazolidines. However, thus far the scope of these reactions is limited to incorporation of two equivalents of the same aryl group (Scheme 42) [74].

Synthesis of Three-, Six-, and Seven-Membered Nitrogen Heterocycles

Although most Pd-catalyzed alkene carboamination reactions that have thus far been developed lead to the formation of five-membered heterocyclic products, the synthesis of smaller and larger-ring compounds has also been accomplished in a few cases. For example, Oshima has employed Pd-catalyzed carboamination reactions between allylic amines and aryl/alkenyl bromides for the generation of substituted aziridines (Scheme 43) [75]. The scope of this method is currently limited to substrates bearing two substituents adjacent to the N-atom of the substrates (R^1, $R^2 \neq$ H). Despite this limitation, substrates that contain two different substituents at this position are converted to trisubstituted aziridines with generally good levels of stereocontrol.

Palladium-catalyzed alkene carboamination reactions have been employed for the stereoselective construction of two classes of six-membered heterocycles, both of which contain two heteroatoms. The synthesis of *cis*-2,6-disubstituted piperazines **64** has been achieved via Pd-catalyzed coupling reactions of *N*-allyl-1,2-diamines **63** with aryl and alkenyl bromides (Scheme 44) [76, 77]. Substrates bearing an aryl group on the cyclizing nitrogen-atom are transformed with high levels of diastereoselectivity. However, analogous cyclizations of boc-protected substrates proceed with low dr. A conceptually related synthesis of *cis*-3,5-disubstituted morpholines **66** from *O*-allyl 1,2-amino alcohols **65** has also been described (Scheme 45) [78]. Both substrates **63** and **65** can be prepared in a few steps from readily available enantiopure

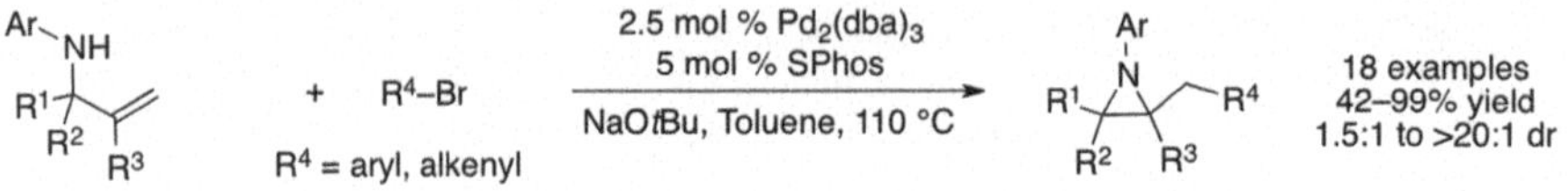

Scheme 43 Synthesis of substituted aziridines

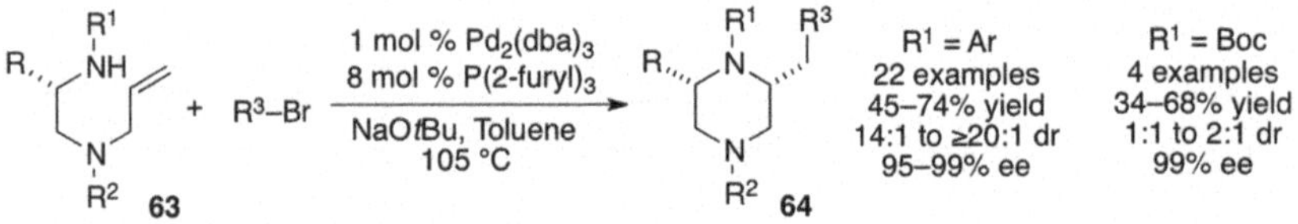

Scheme 44 Synthesis of disubstituted piperazines

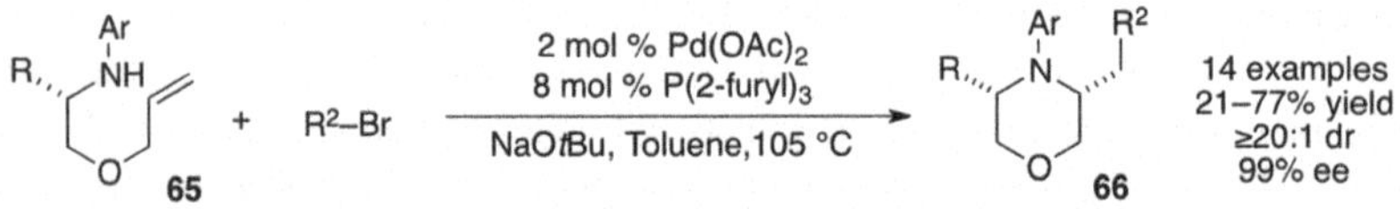

Scheme 45 Synthesis of disubstituted morpholines

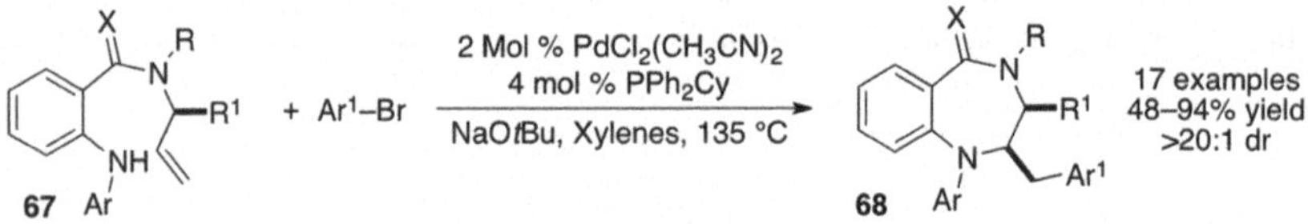

Scheme 46 Synthesis of saturated 1,4-benzodiazepines

amino alcohols, and the cyclizations proceed without erosion of enantiopurity. The generation of benzo-fused or cycloalkyl-fused derivatives is also feasible. Efforts to extend these transformations to the generation of 2,3- or 2,5-disubstituted products resulted in modest diastereoselectivity.

The preparation of one class of seven-membered heterocycles, saturated 1,4-benzodiazepines **68**, has been achieved by Pd-catalyzed carboamination of substrates **67** which are derived from 2-aminobenzylamine (Scheme 46) [79]. In most instances the reactions proceed in good chemical yield, and 2,3-disubstituted products are generated with excellent diastereoselectivity. In contrast with analogous five-membered ring-forming transformations, which proceed through chair-like transition states, the six- and seven-membered ring-forming reactions shown in Schemes 44, 45, and 46 appear to proceed through boat-like transition states.

Synthesis of Nitrogen Heterocycles Bearing 2-Alkynylmethyl Groups

The synthesis of nitrogen heterocycles such as pyrrolidines, lactams, cyclic ureas, and cyclic carbamates bearing 2-alkynylmethyl groups has been achieved by

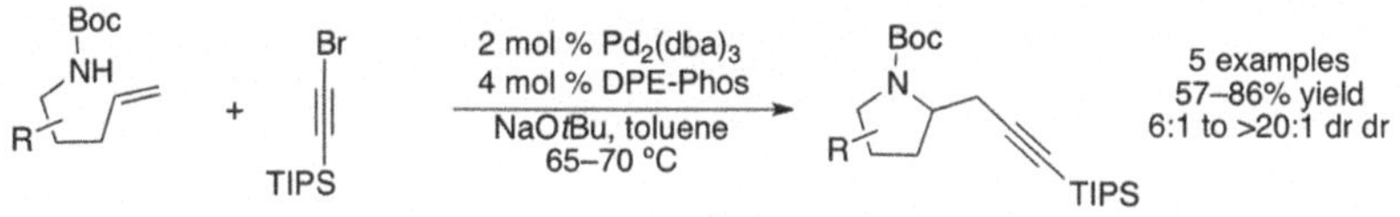

Scheme 47 Pd-catalyzed coupling of γ-(*N*-boc)aminoalkenes with triisopropylsilyl ethynyl bromide

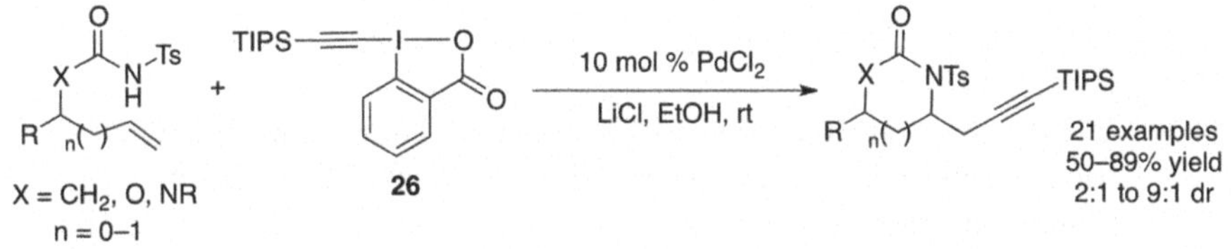

Scheme 48 Pd-catalyzed coupling of unsaturated amides, ureas, and carbamates with hypervalent iodine reagent **26**

Pd-catalyzed carboamination reactions between *N*-boc-pentenylamine derivatives and either triisopropylsily ethynyl bromide or hypervalent iodine reagent **26** (Schemes 47 and 48) [16, 80]. The mechanisms of these reactions are analogous to those described above for related syntheses of 2-alkynylmethyl tetrahydrofurans and lactones (Scheme 15). This method was employed as a key step in the synthesis of the pyrrolizidine natural product trachelanthamidine [80].

3.1.2 Metal-Catalyzed Reactions of Alkene-Appended Amines and Related Nucleophiles with Arylboronic Acids

Zhang and Toste have independently developed Au-catalyzed cross-coupling reactions between aryl boronic acids and sulfonamides bearing pendant alkenes. These transformations afford five- and six-membered nitrogen heterocycles in good yield, although diastereoselectivities are modest in reactions of substrates that contain substituents on the alkyl tether between the sulfonamide and the alkene (Scheme 49) [19, 21, 81]. The reactions appear to proceed through a mechanism that is similar to the one described above for related Au-catalyzed alkene carboalkoxylations between unsaturated alcohols and boronic acids (Scheme 18). However, mechanistic studies performed by Toste indicate that in the carboamination reactions the C–N bond is formed through *syn*-aminoauration rather than *anti*-addition of the sulfonamide and the gold complex to the alkene. Moreover, the reductive elimination step in the reaction of sulfonamide derivatives likely involves a dinuclear gold complex [81].

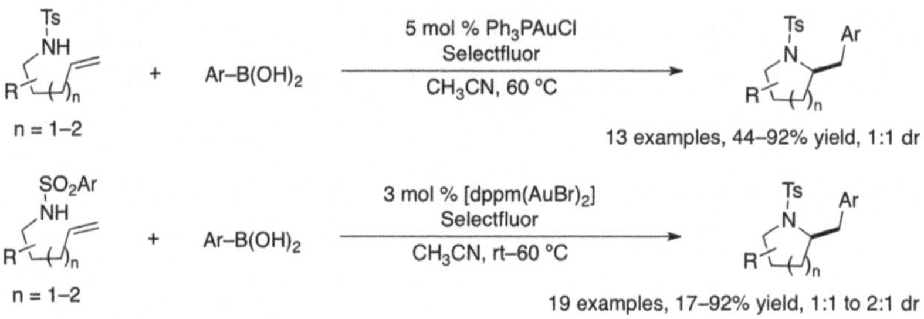

Scheme 49 Au-catalyzed coupling of unsaturated sulfonamides with arylboronic acids

3.1.3 Metal-Catalyzed Reactions of Unsaturated Amines and Related Nucleophiles with Arenes or Alkenes Via Formal C–H Bond Functionalization

Copper-Catalyzed Reactions

Chemler has developed a new approach to the carboamination of alkenes that effects formal C–H functionalization of an arylsulfonamide in a C–C bond-forming event [82, 83]. This method results in the conversion of alkene-tethered arylsulfonamide substrates **69** to tricyclic sulfonamide products **70** in good yield (Scheme 50). Substrates bearing substituents adjacent to the amino group are converted to *cis*-2,5-disubstituted pyrrolidine derivatives with excellent diastereoselectivity ($\geq$20:1 dr). However stereoselectivities are lower in reactions that provide 2,3- or 2,4-disubstituted products. The reactions are effective with terminal alkenes or alkenes bearing a substituent at the internal carbon (1,1-disubstituted). However, 1,2-disubstitued alkenes do not undergo effective cyclization. The polycyclic products of the carboamination reactions can be transformed to 2-benzylpyrrolidine or -indoline derivatives via reductive cleavage of the sulfonyl group.

The carboamination reactions can also be conducted using catalytic amounts (20 mol%) of $Cu(OTf)_2$ and bipy in place of the copper carboxylate complex [84]. In these cases stoichiometric amounts of MnO_2 are added to the reaction mixture to re-oxidize the copper complex and complete the catalytic cycle. The mechanism of the Cu-mediated or -catalyzed carboamination is similar to that of the related Cu-catalyzed carboalkoxylation reactions described above (Scheme 22) and involves *syn*-heterocupration (aminocupration in this case) to afford an intermediate alkylcopper species. This intermediate undergoes C–Cu bond homolysis to generate an alkyl radical, which is then captured by the pendant aryl group of the sulfonamide moiety. The use of amides or imides as substrates in the stoichiometric Cu-mediated carboamination reactions is also feasible (Scheme 51) [85].

The use of a catalyst composed of $Cu(OTf)_2$ and (*R*,*R*)-Ph-Box allows for the enantioselective intramolecular carboamination of *N*-(arylsulfonyl)-pent-4-enylamines (Scheme 52) [86]. The transformations provide tricyclic products **71** with enantioselectivities of up to 94% ee. The use of *N*-tosyl-2-allylaniline or *N*-tosyl-2-allylbenzylamine

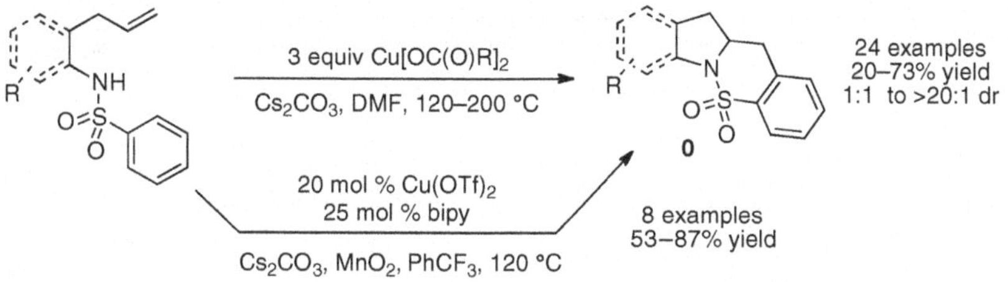

Scheme 50 Cu-catalyzed or Cu-mediated intramolecular carboamination reactions of unsaturated sulfonamides

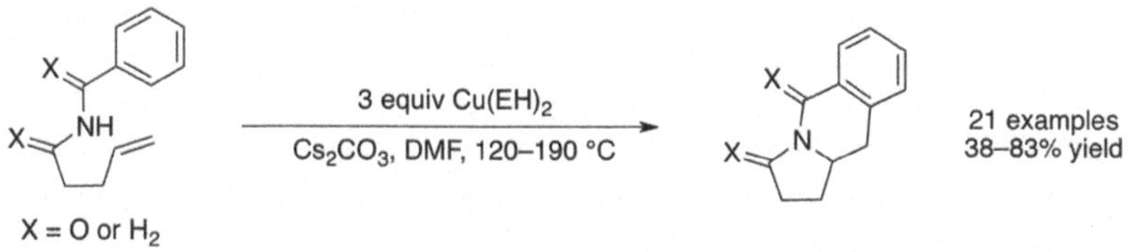

Scheme 51 Cu-catalyzed C–H activation/aminoarylation of unsaturated imides and amides

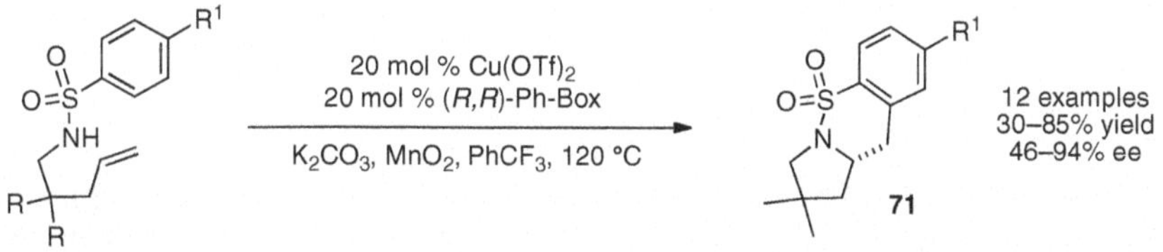

Scheme 52 Cu-catalyzed asymmetric intramolecular carboamination reactions of unsaturated sulfonamides

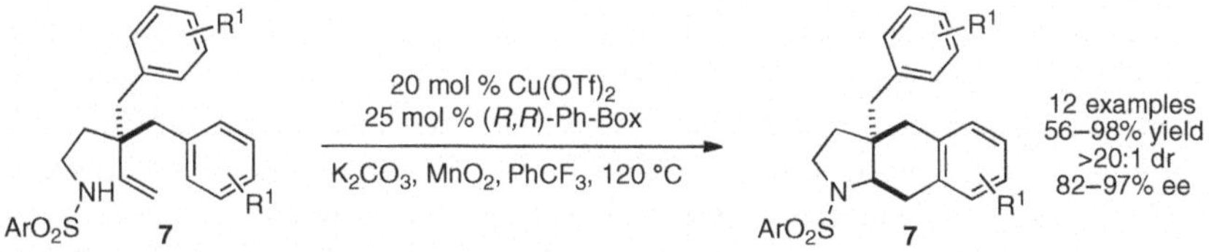

Scheme 53 Cu-catalyzed asymmetric desymmetrization of unsaturated sulfonamides

as substrates leads to polycyclic indoline or tetrahydroisoquinoline products, although enantioselectivities are diminished in these cases. The sulfonamide products **71** can be converted to 2-benzylpyrrolidine derivatives with no loss of enantiopurity via dissolving-metal reductive cleavage of the SO_2 group. This method has been applied to the asymmetric total synthesis of the alkaloid natural product (*S*)-tylophorine [87].

The Cu-catalyzed alkene carboamination methodology developed by Chemler has also been used to effect asymmetric desymmetrization of sulfonamide substrates **72** (Scheme 53) [88]. These reactions lead to the formation of benzo-fused indolizidines

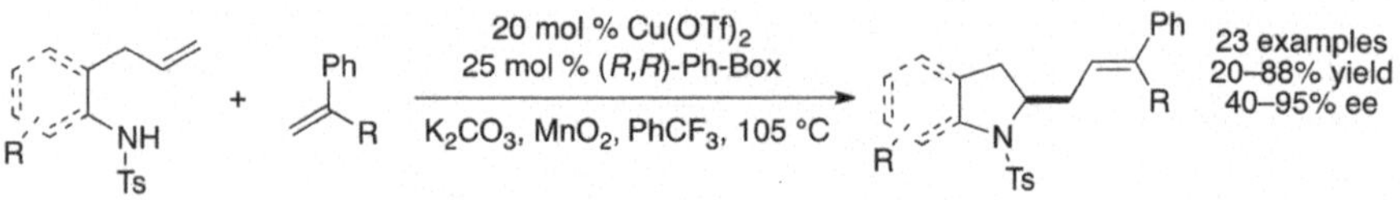

Scheme 54 Cu-catalyzed asymmetric carboamination reactions between unsaturated sulfonamides and alkenes

73 that bear quaternary carbon stereocenters. The products are generated with good to excellent levels of asymmetric induction and essentially complete diastereoselectivity.

Chemler has recently developed a new asymmetric cross-coupling reaction between styrene derivatives and sulfonamides bearing pendant alkenes that afford 2-allypyrolidines or 2-allylindolines in good yield with good to excellent enantioselectivity (Scheme 54) [89]. The mechanism of these reactions is similar to other Cu-catalyzed carboaminations and involves *syn*-aminocupration followed by homolysis of the resulting Cu–C bond to generate an alkyl radical. This intermediate is captured by the styrene derivative, and the resulting radical then undergoes oxidative loss of a hydrogen atom to provide the alkene product. The utility of this method was illustrated through a concise synthesis of a 5-HT_7 receptor antagonist.

Palladium-Catalyzed Reactions

Palladium catalysts have also shown good utility in alkene carboamination reactions that involve C–H functionalization. For example, Yang (Scheme 55) [90] and Zhu (Scheme 56) [91] have independently described Pd-catalyzed oxidative intramolecular alkene aminoarylation reactions of *N*-aryl amide substrates that effect C–H functionalization of the *N*-aryl moiety. The former transformations lead to the generation of benzo-fused pyrrolizidine derivatives **74** in good yield with excellent diastereoselectivity and a broad substrate scope. Mechanistic studies by the Yang group suggest the reactions illustrated in Scheme 55 proceed via *syn*-aminopalladation of the alkene followed by a subsequent C–H functionalization reaction of the resulting Pd(II) complex. The transformations developed by Zhu (Scheme 56) lead to the formation of spirooxindole products **75** in moderate to good yield, although diastereoselectivities are low. In contrast to Yang's transformations, the mechanism of Zhu's reaction is not entirely clear, but the use of the relatively strong oxidant $PhI(OAc)_2$ may facilitate the generation of reactive intermediate Pd(IV) complexes.

Michael has reported the cross-coupling of unsaturated carbamates with aromatic solvents such as benzene or toluene to afford 2-benzylpyrrolidines and related six- and seven-membered heterocyclic products (Scheme 57) [92, 93]. These reactions proceed via an interesting mechanism involving initial *anti*-aminopalladation to generate an alkylpalladium(II) complex **76**, which is then oxidized to the analogous Pd(IV) complex **77** by NFBS (*N*-flurobenzenesulfonamide). The highly reactive Pd(IV) species effects the C–H activation of the arene solvent to provide **78**, which undergoes

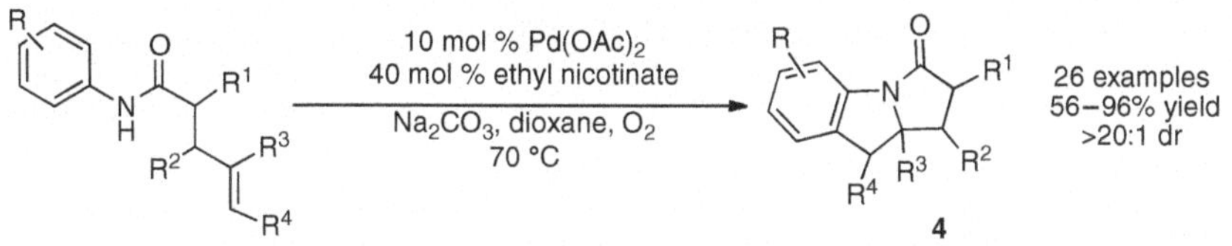

Scheme 55 Pd-catalyzed intramolecular aminoarylation of unsaturated *N*-aryl amides

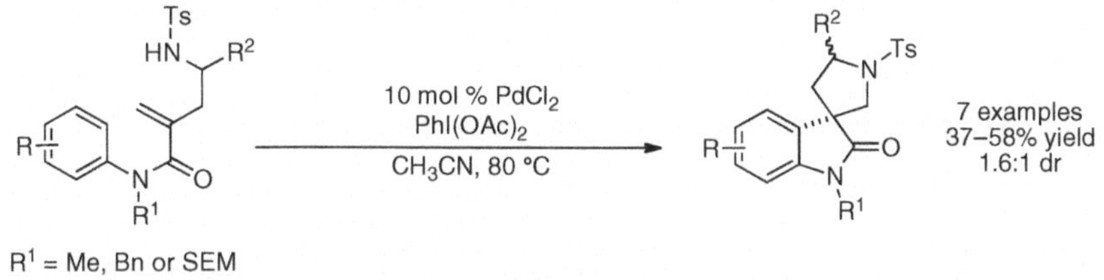

Scheme 56 Synthesis of spirooxindoles via intramolecular alkene aminoarylation

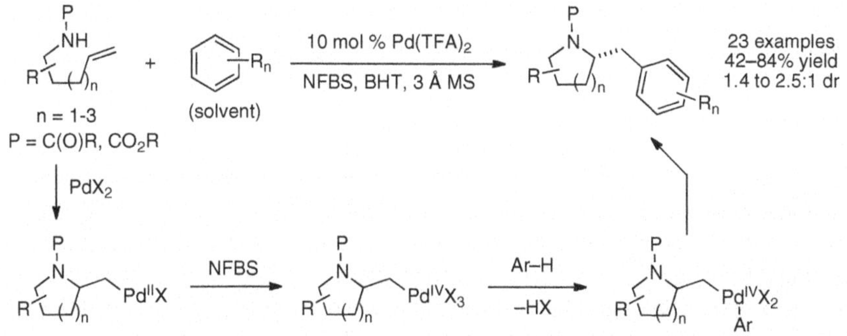

Scheme 57 Pd-catalyzed C–H activation/aminoarylation of unsaturated amides and carbamates

reductive elimination to generate the observed product. The transformations proceed with generally good chemical yields, but diastereoselectivities are modest in reactions of starting materials bearing substituents on the tether between the heteroatom and the alkene.

Sigman has reported mechanistically distinct alkene carboamination reactions between substrates **79** and nucleophilic heteroarenes such as indoles, indolizines, and furans (Scheme 58) [94]. The reactions proceed via initial *anti*-aminopalladation of the alkene to afford an intermediate benzylpalladium complex **81**, which then undergoes elimination to provide reactive quinone methide intermediate **82**. The quinone methide is then captured by the nucleophilic arene component to afford the observed product. These transformations provide substituted pyrrolidine derivatives **80** in good yield with excellent diastereoselectivity and high enantioselectivity. However, due to the mechanistic requirement for the formation of a reactive quinone methide intermediate, these transformations appear to be limited to phenol-derived substrates.

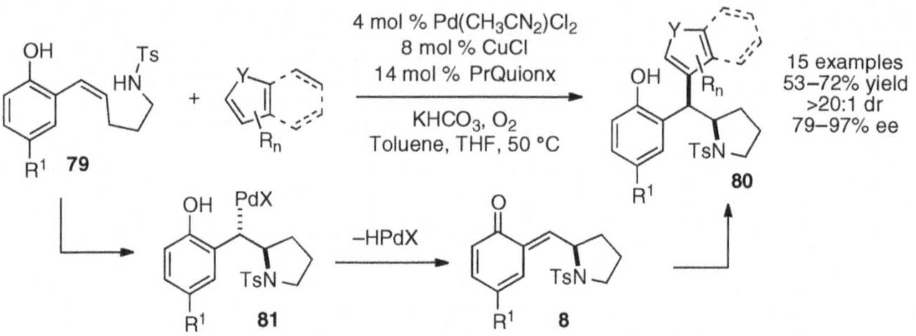

Scheme 58 Alkene carboamination via Pd-catalyzed aminoarylation/quinone methide capture

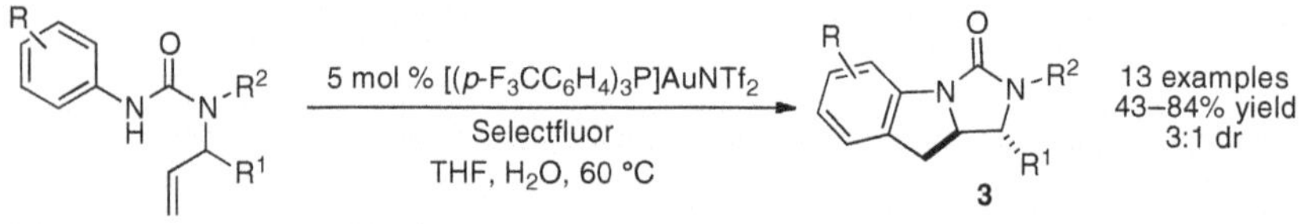

Scheme 59 Au-catalyzed intramolecular aminoarylation of *N*-allylureas

Gold-Catalyzed Reactions

An intramolecular alkene aminoarylation sequence of *N*-allylureas that effects C-H functionalization in the C–C bond-forming event has been described by Zhang and coworkers (Scheme 59) [95]. The combination of an Au(I) phosphine complex and the oxidant selectfluor is utilized to effect these transformations, which provide tricyclic products **83** in good yield with moderate diastereoselectivity. Although the reactions employ a gold catalyst system, their mechanism is quite similar to that of Michael's reactions that are described above (Scheme 57) and appears to involve sequential *anti*-aminoauration, oxidation, and C–H functionalization.

3.1.4 Metal-Catalyzed Cascade Cyclization Reactions of Polyunsaturated Amines and Related Nucleophiles

Cascade cyclization reactions of amines bearing two alkenes provide a concise approach to the construction of polycyclic heterocycles. For example, the Pd-catalyzed reaction of aryl bromides with *N*-allyl-2-allylaniline and related substrates **84** provides benzo-fused heterocycles **85** in moderate to good yield (Scheme 60) [96]. Reactions that generate pyrrolizidine derivatives proceed with good diastereoselectivity, although stereocontrol is less efficient in transformations that yield indolizidine derivatives. These reactions proceed via aminopalladation of the *o*-allyl group to provide **86** followed by subsequent carbopalladation of the *N*-allyl group to yield **87**. Reductive elimination from this latter intermediate then generates the aryl-carbon bond and affords the product **85**. The nature of the

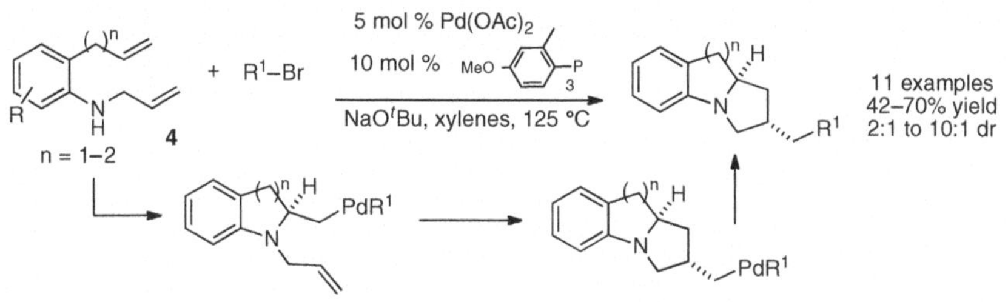

Scheme 60 Cascade cyclization of *N*-allyl-2-allylaniline derivatives

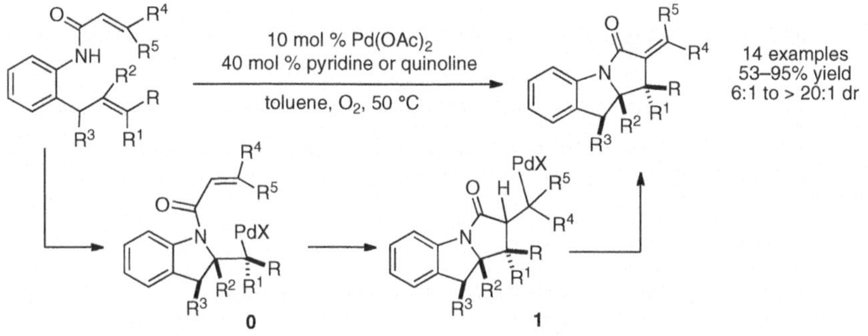

Scheme 61 Cascade oxidative cyclization reactions

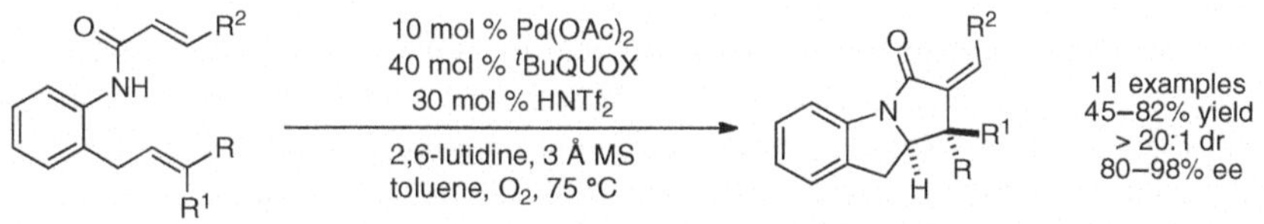

Scheme 62 Cascade asymmetric oxidative cyclization reactions

phosphine ligand is extremely important, as many other catalysts lead to formation of *N*-allyl-2-benzylindoline side products that result from competing reductive elimination from intermediate **86** prior to the carbopalladation of **86–87**.

A cascade oxidative cyclization reaction of *N*-(2-allylphenyl)acrylamide derivatives **88** has been developed by Yang that provides benzo-fused pyrrolizidine derivatives **89** via intramolecular alkene carboamination (Scheme 61) [97]. The key alkene insertion steps of these reactions are related to those of the transformations illustrated above in Scheme 60, although oxidative addition and reductive elimination steps are not involved in the conversion of **88–89**. Instead, this process is initiated by *syn*-aminopalladation of one alkene to generate **90** followed by carbopalladation of the second alkene to yield **91**. The alkylpalladium intermediate **91** then undergoes β-hydride elimination to afford the product **89**. Substitution is tolerated on both alkenes as well as at the allylic position of the substrate. High levels of asymmetric induction can be obtained in these transformations when ^{t}BuQUOX is employed as a chiral ligand for the palladium catalyst. Under these conditions products are formed as single diastereomers with 80–98% ee (Scheme 61 and 62) [98]. Substitution on the

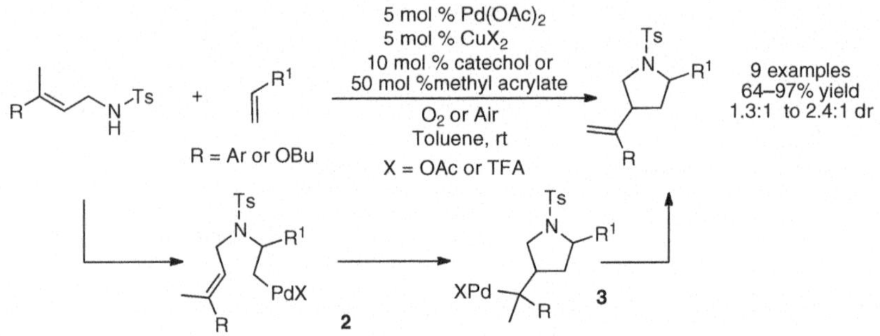

Scheme 63 Pd-catalyzed oxidative carboamination of alkenes with allylic sulfonamides

allyl group is well tolerated in the Pd/^{t}BuQUOX-catalyzed reactions. However, asymmetric reactions of substrates bearing substitution on the acrylamide moiety ($R^2 \neq H$) proceed in modest yield (45–55%).

3.1.5 Palladium-Catalyzed Oxidative Carboamination Reactions of Alkenes with Allylic Sulfonamides

Stahl has developed an interesting carboamination reaction between alkenes and allylic sulfonamides that affords substituted pyrrolidine derivatives (Scheme 63) [99]. These reactions proceed via initial intermolecular aminopalladation of the alkene with the sulfonamide to provide **92** followed by intramolecular carbopalladation to give **93**. A subsequent β-hydride elimination from intermediate **93** then affords the pyrrolidine product. The reactions are effective either with styrene derivatives or with butyl vinyl ether as the alkene component. Chemical yields in these transformations are good, although diastereoselectivities are modest.

3.2 Palladium-Catalyzed Alkene Aminocarboalkoxylation and Aminocarboamidation

The first efficient and chemoselective Pd-catalyzed alkene aminocarbonylation reactions were reported by Tamaru in 1985 [100] and have been employed for the construction of a variety of nitrogen heterocycles bearing pendant carbonyl-containing functional groups (Scheme 64) [101–106]. Reactions of *O*-allyltosylcarbamates and related substrates **94** that contain a single nucleophilic heteroatom lead to the formation of monocyclic five- or six-membered ring products such as **95** that contain an ester functional groups. In contrast, substrates bearing two nucleophilic components, such as ureas **96**, are transformed to bicyclic products such as **97**. Most reactions proceed with good chemical yield and provide products resulting

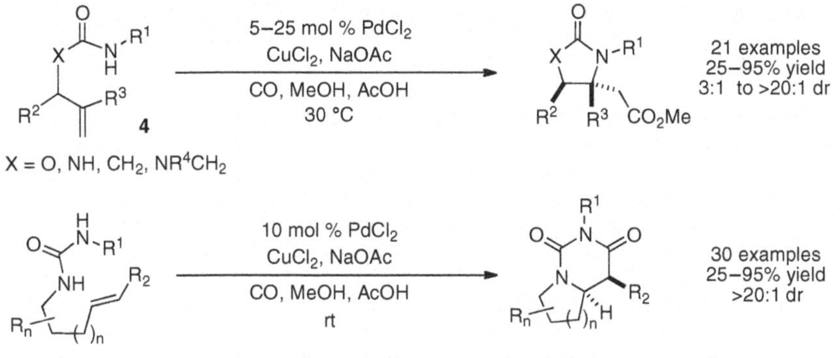

Scheme 64 Pd-catalyzed alkene aminocarboalkoxylation and aminocarboamidation

from *anti*-addition to internal alkenes with high diastereoselectivity. The mechanism of these reactions is similar to the related Pd-catalyzed alkene alkoxycarboalkoxylations described above (Scheme 25). Enantioselective variants of these reactions have also been described, although enantioselectivities in most cases are modest [107–109]. This method has been applied to the synthesis of a number of natural products, including anatoxin-A [110], calvine [111], pinidinone [112], and C6 homologues of 1-deoxynojirimycin [113]. Further details on these types of reactions are described in a recent review [41].

4 Conclusion and Future Outlook

Over the past 10 years great advances have been made in the field of heterocycle synthesis via metal-catalyzed alkene carboamination and carboalkoxylation reactions. The scope of these transformations has expanded considerably since initial reports of these reactions, and good progress has been made in the development of enantioselective variants. Nonetheless, there remain many challenges for the future, including the development of improved methods to access small (three to four membered) or large (six to eight membered) rings, and the invention of improved chiral catalysts with broad substrate scope. Additional development of methods that involve cascade reactions or C–H bond functionalization processes are also likely to continue to attract interest and research efforts, as are applications of these transformations to complex molecule synthesis.

References

1. Wolfe JP, Rossi MA (2004) Stereoselective synthesis of tetrahydrofurans via the palladium-catalyzed reaction of aryl bromides with γ-hydroxy alkenes: evidence for an unusual intramolecular olefin insertion into a Pd(Ar)(OR) intermediate. J Am Chem Soc 126:1620–1621
2. Hay MB, Hardin AR, Wolfe JP (2005) Palladium-catalyzed synthesis of tetrahydrofurans from γ-hydroxy terminal alkenes: scope, limitations, and stereoselectivity. J Org Chem 70: 3099–3107

3. Lasikova A, Dohanosova J, Hlavinova L, Toffano M, Vo-Thanh G, Kozisek J, Gracza T (2012) Domino reaction: Pd(II)-catalyzed cyclization of unsaturated polyols and cross-coupling. Tetrahedron Asymmetry 23:818–827
4. Rosen BR, Ney JE, Wolfe JP (2010) Use of aryl chlorides as electrophiles in Pd-catalyzed alkene difunctionalization reactions. J Org Chem 75:2756–2759
5. Hay MB, Wolfe JP (2005) Palladium-catalyzed synthesis of 2,1′-disubstituted tetrahydrofurans from γ-hydroxy internal alkenes. Evidence for alkene insertion into a Pd–O bond and stereochemical scrambling via β-hydride elimination. J Am Chem Soc 127:16468–16476
6. Ward AF, Wolfe JP (2010) Highly diastereoselective Pd-catalyzed carboetherification reactions of acyclic internal alkenes. Stereoselective synthesis of polysubstituted tetrahydrofurans. Org Lett 12:1268–1271
7. Yeh MCP, Tsao WC, Tu LH (2005) Palladium-catalyzed reaction of aryl bromides with 7-hydroxy-1,3-dienes. Organometallics 24:5909–5915
8. Ward AF, Wolfe JP (2009) Synthesis of fused-ring and attached-ring bis-tetrahydrofurans via Pd-catalyzed carboetherification. Org Lett 11:2209–2212
9. Nakhla JS, Kampf JW, Wolfe JP (2006) Intramolecular Pd-catalyzed carboetherification and carboamination. Influence of catalyst structure on reaction mechanism and product stereochemistry. J Am Chem Soc 128:2893–2901
10. Cacchi S, Fabrizi G, Goggiamani A, Iazzetti A, Madec D, Poli G, Prestat G (2011) Functionalized 2,3-dihydropyrans via palladium-catalyzed oxyarylation of α-allyl-β-ketoesters. Org Biomol Chem 9:8233–8236
11. Kim ES, Kim KH, Park S, Kim JN (2010) Synthesis of dihydroindenofuran scaffold via a Pd-catalyzed 5-*endo-trig* cyclization/enolate *O*-alkylation cascade. Tetrahedron Lett 51: 4648–4652
12. Hay MB, Wolfe JP (2007) Stereoselective synthesis of isoxazolidines through Pd-catalyzed carboetherification of *N*-butenylhydroxylamines. Angew Chem Int Ed 46:6492–6494
13. Jiang D, Peng J, Chen Y (2008) Pd-catalyzed carboetherification of β, γ-unsaturated oximes: a novel approach to Δ2-isoxazolines. Org Lett 10:1695–1698
14. Hayashi S, Yorimitsu H, Oshima K (2009) Synthesis of epoxides by palladium-catalyzed reactions of tertiary allylic alcohols with aryl or alkenyl halides. J Am Chem Soc 131: 2052–2053
15. Ward AF, Xu Y, Wolfe JP (2012) Synthesis of chromans via Pd-catalyzed alkene carboetherification reactions. Chem Commun 48:609–611
16. Waser J, Nicolai S (2011) Pd(0)-catalyzed oxy- and aminoalkynylation of olefins for the synthesis of tetrahydrofurans and pyrrolidines. Org Lett 13:6324–6327
17. Nicolai S, Erard S, Fernandez Gonzalez D, Waser J (2010) Pd-catalyzed intramolecular oxyalkynylation of alkenes with hypervalent iodine. Org Lett 12:384–387
18. Zhu R, Buchwald SL (2012) Copper-catalyzed oxytrifluoromethylation of unactivated alkenes. J Am Chem Soc 134:12462–12465
19. Zhang G, Cui L, Wang Y, Zhang L (2010) Homogeneous gold-catalyzed oxidative carboheterofunctionalization of alkenes. J Am Chem Soc 132:1474–1475
20. Ball LT, Green M, Lloyd-Jones GC, Russell CA (2010) Arylsilanes: application to gold-catalyzed oxyarylation of alkenes. Org Lett 12:4724–4727
21. Tkatchouk E, Mankad NP, Benitez D, Goddard WA III, Toste FD (2011) Two metals are better than one in the gold catalyzed oxidative heteroarylation of alkenes. J Am Chem Soc 133:14293–14300
22. Zhu C, Falck JR (2011) Alternative pathways for heck intermediates: palladium-catalyzed oxyarylation of homoallylic alcohols. Angew Chem Int Ed 50:6626–6629
23. Zhu R, Buchwald SL (2012) Combined oxypalladation/C-H functionalization: palladium(II)-catalyzed intramolecular oxidative oxyarylation of hydroxyalkenes. Angew Chem Int Ed 51: 1926–1929
24. Matsuura BS, Condie AG, McBee IA, Buff RC, Karahalis GJ, Stephenson CRJ (2011) Intercepting Wacker intermediates with arenes: C-H functionalization and dearomatization. Org Lett 13:6320–6323

25. Miller Y, Miao L, Hosseini AS, Chemler SR (2012) Copper-catalyzed intramolecular alkene carboetherification: synthesis of fused-ring and bridged-ring tetrahydrofurans. J Am Chem Soc 134:12149–12156
26. Fries P, Halter D, Kleinschek A, Hartung J (2011) Functionalized tetrahydrofurans from alkenols and olefins/alkynes via aerobic oxidation-radical addition cascades. J Am Chem Soc 133:3906–3912
27. Semmelhack MF, Bodurow C (1984) Intramolecular alkoxypalladation/carbonylation of alkenes. J Am Chem Soc 106:1496–1498
28. Semmelhack MF, Zhang N (1989) Stereoselective formation of tetrahydrofuran rings via intramolecular alkoxycarbonylation of hydroxyalkenes. J Org Chem 54:4483–4485
29. McCormick M, Monahan R III, Soria J, Goldsmith D, Liotta D (1989) Effects of substitution on intramolecular alkoxypalladation carbonylation reactions. J Org Chem 54:4485–4487
30. Semmelhack MF, Kim C, Zhang N, Bodurow C, Sanner M, Dobler W, Meier M (1990) Intramolecular alkoxy-carbonylation of hydroxy alkenes promoted by Pd(II). Pure Appl Chem 62:2035–2040
31. Tamaru Y, Kobayashi T, Kawamura S, Ochiai H, Hojo M, Yoshida Z (1985) Palladium catalyzed oxycarbonylation of 4-penten-1,3-diols: efficient stereoselective synthesis of cis-3-hydroxytetrahydrofuran-2-acetic acid lactones. Tetrahedron Lett 26:3207–3210
32. Tietze LF, Zinngrebe J, Spiegl DA, Stecker F (2007) Palladium-catalyzed Domino Wacker carbonylation reaction for the enantioselective synthesis of chromans and benzodioxins. Heterocycles 74:473–489
33. Semmelhack MF, Epa WR, Cheung AWH, Gu Y, Kim C, Zhang N, Lew W (1994) Palladium-promoted synthesis of ionophore antibiotics. Strategy and assembly of the homochiral tetrahydrofuran and tetrahydropyran portions of tetronomycin. J Am Chem Soc 116:7455–7456
34. Paddon-Jones GC, Hungerford NL, Hayes P, Kitching W (1999) Efficient palladium(II)-mediated construction of functionalized plakortone cores. Org Lett 1:1905–1907
35. Semmelhack MF, Shanmugam P (2000) Development of an approach to the synthesis of the plakortones. Tetrahedron Lett 41:3567–3571
36. Hayes PY, Kitching W (2002) Total synthesis and absolute stereochemistry of plakortone D. J Am Chem Soc 124:9718–9719
37. Paddon-Jones GC, McErlean CSP, Hayes P, Moore CJ, Konig WA, Kitching W (2001) Synthesis and stereochemistry of some bicyclic γ-lactones from parasitic wasps (Hymenoptera: Braconidae). Utility of hydrolytic kinetic resolution of epoxides and palladium(II)-catalyzed hydroxycyclization-carbonylation-lactonization of ene-diols. J Org Chem 66:7487–7495
38. Babjak M, Kapitan P, Gracza T (2005) Synthesis of (+)-goniothalesdiol and (+)-7-epi-goniothalesdiol. Tetrahedron 61:2471–2479
39. Hayes PY, Chow S, Rahm F, Bernhardt PV, DeVoss JJ, Kitching W (2010) Synthesis of the sponge-derived plakortone series of bioactive compounds. J Org Chem 75:6489–6501
40. Karlubikova O, Babjak M, Gracza T (2011) Tetrahydropyran synthesis by palladium(II)-catalyzed hydroxycarbonylation of hexenols: synthesis of (±)-diospongin A and (+)-civet cat compound. Tetrahedron 67:4980–4987
41. Wu XF, Neumann H, Beller M (2013) Synthesis of heterocycles via palladium-catalyzed carbonylations. Chem Rev 113:1–35
42. Hoang GT, Reddy VJ, Nguyen HHK, Douglas CJ (2011) Insertion of an alkene into an ester: intramolecular oxyacylation reaction of alkenes through acyl C–O bond activation. Angew Chem Int Ed 50:1882–1884
43. Koester DC, Kobayashi M, Werz DB, Nakao Y (2012) Intramolecular oxycyanation of alkenes by cooperative Pd/BPh_3 catalysis. J Am Chem Soc 134:6544–6547
44. Ney JE, Wolfe JP (2004) Palladium-catalyzed synthesis of *N*-aryl pyrrolidines from γ-(*N*-arylamino)alkenes: evidence for chemoselective alkene insertion into Pd–N bonds. Angew Chem Int Ed 43:3605–3608
45. Ney JE, Hay MB, Yang Q, Wolfe JP (2005) Synthesis of *N*-aryl-2-allyl pyrrolidines via palladium-catalyzed carboamination reactions of γ-(*N*-arylamino)alkenes with vinyl bromides. Adv Synth Catal 347:1614–1620

46. Neukom JD, Perch NS, Wolfe JP (2010) Intramolecular alkene aminopalladation reactions of (dppf)Pd(Ar)[N(Ar^1)(CH_2)$_3$CH=CH_2] complexes. Insertion of unactivated alkenes into Pd–N bonds. J Am Chem Soc 132:6276–6277
47. Neukom JD, Perch NS, Wolfe JP (2011) Intramolecular insertion of alkenes into Pd–N bonds. Effects of substrate and ligand structure on the reactivity of (P–P)Pd(Ar)[N(Ar^1)(CH_2)$_3$CR=CHR′] complexes. Organometallics 30:1269–1277
48. Hanley PS, Markovic D, Hartwig JF (2010) Intermolecular insertion of ethylene and octene into a palladium-amide bond. Spectroscopic evidence for an ethylene amido intermediate. J Am Chem Soc 132:6302–6303
49. Hanley PS, Hartwig JF (2011) Intermolecular migratory insertion of unactivated olefins into palladium-nitrogen bonds. Steric and electronic effects on the rate of migratory insertion. J Am Chem Soc 133:15661–15673
50. White PB, Stahl SS (2011) Reversible alkene insertion into the Pd–N bond of Pd(II)-sulfonamidates and implications for catalytic amidation reactions. J Am Chem Soc 133: 18594–18597
51. Ney JE, Wolfe JP (2005) Selective synthesis of 5- or 6-aryl octahydrocyclopenta[*b*]pyrroles from a common precursor through control of competing pathways in a Pd-catalyzed reaction. J Am Chem Soc 127:8644–8651
52. Schultz DM, Wolfe JP (2011) Intramolecular alkene carboamination reactions for the synthesis of enantiomerically enriched tropane derivatives. Org Lett 13:2962–2965
53. Lira R, Wolfe JP (2004) Palladium-catalyzed synthesis of *N*-aryl-2-benzylindolines via tandem arylation of 2-allylaniline: control of selectivity through in situ catalyst modification. J Am Chem Soc 126:13906–13907
54. Yang Q, Ney JE, Wolfe JP (2005) Palladium-catalyzed tandem *n*-arylation/carboamination reactions for the stereoselective synthesis of *N*-aryl-2-benzylpyrrolidines. Org Lett 7: 2575–2578
55. Lemen GS, Wolfe JP (2011) Cascade intramolecular *N*-arylation/intermolecular carboamination reactions for the construction of tricyclic heterocycles. Org Lett 13:3218–3221
56. Bertrand MB, Wolfe JP (2005) Stereoselective synthesis of *N*-protected pyrrolidines via Pd-catalyzed reactions of γ-(*N*-acylamino) and γ-(*N*-boc-amino)alkenes with aryl bromides. Tetrahedron 61:6447–6459
57. Tepsen TH, Larsen M, Nielsen MB (2010) On the scope of Pd-catalyzed carboamination reactions-synthesis of 2,4-disubstituted pyrrolidines and 2-substituted piperidines and morpholines. Tetrahedron 66:6133–6137
58. Bertrand MB, Leathen ML, Wolfe JP (2007) Mild conditions for the synthesis of functionalized pyrrolidines via Pd-catalyzed carboamination reactions. Org Lett 9:457–460
59. Bertrand MB, Neukom JD, Wolfe JP (2008) Mild conditions for Pd-catalyzed carboamination of *N*-protected hex-4-enylamines and 1-, 3-, and 4-substituted pent-4-enylamines. Scope, limitations, and mechanism of pyrrolidine formation. J Org Chem 73: 8851–8860
60. Bertrand MB, Wolfe JP (2006) A concise stereoselective synthesis of preussin, 3-epi-preussin, and analogues. Org Lett 8:2353–2356
61. Rossiter LM, Slater ML, Giessert RE, Sakwa SA, Herr RJ (2009) A concise palladium-catalyzed carboamination route to (±)-tylophorine. J Org Chem 74:9554–9557
62. Mai DN, Wolfe JP (2010) Asymmetric palladium-catalyzed carboamination reactions for the synthesis of enantiomerically enriched 2-(arylmethyl) and 2-(alkenylmethyl)pyrrolidines. J Am Chem Soc 132:12157–12159
63. Mai DN, Rosen BR, Wolfe JP (2011) Enantioconvergent synthesis of (+)-aphanorphine via asymmetric Pd-catalyzed alkene carboamination. Org Lett 13:2932–2935
64. Lemen GS, Wolfe JP (2010) Pd-catalyzed carboamination of oxazolidin-2-ones: a stereoselective route to *trans*-2,5-disubstituted pyrrolidines. Org Lett 12:2322–2325
65. Bagnoli L, Cacchi S, Fabrizi G, Goggiamani A, Scarponi C, Tiecco M (2010) Diastereoselective synthesis of hexahydro-3H-pyrrolizidin-3-ones through Pd-catalyzed carboamination. J Org Chem 75:2134–2137

66. Jaegli S, Erb W, Retailleau P, Vors JP, Neuville L, Zhu J (2010) Palladium-catalyzed domino process to spirooxindoles: ligand effect on aminopalladation versus carbopalladation. Chemistry 16:5863–5867
67. Ward AF, Wolfe JP (2011) Stereoselective synthesis of 1,3-oxazolidines via Pd-catalyzed carboamination reactions of *O*-vinyl-1,2-amino alcohols. Org Lett 13:4728–4731
68. Fritz JA, Nakhla JS, Wolfe JP (2006) A new synthesis of imidazolidin-2-ones via Pd-catalyzed carboamination of *N*-allylureas. Org Lett 8:2531–2534
69. Fritz JA, Wolfe JP (2008) Stereoselective synthesis of imidazolidin-2-ones via Pd-catalyzed alkene carboamination. Scope and limitations. Tetrahedron 64:6838–6852
70. Babij NR, Wolfe JP (2012) Asymmetric total synthesis of (+)-merobatzelladine B. Angew Chem Int Ed 51:4128–4130
71. Hopkins BA, Wolfe JP (2012) Synthesis of enantiomerically enriched imidazolidin-2-ones through asymmetric Pd-catalyzed carboamination reactions. Angew Chem Int Ed 51: 9886–9890
72. Giampietro NC, Wolfe JP (2008) Stereoselective synthesis of *cis*- or *trans*-3,5-disubstituted pyrazolidines via Pd-catalyzed carboamination reactions. Use of allylic strain to control product stereochemistry through N-substituent manipulation. J Am Chem Soc 130:12907–12911
73. Lemen GS, Giampietro NC, Hay MB, Wolfe JP (2009) Influence of hydroxylamine conformation on stereocontrol in Pd-catalyzed isoxazolidine-forming reactions. J Org Chem 74: 2533–2540
74. Peng J, Lin W, Yuan S, Chen Y (2007) J Org Chem 72:3145–3148
75. Hayashi S, Yorimitsu H, Oshima K (2009) Synthesis of aziridines by palladium-catalyzed reactions of allylamines with aryl and alkenyl halides. Evidence of a syn-carboamination pathway. Angew Chem Int Ed 48:7224–7226
76. Nakhla JS, Wolfe JP (2007) A concise asymmetric synthesis of *cis*-2,6-disubstituted *N*-aryl piperazines via Pd-catalyzed carboamination reactions. Org Lett 9:3279–3282
77. Nakhla JS, Schultz DM, Wolfe JP (2009) Palladium-catalyzed alkene carboamination reactions for the synthesis of substituted piperazines. Tetrahedron 65:6549–6570
78. Leathen ML, Rosen BR, Wolfe JP (2009) New strategy for the synthesis of substituted morpholines. J Org Chem 74:5107–5110
79. Neukom JD, Aquino AS, Wolfe JP (2011) Synthesis of saturated 1,4-benzodiazepines via Pd-catalyzed carboamination reactions. Org Lett 13:2196–2199
80. Nicolai S, Piemontesi C, Waser J (2011) A palladium-catalyzed aminoalkynylation strategy towards bicyclic heterocycles: synthesis of (±)-trachelanthamidine. Angew Chem Int Ed 50: 4680–4683
81. Brenzovich WE Jr, Benitez D, Lackner AD, Shunatona HP, Tkatchouk E, Goddard WA III, Toste FD (2010) Gold-catalyzed intramolecular aminoarylation of alkenes: C–C bond formation through bimolecular reductive elimination. Angew Chem Int Ed 49:5519–5522
82. Sherman ES, Chemler SR, Tan TB, Gerlits O (2004) Copper(II) acetate-promoted oxidative cyclization of arylsulfonyl-*o*-allylanilines. Org Lett 6:1573–1575
83. Sherman ES, Fuller PH, Kasi D, Chemler SR (2007) Pyrrolidine and piperidine formation via copper(II) carboxylate-promoted intramolecular carboamination of unactivated olefins: diastereoselectivity and mechanism. J Org Chem 72:3896–3905
84. Sherman ES, Chemler SR (2009) Copper(II)-catalyzed aminooxygenation and carboamination of *N*-aryl-2-allylanilines. Adv Synth Catal 351:467–471
85. Fuller PH, Chemler SR (2007) Copper(II) carboxylate-promoted intramolecular carboamination of alkenes for the synthesis of polycyclic lactams. Org Lett 9:5477–5480
86. Zeng W, Chemler SR (2007) Copper(II)-catalyzed enantioselective intramolecular carboamination of alkenes. J Am Chem Soc 129:12948–12949
87. Zheng W, Chemler SR (2008) Total synthesis of (*S*)-(+)-tylophorine via enantioselective intramolecular alkene carboamination. J Org Chem 73:6045–6047
88. Miao L, Haque I, Manzoni MR, Tham WS, Chemler SR (2010) Diastereo- and enantioselective copper-catalyzed intramolecular carboamination of alkenes for the synthesis of hexahydro-1*H*-benz[*f*]indoles. Org Lett 12:4739–4741

89. Liwosz TW, Chemler SR (2012) Copper-catalyzed enantioselective intramolecular alkene amination/intermolecular Heck-type coupling cascade. J Am Chem Soc 134:2020–2023
90. Yip KT, Yang D (2011) Pd(II)-catalyzed intramolecular amidoarylation of alkenes with molecular oxygen as the sole oxidant. Org Lett 13:2134–2137
91. Jaegli S, Dufour J, Wei HI, Piou T, Duan XH, Vors JP, Neuville L, Zhu J (2010) Palladium-catalyzed carbo-heterofunctionalization of alkenes for the synthesis of oxindoles and spirooxindoles. Org Lett 12:4498–4501
92. Rosewall CF, Sibbald PA, Liskin DV, Michael FE (2009) Palladium-catalyzed carboamination of alkenes promoted by *N*-fluorobenzenesulfonimide via C–H activation of arenes. J Am Chem Soc 131:9488–9489
93. Sibbald PA, Rosewall CF, Swartz RD, Michael FE (2009) Mechanism of *N*-fluorobenzene-sulfonimide promoted diamination and carboamination reactions: divergent reactivity of a Pd(IV) species. J Am Chem Soc 131:15945–15951
94. Jana R, Pathak TP, Jensen KH, Sigman MS (2012) Palladium(II)-catalyzed enantio- and diastereoselective synthesis of pyrrolidine derivatives. Org Lett 14:4074–4077
95. Zhang G, Luo Y, Wang Y, Zhang L (2011) Combining gold(I)/Gold(III) catalysis and C–H functionalization: a formal intramolecular [3+2] annulation towards tricyclic indolines and mechanistic studies. Angew Chem Int Ed 50:4450–4454
96. Schultz DM, Wolfe JP (2010) Synthesis of polycyclic nitrogen heterocycles via alkene aminopalladation/carbopalladation cascade reactions. Org Lett 12:1028–1031
97. Yip KT, Yang D (2011) Palladium(II)-catalyzed oxidative cascade cyclization reactions of anilides and anilines: scope and mechanistic investigations. Chem Asian J 6:2166–2175
98. He W, Yip KT, Zhu NY, Yang D (2009) Pd(II)/tBu-quinolineoxazoline: an air-stable and modular chiral catalyst system for enantioselective oxidative cascade cyclization. Org Lett 11:5626–5628
99. Scarborough CC, Stahl SS (2006) Synthesis of pyrrolidines via palladium(II)-catalyzed aerobic oxidative carboamination of butyl vinyl ether and styrenes with allyl tosylamides. Org Lett 8:3251–3254
100. Tamaru Y, Kobayashi T, Kawamura S, Ochiai H, Yoshida Z (1985) Stereoselective intramolecular aminocarbonylation of 3-hydroxypent-4-enylamides catalyzed by palladium. Tetrahedron Lett 26:4479–4482
101. Tamaru Y, Hojo M, Higashimura H, Yoshida Z (1988) Urea as the most reactive and versatile nitrogen nucleophile for the palladium(2+)-catalyzed cyclization of unsaturated amines. J Am Chem Soc 110:3994–4002
102. Tamaru Y, Hojo M, Yoshida Z (1988) Palladium(2+)-catalyzed intramolecular aminocarbonylation of 3-hydroxy-4-pentenylamines and 4-hydroxy-5-hexenylamines. J Org Chem 53: 5731–5741
103. Tamaru Y, Tanigawa H, Ito S, Kimura M, Tanaka S, Fugami K, Sekiyama T, Yoshida Z (1992) Palladium(II)-catalyzed oxidative aminocarbonylation of unsaturated carbamates. Tetrahedron Lett 33:631–634
104. Harayama H, Okuno H, Takahashi Y, Kimura M, Fugami K, Tanaka S, Tamaru Y (1996) Chemoselective intramolecular aminocarbonylation of unsaturated amides under Wacker-type conditions. Tetrahedron Lett 37:7287–7290
105. Harayama H, Abe A, Sakado T, Kimura M, Fugami K, Tanaka S, Tamaru Y (1997) Palladium (II)-catalyzed intramolecular aminocarbonylation of endo-carbamates under Wacker-type conditions. J Org Chem 62:2113–2122
106. Borsini E, Broggini G, Fasana A, Galli S, Khansaa M, Piarulli U, Rigamonti M (2011) Intramolecular palladium-catalyzed aminocarboxylation of olefins as a direct rouge to bicyclic oxazolidinones. Adv Synth Catal 353:985–994
107. Shinohara T, Arai MA, Wakita K, Arai T, Sasai H (2003) The first enantioselective intramolecular aminocarbonylation of alkenes promoted by Pd(II)-spiro bis(isoxazoline) catalyst. Tetrahedron Lett 44:711–714

108. Tsujihara T, Shinohara T, Takenaka K, Takizawa S, Onitsuka K, Hatanaka M, Sasai H (2009) Enantioselective intramolecular oxidative aminocarbonylation of alkenylureas catalyzed by palladium-spiro bis(isoxazoline) complexes. J Org Chem 74:9274–9279
109. Koos P, Spanik I, Gracza T (2009) Asymmetric Pd(II)-catalyzed amidocarbonylation of unsaturated amino alcohols. Tetrahedron Asymmetry 20:2720–2723
110. Oh CY, Kim KS, Ham WH (1998) A formal synthesis of (±)-anatoxin A by an intramolecular Pd-catalyzed aminocarbonylation reaction. Tetrahedron Lett 39:2133–2136
111. Kubizna P, Spanik I, Kozisek J, Szolcsanyi P (2010) Synthesis of 2,6-disubstituted piperidine alkaloids from ladybird beetles Calvia 10-guttata and Calvia 14-guttata. Tetrahedron 66: 2351–2355
112. Csatayova K, Spanik I, Durisova V, Szolcsanyi P (2010) Synthesis of (−)-pinidinone. Tetrahedron Lett 51:6611–6614
113. Szolcsanyi P, Gracza T, Koman M, Pronayova N, Liptaj T (2000) Pd(II)-catalyzed aminocarbonylation as a key step in the total synthesis of C-6 homologues of 1-deoxynojirimycin and 1-deoxy-l-idonojirimycin. Tetrahedron Asymmetry 11:2579–2597

Top Heterocycl Chem (2013) 32: 39–76
DOI: 10.1007/7081_2013_104

Published online: 19 April 2013

Synthesis of Saturated Heterocycles via Metal-Catalyzed Alkene Diamination, Aminoalkoxylation, or Dialkoxylation Reactions

Sherry R. Chemler and David A. Copeland

Abstract The development of metal-catalyzed additions of nitrogen and oxygen moieties across alkenes to form saturated nitrogen and oxygen heterocycles is described herein. This chapter covers the most recent advances in osmium and palladium-catalyzed alkene oxidation and amination reactions and also summarizes the emerging areas of copper, iron, and gold-catalyzed alkene oxidations and aminations. In most examples, moderate to excellent levels of diastereoselectivity, either by stereospecific addition across the alkene or substrate-directed diastereocontrol, have been achieved. This enables the synthesis of nitrogen and oxygen-containing heterocycles with predictable control of stereogenic centers. In a few cases, asymmetric catalysis has been achieved, allowing for the synthesis of chiral nitrogen and oxygen-containing heterocycles from achiral substrates. In many of these oxidation reactions, use of pre-oxidized substrates or stoichiometric amounts of added oxidants are required to achieve the catalytic cycles, which frequently involve higher oxidation states of the metal catalysts.

Keywords Alkenes · Aminohydroxylation · Asymmetric catalysis · Copper · Diamination · Dihydroxylation · Gold · Indolines · Iron · Osmium · Palladium · Pyrolidines · Saturated heterocycles · Tetrahydrofurans

Contents

S.R. Chemler (✉) and D.A. Copeland
Department of Chemistry, The State University of New York at Buffalo, Buffalo, NY 14260, USA
e-mail: schemler@buffAlo.edu

1 Alkene Diamination

The synthesis of a saturated nitrogen heterocycle and concomitant introduction of two nitrogen functionalities across an alkene can be enabled in a very direct and efficient manner using transition metal catalysis. Alkene diamination, initially explored in the 1970s with stoichiometric metal promoters, has experienced a resurgence of effort in the last decade [1–4]. Transition metals employed to catalyze olefin diamination reactions for the synthesis of saturated heterocyclic compounds include palladium, copper, nickel, and gold (vide infra). Methods for alkene diamination that do not use metals have also been recently developed [5–13], but these reactions fall out of the scope of which will be covered in this chapter. This review will focus on contributions made in the last decade, with emphasis on stereoselective metal-catalyzed alkene diamination protocols.

1.1 Palladium-Catalyzed Alkene Diaminations

The first Pd-catalyzed alkene diamination was reported in 2005 and enabled the regio- and diasteroselective synthesis of cyclic ureas from conjugated dienes (Scheme 1) [14]. The regioselectivity is thought to result from the required formation of a π-allyl palladium intermediate (Scheme 1). Displacement of [Pd(0)] with the second amine generates the product and oxidation of [Pd(0)] with benzoquinone regenerates the [Pd(II)] catalyst.

That same year, a Pd-catalyzed intramolecular diamination of unactivated, isolated alkenes was reported to occur in the presence of stoichiometric $PhI(OAc)_2$ (2.2 equiv.) [15]. This reaction generated fused 5,5-, 6,5-, and 7,5-bicyclic ureas in high yields from unsaturated *N*-tosylureas (Scheme 2).

A mechanism involving a [Pd(II)]/[Pd(IV)] catalytic cycle was proposed (Scheme 3) [15, 16]. The authors proposed a sequence involving *syn*-aminopalladation, oxidation of [Pd(II)] to [Pd(IV)], N–Pd disassociation and C–N bond formation via S_N2 substitution at carbon. An alternative mechanism that would give the same

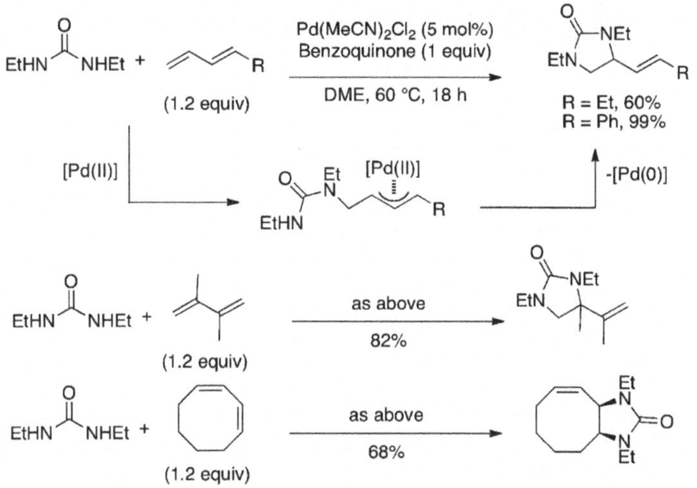

Scheme 1 Pd-catalyzed diamination of conjugated dienes [14]

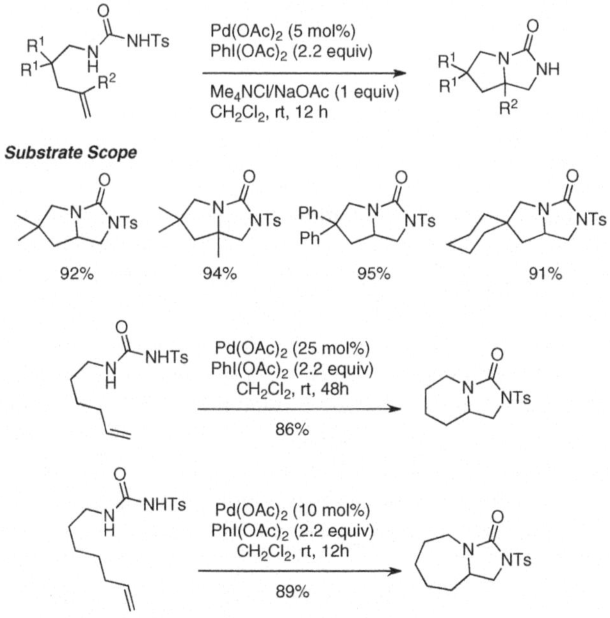

Scheme 2 Pd-catalyzed diamination of isolated alkenes [15]

stereochemical result has been supported by density functional theory (DFT) calculations and entails anti-aminopalladation, N–Pd association, oxidation to Pd(IV) and reductive elimination to give the N–C bond [17].

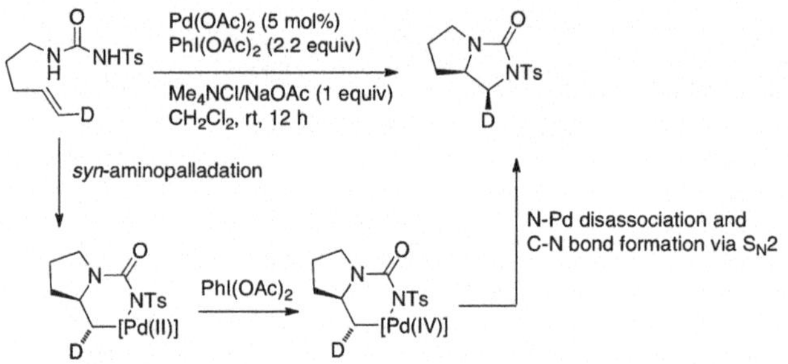

Scheme 3 Proposed Pd-catalyzed diamination mechanism [15, 16]

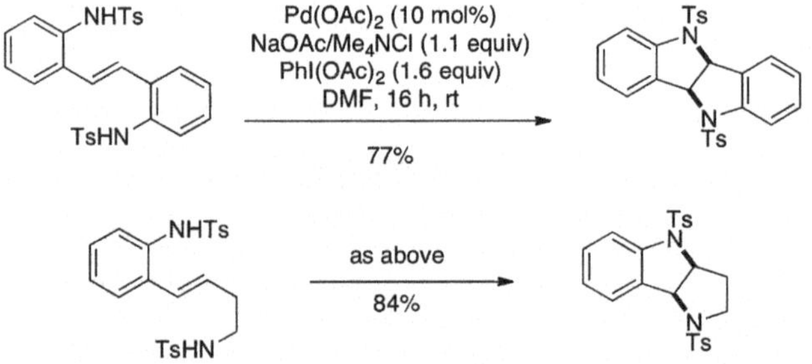

Scheme 4 Diamination of internal alkenes [18]

Using a similar protocol, bis-amination of internal alkenes for the synthesis of bisindoles was also achieved (Scheme 4) [18]. It is noteworthy that a metal-free alkene diamination has been reported to occur with similar substrates and reagents to give largely identical products, but in the absence of a palladium catalyst [8].

Other oxidants were subsequently explored to enable the intramolecular Pd-catalyzed alkene diamination. In 2009, a Pd-catalyzed intra/intermolecular diamination that used *N*-fluorobenzenesulfonamide (NFBS) as both the oxidant and external amine source was reported [19]. Both γ-unsaturated amides and carbamates underwent the *exo*-selective reaction in good to excellent yield. A Pd(II)/Pd(IV) catalytic cycle involving *trans*-aminopalladation and C–N formation via S_N2 substitution was also proposed for this alkene diamination sequence (Scheme 5) [19, 20].

Copper(II) bromide has also been used as the stoichiometric oxidant for the Pd-catalyzed diamination of internal acrylates (Scheme 6) [21–23]. Complementary diastereoselectivities were obtained based upon the substrate structure: ureas gave *cis*-substituted cyclic urea products from *trans*-acrylates [23] and sulfamides gave *trans*-substituted cyclic sulfamide products from *trans*-acrylates [22]. In the urea substrate case, the proposed mechanism is *cis*-aminopalladation followed by S_N2 displacement of [Pd], activated by $CuBr_2$. In the sulfamide case, the proposed

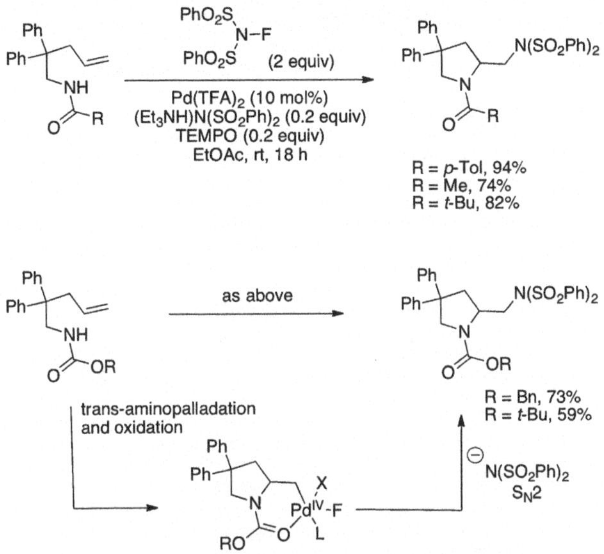

Scheme 5 Pd-catalyzed diamination with NFBS as oxidant [19, 20]

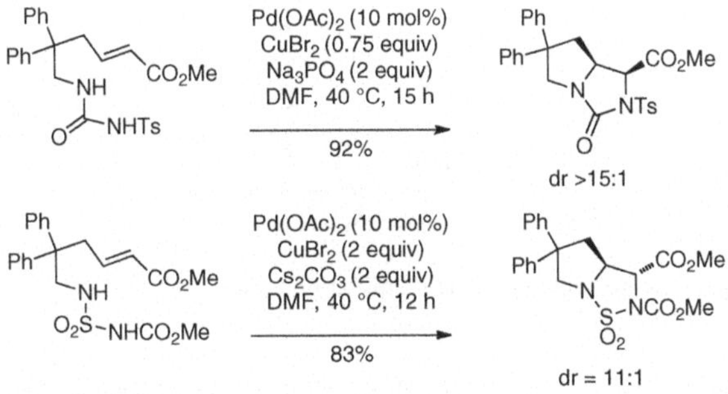

Scheme 6 Pd-catalyzed diamination of acrylates [22, 23]

mechanism is *cis*-aminopalladation, displacement of [Pd] with bromide and subsequent C–N bond formation via bromide displacement.

The catalytic asymmetric alkene diamination has been a long sought-after goal in asymmetric catalysis. The first catalytic enantioselective diamination was reported in 2007. This reaction forms cyclic ureas via Pd-catalyzed intermolecular diamination of conjugated dienes using di-*tert*-butyldiaziridinone as a pre-oxidized diamine source [24]. The reaction is general for alkyl and aryl-substituted dienes and was regioselective for diamination at the more substituted, internal alkene of the diene (Scheme 7). Yields were good to excellent and enantioselectivity levels were generally high. Chiral phosphoramidite ligands proved superior in imparting enantioselectivity to the products. The mechanism involves oxidative addition of

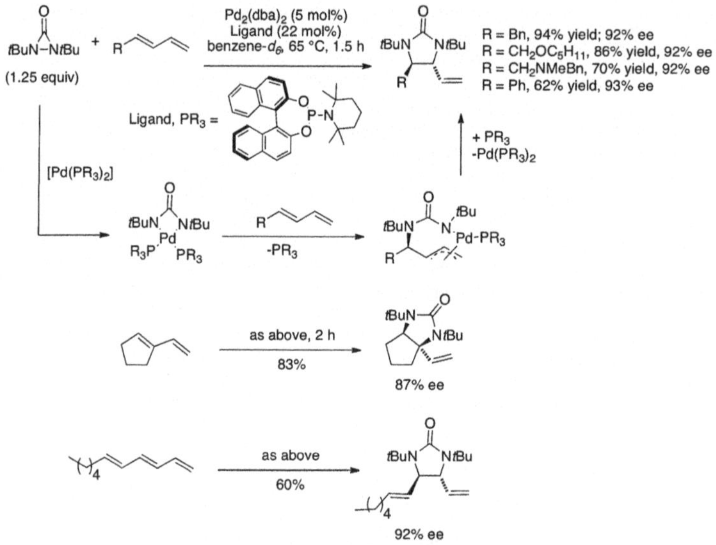

Scheme 7 Pd-catalyzed enantioselective diamination of dienes [24, 25]

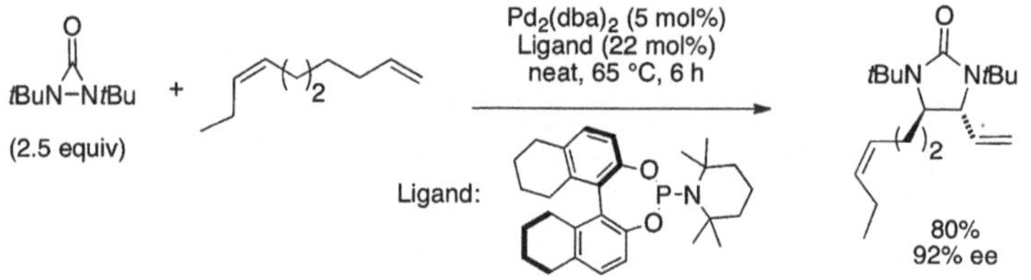

Scheme 8 Catalytic enantioselective C–H diamination [26]

Pd(0) into the N–N bond of the diaziridinone, addition of the N–Pd(II) complex to the internal alkene of the diene and concomitant π-allyl Pd formation and finally C–N bond formation via Pd(0) displacement (Scheme 7) [25].

This method was further advanced by the demonstration that the diene could be formed in situ from terminal alkenes (Scheme 8) [26]. This C–H diamination reaction could be performed neat and provided similar products to those shown in Scheme 7 (vide supra).

1.2 Copper-Catalyzed Alkene Diaminations

Since 2005, the diamination of alkenes has been similarly pursued using less expensive copper complexes as reaction promoters and catalysts [27]. Highly diastereoselective intra/intramolecular and intra/intermolecular copper(II)-promoted alkene diaminations have enabled the synthesis of pyrrolidines and indolines from

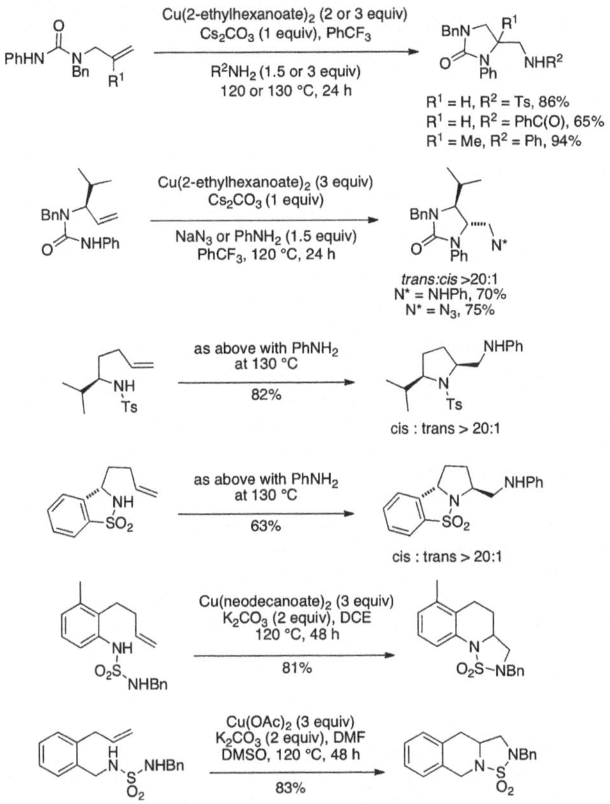

Scheme 9 Copper(II)-promoted alkene diaminations [27–29]

unsaturated sulfamides, sulfonamides, amides, and ureas (Scheme 9) [27–29]. Some examples of tetrahydroisoquinoline-forming diaminations have also been reported (Scheme 9). The reactions were performed with $Cu(OAc)_2$ [27], Cu(neodecanoate)$_2$ [28], and Cu(2-ethylhexanoate)$_2$ [29] as reaction promoter. Reactions with Cu(2-ethylhexanoate)$_2$ in $PhCF_3$ generally proved most efficient [29].

The intra/intermolecular alkene diamination was rendered catalytic in [Cu(II)] with sulfonamides as the intermolecular amine component and MnO_2 (3 equiv.) as the stoichiometric oxidant (Scheme 10) [29]. A promising catalytic enantioselective intra/intermolecular alkene diamination using [Cu((*R*,*R*)-Ph-box)](OTf)$_2$ as the catalyst was also reported (Scheme 10).

The reaction mechanism, based on reaction diastereoselectivity and isotopic labeling studies [29], is thought to involve *cis*-aminocupration, homolysis of the resulting C–[Cu(II)] bond, addition of the resulting carbon radical to [Cu(II)], amine coordination and reductive elimination of the [Cu(III)] intermediate to form the new C–N bond and [Cu(I)] (Scheme 11). In the catalytic reactions, MnO_2 is thought to oxidize the extruded [Cu(I)] back to [Cu(II)] [29].

In 2012, more electron-rich unsaturated amidine substrates were shown to undergo copper-catalyzed intramolecular alkene diamination to form bi- and

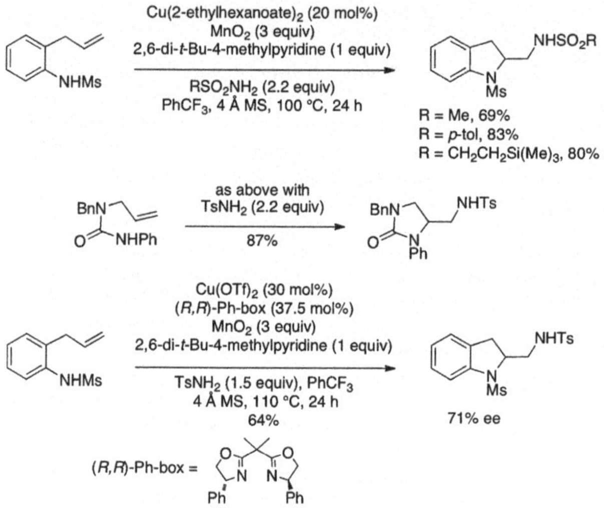

Scheme 10 Cu(II)-catalyzed intra/intermolecular diamination [29]

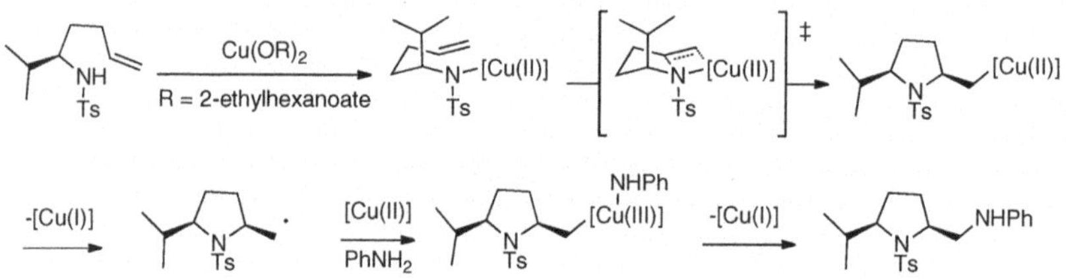

Scheme 11 Proposed mechanism of the Cu(II)-promoted alkene diamination [29]

tricyclic amidines (Scheme 12) [30]. Both terminal and internal alkenes underwent the reaction efficiently.

The proposed mechanism, consistent with product diastereoselectivity, involves two-electron oxidation of the amidine followed by a concerted [3+2]-type annulation (Scheme 13) [30].

The copper-catalyzed intermolecular diamination of terminal alkenes and conjugated dienes has been reported for the synthesis of cyclic ureas, sulfamides, and guanidines, where diaziridinones, thiadiaziridines, and (cyanimino)-diaziridines, respectively, were used as both diamine source and oxidant (Scheme 14) [31–34]. A catalytic, enantioselective diamination (up to 74% ee) of the terminal alkene of conjugated dienes was achieved [35, 36], making the method complementary to analogous Pd-catalyzed diene diaminations (vide supra, Sect. 1.1).

It was further found that depending upon the substrate and catalyst structure, the regioselectivity in the diamination of conjugated dienes can be tuned for either the internal or terminal alkene of the diene (compare Schemes 14 and 15) [33, 38]. Mechanistically, it was determined that the diamination can occur via a more

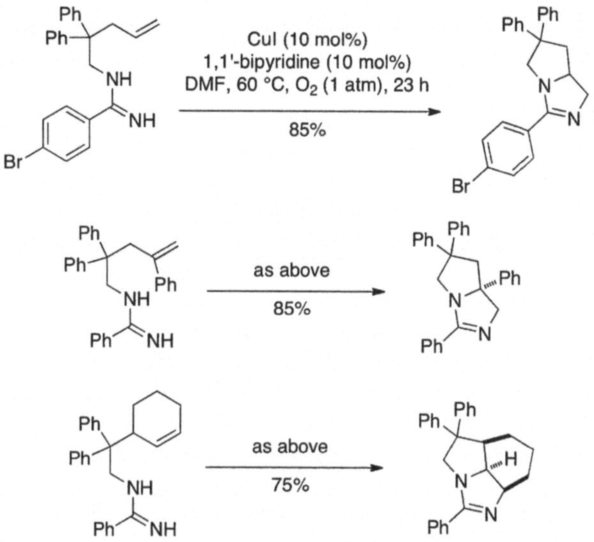

Scheme 12 Cu-catalyzed diamination of unsaturated amidines [30]

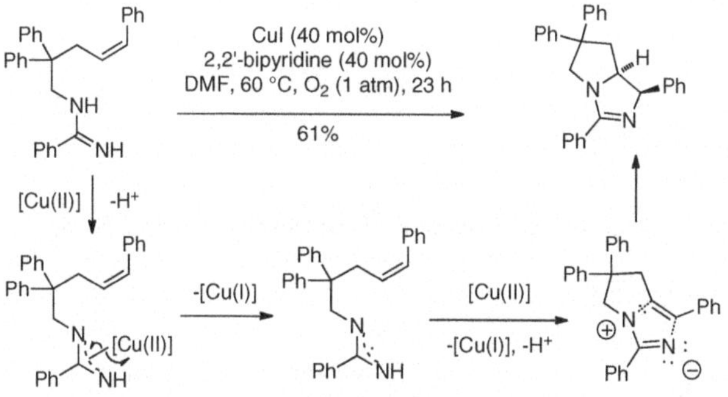

Scheme 13 Proposed mechanism for the unsaturated amidine diamination [30]

radical-type mechanism, selective for the terminal alkene, or via a more electrophilic, Cu(III)-type mechanism, selective for the internal alkene (Scheme 15) [33].

1.3 Nickel-Catalyzed Alkene Diaminations

Nickel complexes have been used to catalyze the intramolecular diamination of unsaturated sulfamides, ureas, and guanidines (Scheme 16) [4, 39]. Terminal and 1,1-disubstituted alkenes underwent the reaction with good efficiency, and $PhI(OAc)_2$ was used as the stoichiometric oxidant.

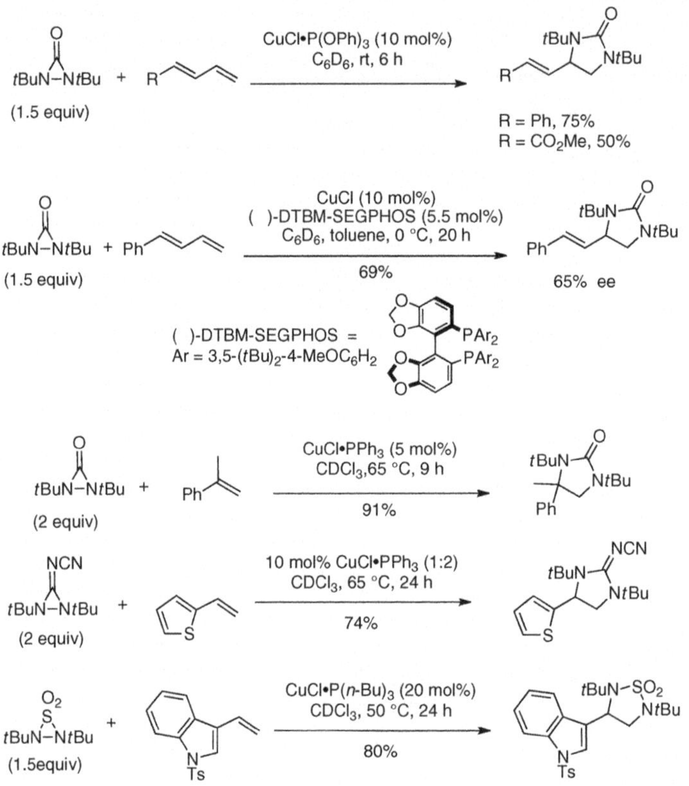

Scheme 14 Copper-catalyzed diamination of alkenes and dienes [31, 32, 34, 35, 37]

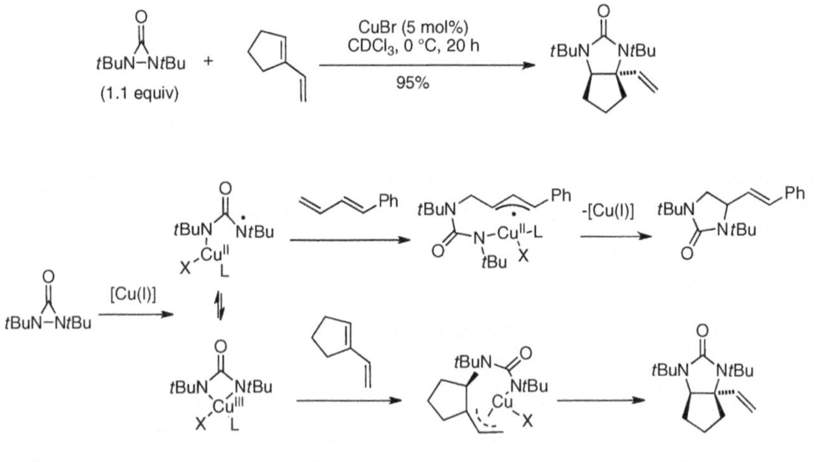

Scheme 15 Regioselectivity and mechanism of the copper-catalyzed diamination of dienes [33, 38]

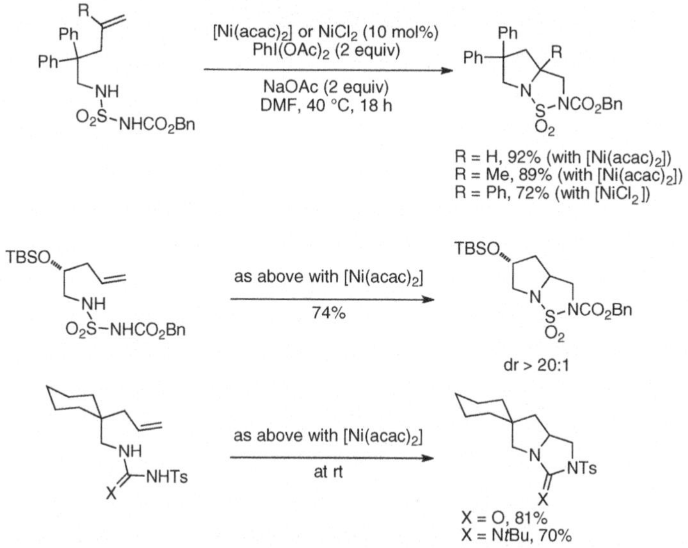

Scheme 16 Nickel-catalyzed intramolecular alkene diamination [4, 39]

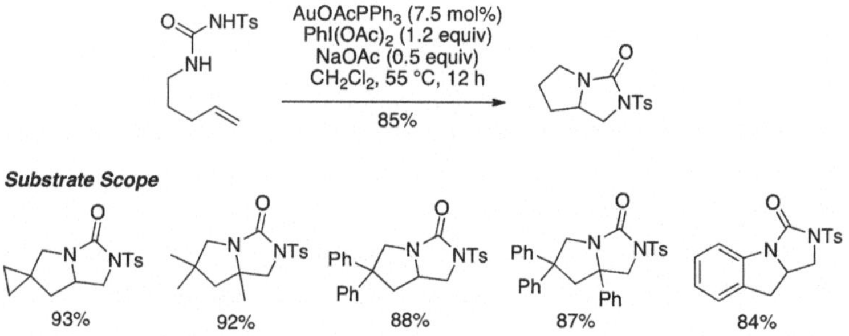

Scheme 17 First Au-catalyzed alkene diamination [40]

1.4 Gold-Catalyzed Alkene Diaminations

Gold-catalyzed alkene diamination was first reported in 2009 [40]. This *exo*-selective reaction provided 5,5-fused bicyclic ureas from terminal and 1,1-disubstituted γ-alkenyl-*N*-tosylureas in high yields (Scheme 17).

One example of a net *cis*-diamination of an internal alkene was also presented (Scheme 18). These reactions require $PhI(OAc)_2$ (1.2 equiv.) as stoichiometric oxidant and an Au(I)/Au(III) catalytic cycle was proposed (Scheme 18). The product stereochemistry is consistent with *trans*-aminoauration followed by S_N2 displacement of [Au(III)] by the second amine moiety.

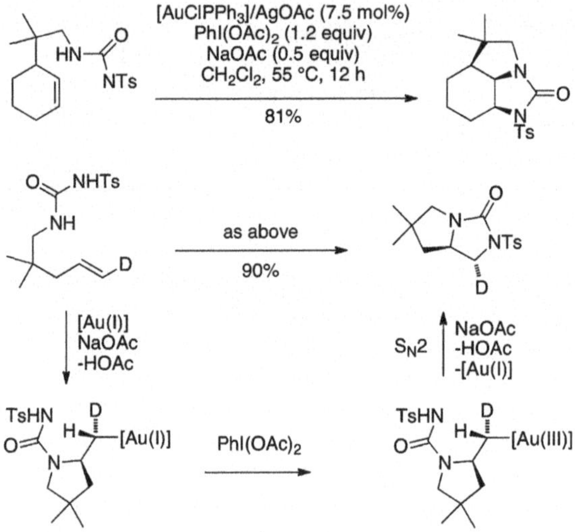

Scheme 18 Proposed Au-catalyzed diamination mechanism [40]

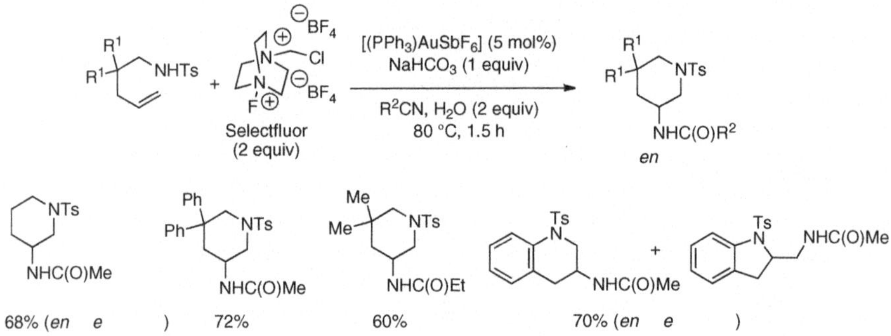

Scheme 19 *Endo*-selective Au-catalyzed diamination [41]

A complementary, largely *endo*-selective Au-catalyzed alkene diamination reaction was reported in 2011 (Scheme 19) [41]. In this reaction, Selectfluor (2 equiv.) was used as the stoichiometric oxidant to enable the Au(I)/Au(III) catalytic cycle. Nitriles served as the source of the second (external) amine nucleophile, providing amide-functionalized piperidine products.

Two possible mechanistic scenarios were proposed (Scheme 20). In the first, *endo*-selective *trans*-aminoauration followed by Au(I)/Au(III) oxidation, nitrile complexation, hydration and reductive elimination provide the piperidine product. In the second, *exo*-selective (or possible *endo*-selective) *trans*-aminoauration followed by Au(I)/Au(III) oxidation and intramolecular S_N2 displacement provide

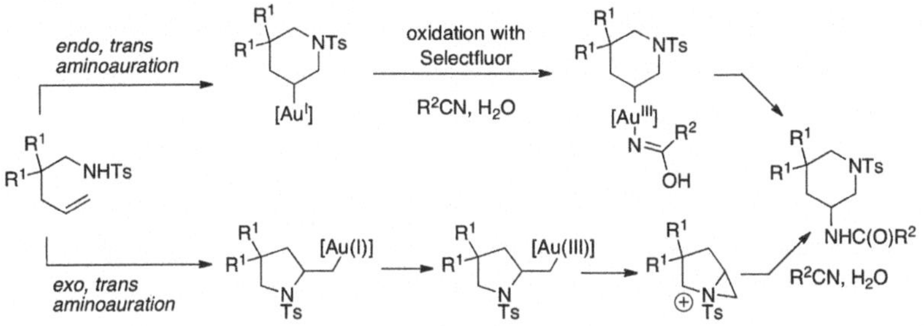

Scheme 20 Mechanistic alternatives for the *endo*-selective Au-catalyzed diamination [41]

an aziridinium ion intermediate that can undergo S_N2 attack by the nitrile via a Ritter-type mechanism [41].

2 Alkene Aminoalkoxylation

Metal-catalyzed alkene aminoalkoxylation reactions furnish various nitrogen and oxygen-containing heterocycles directly and oftentimes stereoselectively. The development of a number of aminoalkoxylation methods has been reviewed [42–45]. While numerous diastereoselective metal-catalyzed ring-forming alkene aminoalkoxylations have been reported (vide infra), catalytic enantioselective alkene aminoalkoxylations are more rare. The synthesis of chiral nitrogen heterocycles with good to excellent levels of enantiomeric excess from achiral alkene substrates is an active and growing topic of asymmetric catalysis (vide infra). It should be noted that a number of aminoalkoxylation reactions promoted by hypervalent iodine species and other non-metal containing compounds have also been reported recently but are outside the scope of this review [6, 46–52].

2.1 *Osmium-Catalyzed Alkene Aminoalkoxylation*

Perhaps the most developed diastereoselective ring-forming alkene aminoalkoxylation method is the osmium-catalyzed tethered aminohydroxylation [43, 53]. In these reactions, both high cyclic and acyclic diastereocontrol has been achieved (vide infra). The tethered alkene aminohydroxylation reaction was introduced in response to a perceived need to better control the regioselectivity of an intermolecular Os-catalyzed alkene aminohydroxylation process (Scheme 21) [44, 54].

The tethered aminohydroxylation reaction involves regiospecific, diastereoselective intramolecular addition of an amine to a pendant olefin and concomitant introduction of an alcohol from an exogenous hydroxyl source, e.g. H_2O. It was initially introduced in 2001 and involved the cyclization/aminohydroxylation of

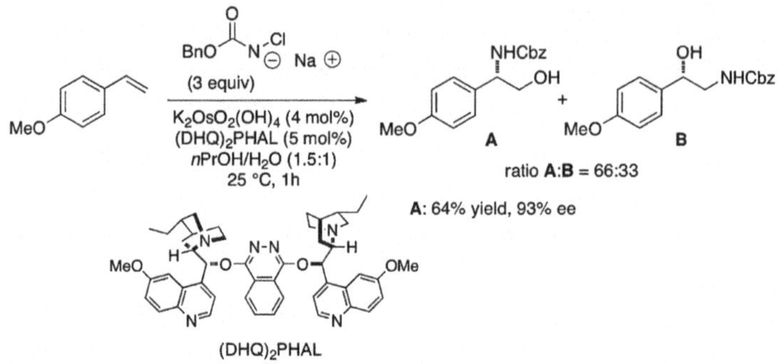

Scheme 21 Intermolecular catalytic enantioselective aminohydroxylation [54]

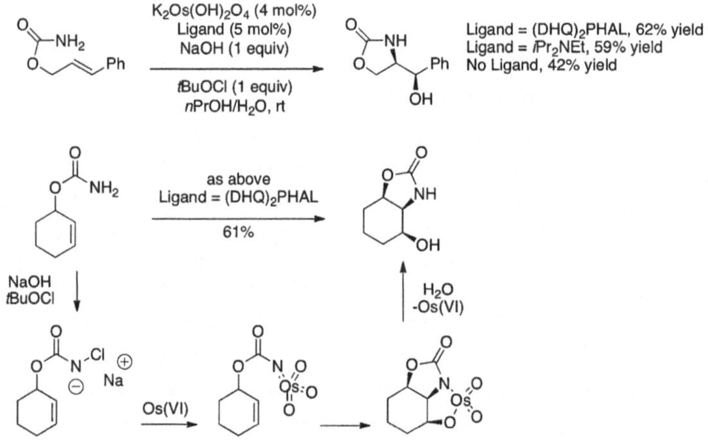

Scheme 22 Tethered aminohydroxylation range and mechanism [55, 56, 58]

allylic carbamates (Scheme 22) [55, 56]. The use of a chiral $(DHQ)_2PHAL$ ligand improved the reaction yield with some substrates but, unlike in the intermolecular aminohydroxylation shown in Scheme 21, did not render the reactions enantioselective. The reoxidant for this reaction is the *N*-chlorocarbamate salt, formed in situ from reaction of *t*-BuOCl and the primary carbamate in the presence of NaOH. The mechanism is thought to involve formation of Os(VIII) from Os(VI) and intramolecular [3+2] cycloaddition followed by osmate ester hydrolysis (Scheme 22). Under these reaction conditions, chlorination of the alkene could become a competing process and the lifetime of the chlorocarbamate could be short, requiring excess of the carbamate and reagents to be used at times [57].

To address the drawbacks presented by the chlorocarbamate intermediate, improvements were made to the reaction. The first improvement involved the use of hydroxycarbamate derivatives that eliminated the need for additional oxidant (Scheme 23) [57, 59]. External ligand was also no longer required in these

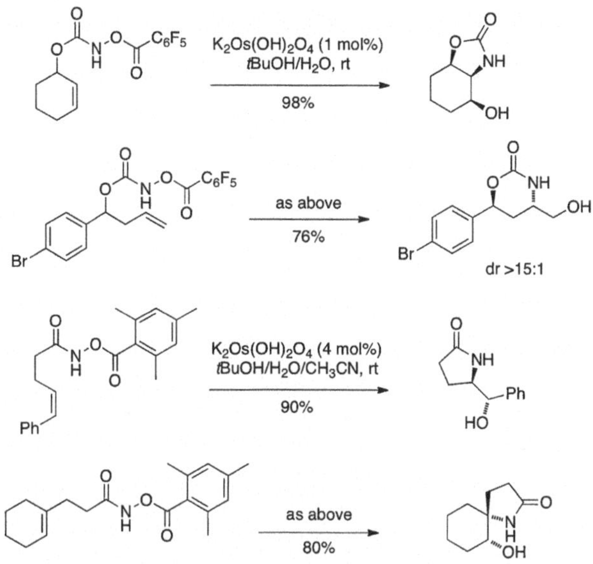

Scheme 23 Pre-oxidized substrates in the tethered aminohydroxylation [57, 59, 60]

reactions, and different hydroxylamine groups were optimal for different kinds of substrates, e.g. carbamates vs amides (Scheme 23).

Subsequently, an experiment using stoichiometric Os(VI) in a tethered alkene aminohydroxylation revealed that the Os(VI) oxidation state is capable of promoting the reaction (Scheme 24) [61]. This enabled the development of a second process improvement involving the use of vicinal hydroxyl-functionalized secondary carbamate and sulfonamide substrates that could employ dual chelation to Os(VI) and the use of more mild oxidants, trimethylamine *N*-oxide (TMO), pyridine *N*-oxide (PNO), and *p*-nitropyridine *N*-oxide (NPNO), used in the presence of Bronsted and Lewis acid catalysts, to enable the reoxidation of Os(IV) to Os(VI) [61–63].

2.2 *Palladium-Catalyzed Alkene Aminoalkoxylation*

A number of Pd-catalyzed alkene aminoalkoxylation reactions that result in the formation of nitrogen and oxygen heterocycles, e.g., pyrrolidines and tetrahydrofurans, were reported from 2005 to 2010 [64–67]. Higher oxidation state organopalladium intermediates were invoked in the majority of the proposed reaction mechanisms (vide infra). In these examples added oxidant was essential to enabling cycles catalytic in Pd(II), and ones that avoided potentially competing β-hydride elimination pathways.

Two novel methods for intramolecular palladium-catalyzed alkene aminoalkoxylation were independently reported in 2005 [64, 65]. The first method involved

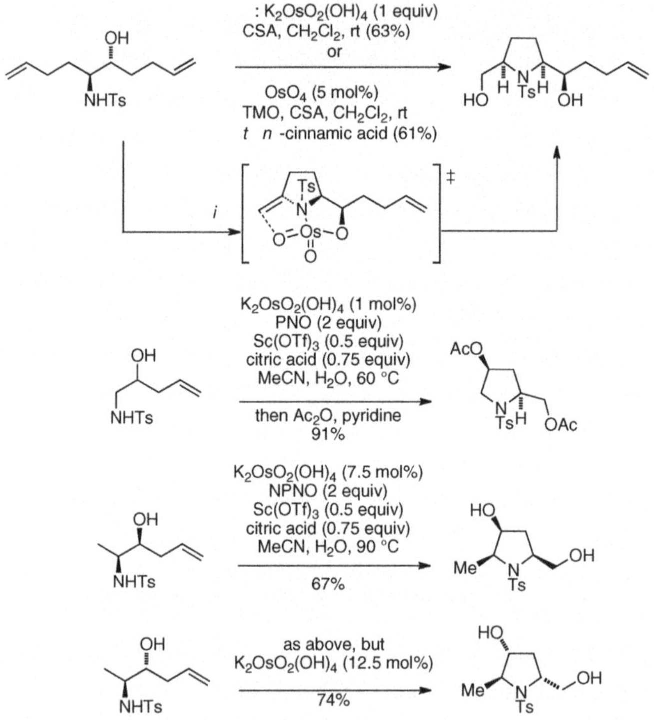

Scheme 24 Diastereoselective pyrrolidine synthesis [61–63]

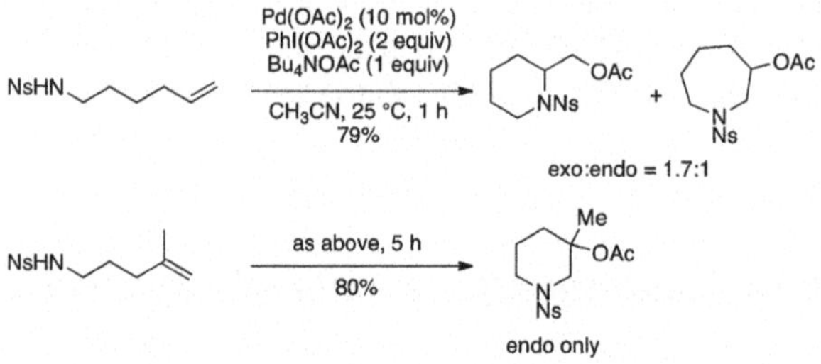

Scheme 25 Regioselectivity of Pd-catalyzed aminoacetoxylation [64]

the use of $PhI(OAc)_2$ (2 equiv.) as the stoichiometric oxidant and the reaction was performed at rt in CH_3CN using catalytic $Pd(OAc)_2$ [64]. Regioselectivity (*exo* vs *endo*) in the cyclization reaction was largely dependent upon the substrate's structure; *N*-tosylamides and *N*-tosylcarbamates cyclized onto pendant alkenes with high *exo*-selectivity, while an *N*-sulfonylalkyl-enes cyclized with poor regioselectivity in the case of monosubstituted alkenes and with *endo* selectivity in the case of 1,1-disubstituted alkenes (Scheme 25).

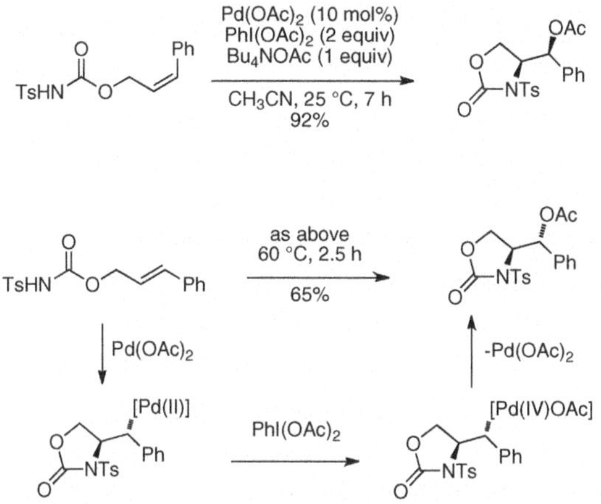

Scheme 26 Diastereoselectivity and mechanism [64]

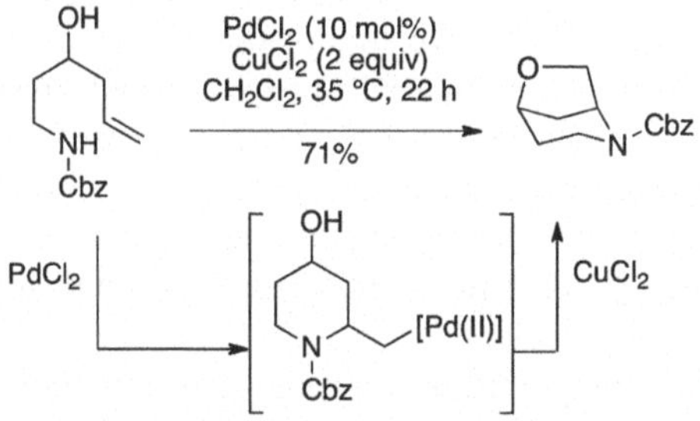

Scheme 27 Doubly intramolecular aminoalkoxylation [65]

E- and *Z*-phenyl-substituted internal alkenes underwent highly *exo*-selective Pd-catalyzed aminoacetoxylations; these reactions also occurred with high stereospecificity (Scheme 26). Based on the observed diastereoselectivity, a catalytic cycle involving *trans*-aminopalladation, oxidation of Pd(II) to Pd(IV) with $PhI(OAc)_2$ and subsequent reductive elimination was proposed (Scheme 26). In 2006, an entirely intermolecular alkene aminoacetoxylation was subsequently reported to occur under similar reaction conditions [68].

A second Pd-catalyzed aminoalkoxylation published in 2005 involved *exo* cyclization of a 3-hydroxy-5-hexenylcarbamate at 35°C in CH_2Cl_2 in the presence of catalytic $PdCl_2$ and $CuCl_2$ (2 equiv.) as the stoichiometric oxidant (Scheme 27) [65]. This aminoalkoxylation is doubly intramolecular since the substrate's hydroxyl group serves as the oxygen source for the terminal alkene carbon. The mechanism is thought to involve aminopalladation followed by $CuCl_2$-assisted oxidative C–O bond formation.

An intra/intermolecular Pd-catalyzed alkene aminoalkoxylation using alcohol solvents as the oxygen source and *N*-fluorobenzenesulfonamide (NFBS) as the oxidant

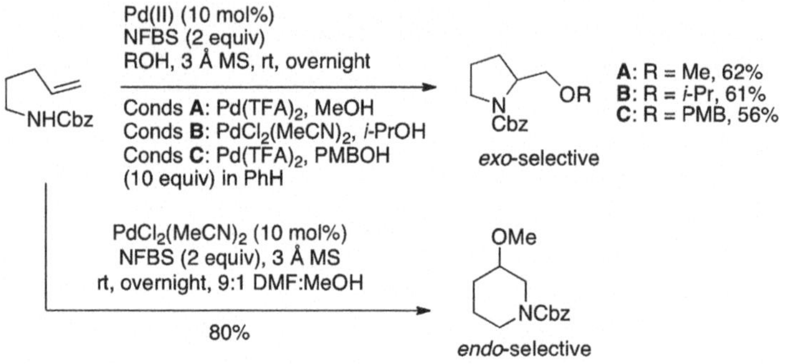

Scheme 28 Complementary regioselective aminoalkoxylations [66]

(2 equiv.) was reported in 2010 [66]. This reaction gave excellent *exo*-selectivity and moderate yields of ether-substituted pyrrolidines using $Pd(TFA)_2$ or $PdCl_2(MeCN)_2$ as catalyst, depending upon the nucleophilic alcohol (Scheme 28). Interestingly, the selectivity could be switched in favor of the *endo* regioisomer when the reaction was performed with $PdCl_2(MeCN)_2$ as catalyst in the polar solvent DMF (Scheme 28).

The authors proposed *exo*-selective aminopalladation, oxidation of the resulting organopalladium(II) intermediate to a Pd(IV) species and subsequent nucleophilic displacement with external alcohol to be the operating mechanism [66]. In the case of the *endo*-selective reactions, the authors speculated an intermediate aziridinium ion is formed by intramolecular displacement (neighboring group participation) of Pd(IV). Preferential attack at the more substituted carbon then provides the piperidine product [66].

A more unusual, tetrahydrofuran-forming inter/intramolecular Pd-catalyzed aminoalkoxylation of homoallylic alcohols was reported in 2007 [67]. In this reaction, $Pd(OAc)_2$ (10 mol%) served as catalyst, $PhI(OAc)_2$ (3 equiv.) was the oxidant and $AgBF_4$ (20 mol%) as additive improved the reaction efficiency. The reaction was generally diastereoselective, favoring formation of 3,4-anti-disubstituted tetrahydrofurans (Scheme 29). The reaction was more efficient with aryl rather than alkyl substituents at the substrate's allylic position.

The proposed reaction mechanism, based upon observed product stereochemistry, involves *cis*-aminopalladation to give a tethered organopalladium(II) intermediate, Pd(II) to Pd(IV) oxidation with $PhI(OAc)_2$, and subsequent reductive elimination to form the C–O bond (Scheme 30).

2.3 Gold-Catalyzed Aminoalkoxylation

The regioselective synthesis of piperidines from 4-pentenylsulfonamides via gold-catalyzed aminoalkoxylation in the presence of Selectfluor as stoichiometric oxidant was reported in 2011 [41]. The reaction occurred with high *endo* regioselectivity and was most efficient for terminal alkenes. Both alcohols and ethers were formed, depending on the reaction solvent (Scheme 31).

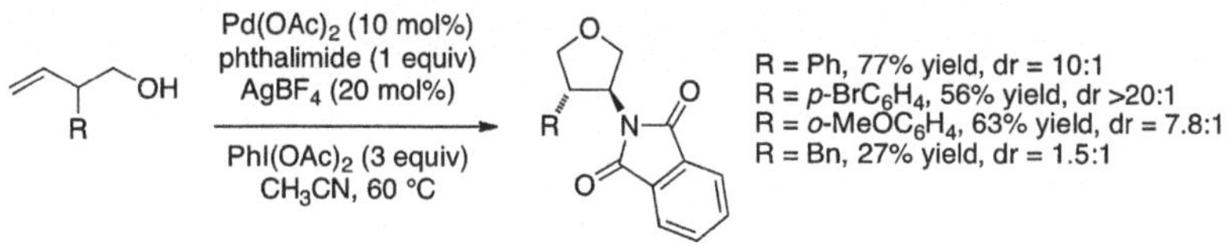

Scheme 29 Tetrahydrofuran-forming aminoalkoxylation [67]

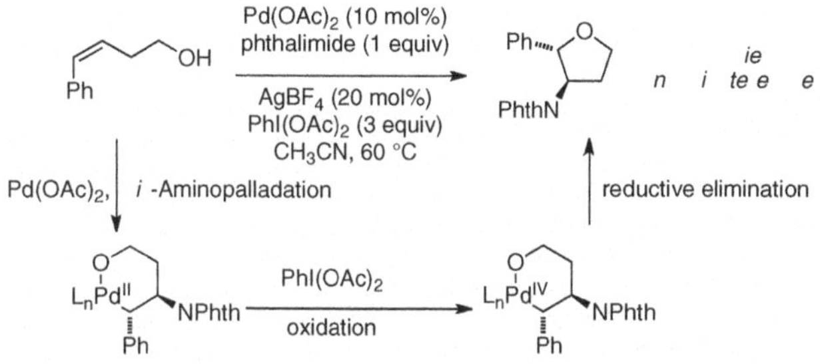

Scheme 30 Proposed mechanism of the Pd-catalyzed aminoalkoxylation [67]

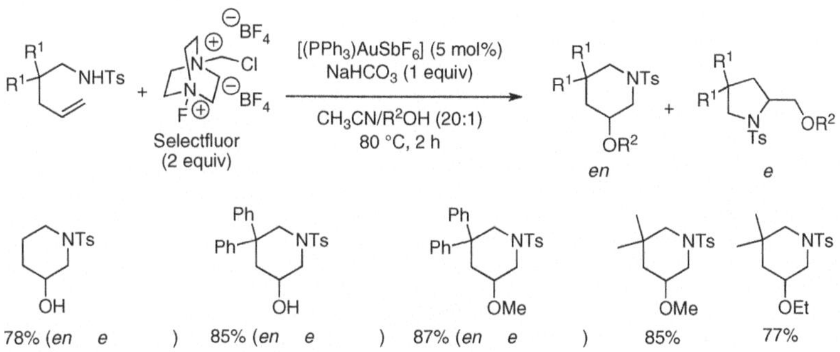

Scheme 31 Au-catalyzed aminoalkoxylation scope [41]

The mechanism, supported by isotopic labeling studies, is thought to involve *endo*-selective anti-aminoauration followed by oxidation of the organo-Au(I) intermediate to an organo-Au(III) intermediate and reductive elimination to secure the C–O bond (Scheme 32).

2.4 Copper-Catalyzed Ring-Forming Alkene Aminoalkoxylation

A number of copper-catalyzed alkene aminoalkoxylations have been reported. A range of reaction mechanisms and copper oxidation states have been invoked in these diverse transformations where the substrate structure, reagents, and copper catalysts largely dictate the reaction pathway.

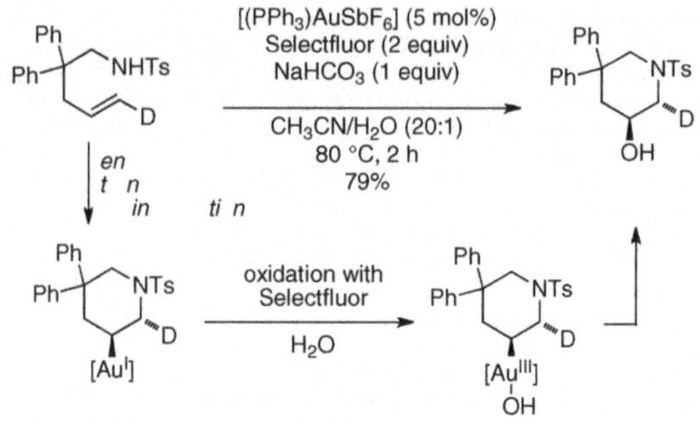

Scheme 32 Proposed mechanism involves an Au(I)/Au(III) cycle [41]

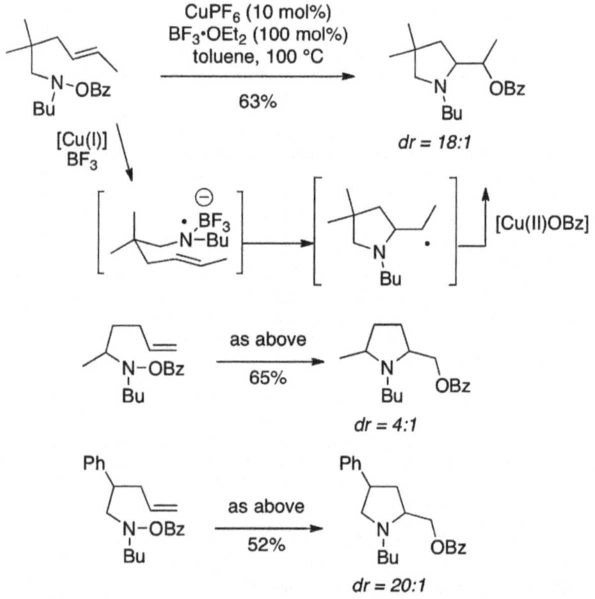

Scheme 33 Regioselective and diastereoselective alkene oxyamination [69]

The first copper-catalyzed alkene aminoalkoxylation for the synthesis of pyrrolidines from 4-pentenyl-*O*-benzoyl-hydroxylamines was reported in 2002 (Scheme 33) [69]. The reaction was largely regioselective (*endo* vs *exo* cyclization) and diastereoselective. A mechanism involving Cu(I)-catalyzed nitrogen radical formation, addition to the alkene and subsequent Cu(II)-assisted benzoylation of the resulting carbon radical was proposed.

The first catalytic enantioselective intramolecular alkene aminoalkoxylation was reported in 2008 [70]. Chiral indolines and pyrrolidines were synthesized from γ-alkenylsulfonamides using catalytic [Cu((4*R*,5*S*)-di-Ph-Box)](OTf)$_2$ in the presence of (2,2,6,6-tetramethylpiperidin-1-yl)oxyl (TEMPO) as alkoxyl source and O_2 (1 atm, balloon) as oxidant (Scheme 34) [70]. Removal of the *N*-sulfonyl group

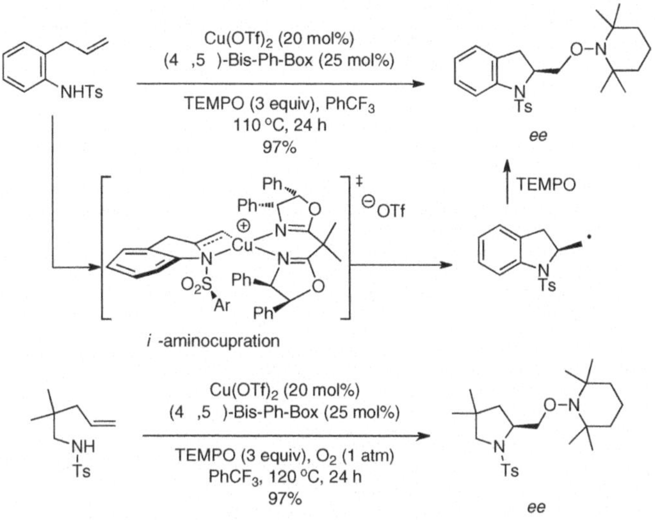

Scheme 34 Enantioselective copper(II)-catalyzed aminoalkoxylation [70]

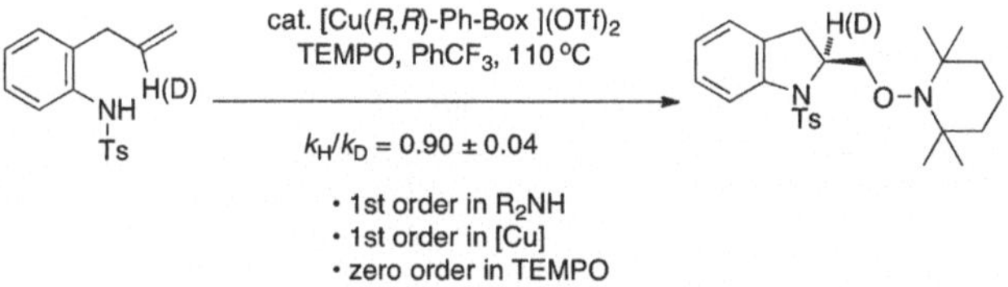

Scheme 35 Kinetic analysis of the enantioselective aminoalkoxylation [71]

and either N–O reduction to the corresponding alcohol or N–O oxidation to the corresponding aldehyde was demonstrated. A mechanism involving *cis*-aminocupration [Cu(II) oxidation state] across the alkene via a chair-like transition state, subsequent C–[Cu(II)] homolysis and direct quenching of the resulting carbon radical with TEMPO was proposed [70, 71].

The copper-catalyzed enantioselective aminooxygenation reaction mechanism was further probed using kinetics and isotope effect studies [71]. The reaction was found to be first order in [Cu], first order in amine substrate and zero order in TEMPO. These data, along with an inverse secondary kinetic isotope effect (see Scheme 35), supported the alkene addition as the rate-determining step of the reaction.

The enantioselective aminoalkoxylation was subsequently optimized for catalyst loading, time and enantioselectivity, and was demonstrated on a multigram scale (Scheme 36) [71, 72].

The copper(II) 2-ethylhexanoate-catalyzed and promoted diastereoselective synthesis of disubstituted pyrrolidines [73], cyclic ureas [74], and isoxazolidines

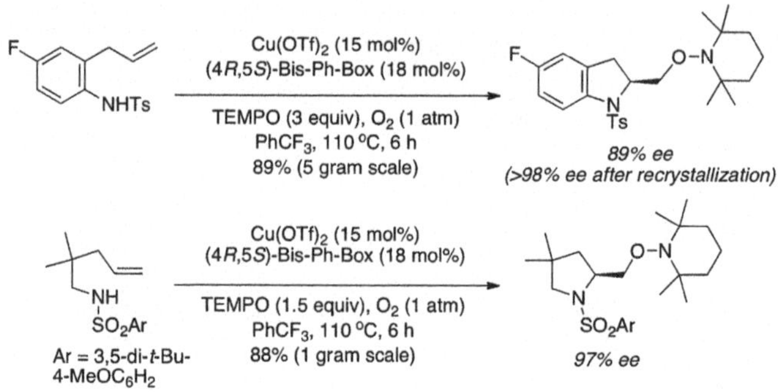

Scheme 36 Multigram scale optimized aminooxygenation reactions [71, 72]

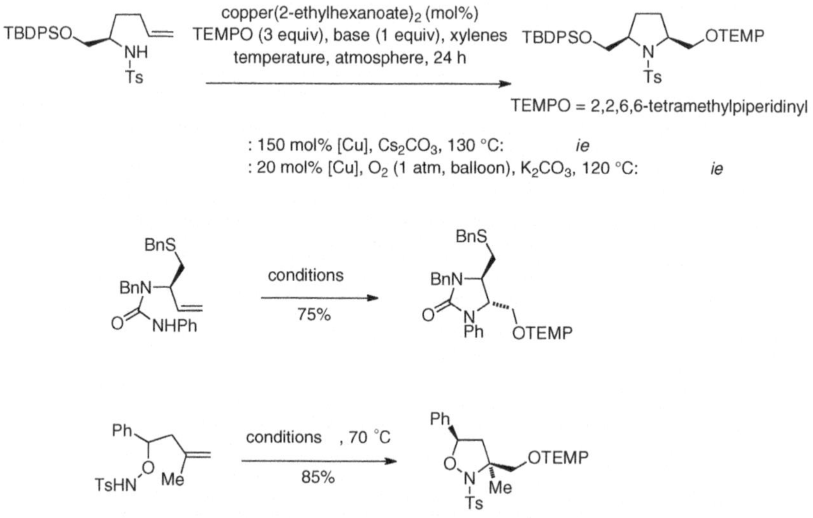

Scheme 37 Diastereoselective copper(II)-catalyzed aminoalkoxylations [73–75]

[75] via alkene aminoalkoxylation has also been reported. The copper(II) acetate-promoted aminoxygenation of alkenylimines and amidines has also been described [76]. Representative examples are shown in Scheme 37.

An in-depth mechanistic analysis of the indoline-forming copper(II) 2-ethylhexanoate-promoted aminooxygenation revealed the reaction is 1/2 order in [Cu], 1/2 order in sulfonamide substrate, and zero order in TEMPO (Scheme 38) [77]. The kinetics are consistent with involvement of a pre-equilibrium step wherein the copper(II) carboxylate dimer is converted to a monomeric species upon complexation with the sulfonamide. An inverse secondary KIE was observed in the alkene addition step, supporting its role as the rate-determining step of the reaction. The existence and viability of an R_2N–[Cu(II)] intermediate along the reaction pathway was supported by reaction kinetics and EPR spectroscopy.

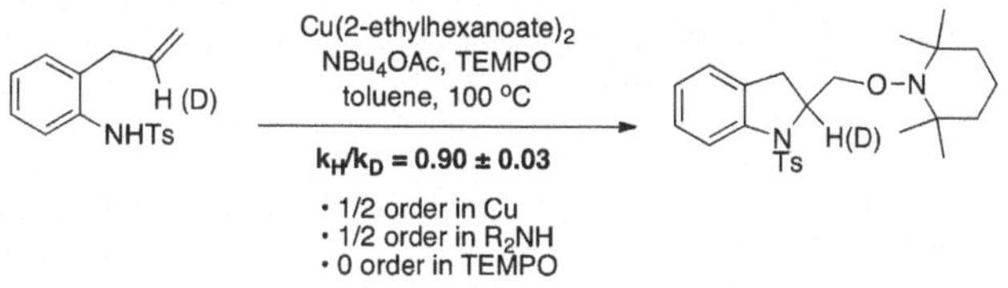

Scheme 38 Kinetic analysis of the copper(II)-promoted aminoalkoxylation [77]

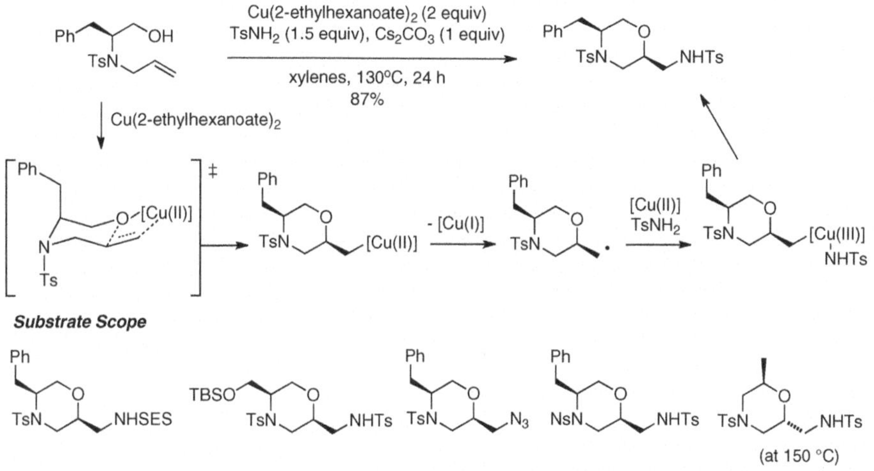

Scheme 39 Morpholine-forming aminoalkoxylation mechanism and scope [78]

The diastereoselective synthesis of morpholines via a copper(2-ethylhexanoate)$_2$-promoted alkene oxyamination was reported in 2012 (Scheme 39) [78]. The oxyamination is a less common transformation; it is thought to initiate with *cis*-oxycupration across the alkene in analogy with a recently reported alkene carboetherication reaction [78, 79]. Homolysis of the carbon–copper(II) bond followed by recombination of the carbon radical with copper(II) in the presence of a primary sulfonamide provides a transient organocopper(III) intermediate that, upon reductive elimination, provides the C–N bond. The reaction was general for a number of alkenol substrates and external amine sources [$TsNH_2$, $MsNH_2$, 2-trimethylsilylethyl-sulfonamide ($SESNH_2$), benzamide and NaN_3].

Copper-catalyzed intramolecular alkene aminoalkoxylation reactions can also be conducted if $PhI(OAc)_2$ is used as the stoichiometric oxidant [80, 81]. These reactions tend to occur at room temperature and oxidation of copper(I)/copper(II) to copper(III) is thought to occur prior to alkene aminocupration. In 2010, 4-pentenyl-sulfonamides were shown to undergo both *endo* and *exo* cyclization pathways where terminal alkenes favored the former and internal alkenes favored the latter pathway (Scheme 40) [80]. Carbocation formation and hydride shift appeared to have occurred in one instance (Scheme 40, Eq. d).

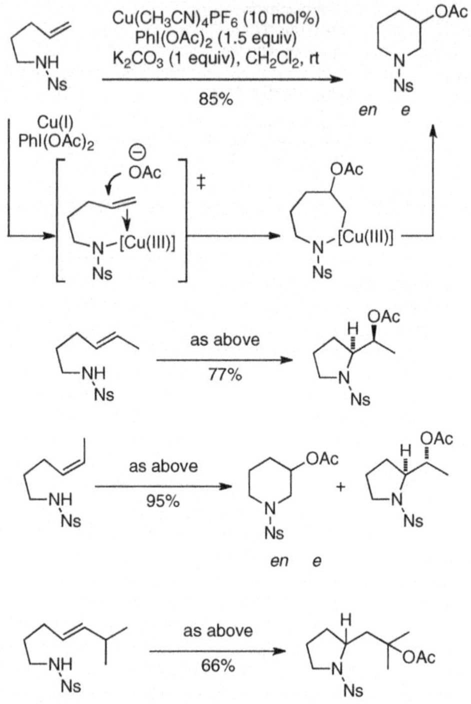

Scheme 40 Aminooxygenation via Cu(III) [80]

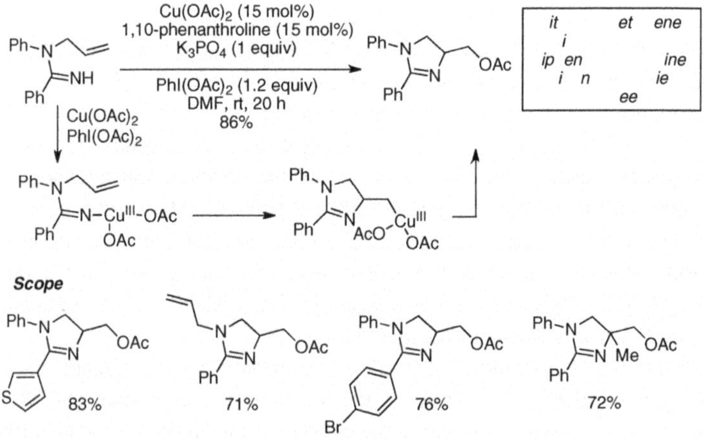

Scheme 41 Aminoacetoxylation of *N*-allylamidines [81]

N-Allylamidines also undergo intramolecular copper-catalyzed aminoacetoxylation in the presence of $PhI(OAc)_2$ as oxidant [81]. Terminal and 1,1-disubstituted alkenes underwent the aminoacetoxylation reaction efficiently (Scheme 41).

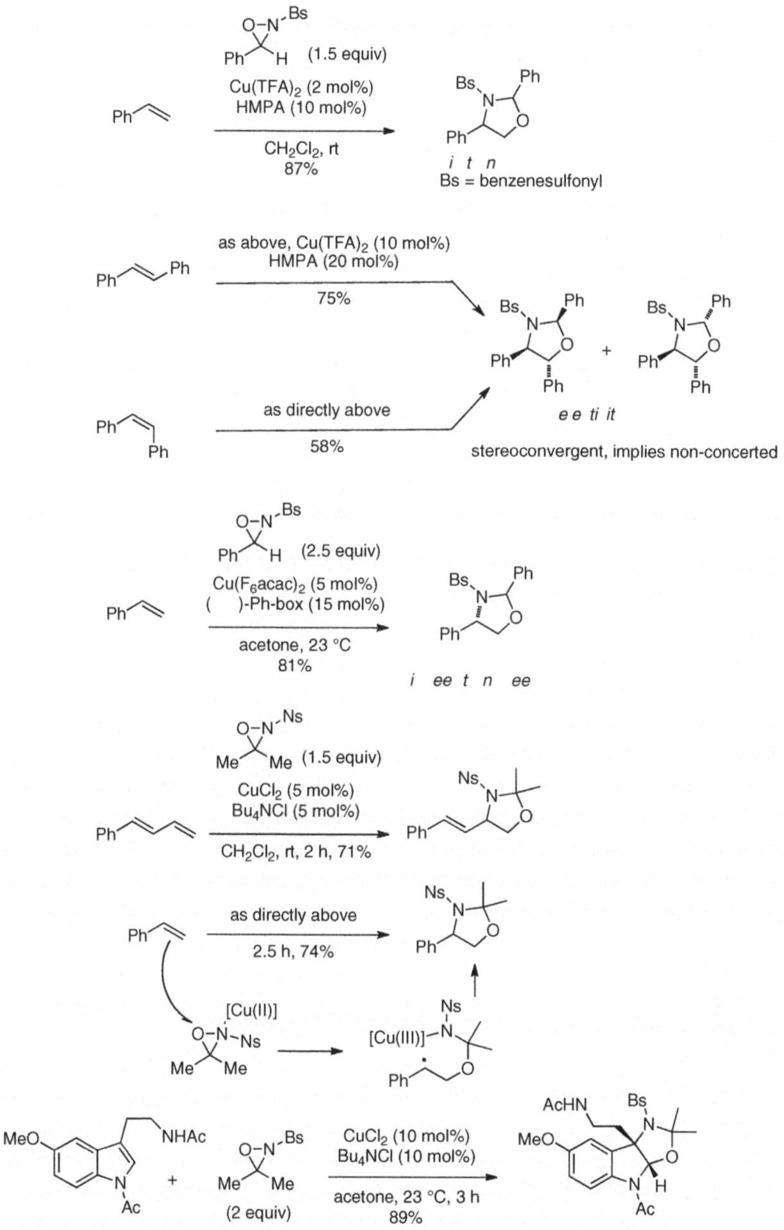

Scheme 42 Copper(II)-catalyzed aminoalkoxylations using oxaziridines [82–84, 86]

Conversion of the cyclic adducts to acyclic diaminoalcohols was demonstrated. The proposed mechanism (Scheme 41) involves copper-amine complexation and copper oxidation with PhI(OAc)$_2$ to give an R$_2$N–copper(III) intermediate.

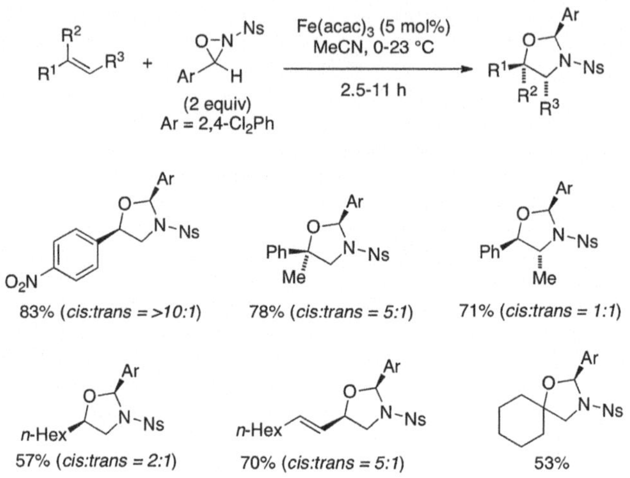

Scheme 43 Iron-catalyzed alkene aminoalkoxylation [87]

Aminocupration and subsequent reductive elimination then secures the C–O bond. When a chiral bis(oxazoline) ligand was used, a promising 48% ee was obtained. The ligand-based asymmetric induction is a strong indication that an aminocupration step is involved in the reaction mechanism.

The intermolecular copper(II)-catalyzed aminoalkoxylation of alkenes via activation of *N*-sulfonyl oxaziridines was first reported in 2006 (Scheme 42) [82]. Styrenes had the highest reactivity and a range of cyclic aminals were synthesized in good yields [82–85]. Hydrolysis of the aminal and removal of the sulfonyl group could be achieved to reveal the respective aminoalcohols with amine-bearing stereocenters. The catalytic enantioselective aminoalkoxylation of styrenes was achieved with moderate enantioselectivities [86]. A non-concerted mechanism involving generation of a benzylic radical was proposed based on stereochemical trends and radical trapping experiments [83].

2.5 *Iron-Catalyzed Alkene Aminoalkoxylation*

An Fe-catalyzed oxyamination of alkenes with *N*-sulfonyloxaziridine was reported in 2010 [87]. This reaction gives complementary regioselectivity to the analogous copper-catalyzed aminooxygenation reaction summarized above in that the reaction generates an oxygen-bearing stereocenter. Both terminal and internal styrenes were reactive and dienes and alkyl-substituted terminal alkenes also underwent oxyamination (Scheme 43). As in the analogous copper-catalyzed reaction (vide supra), mixtures of diastereomers epimeric at the aminal carbon were obtained.

The Fe-catalyzed reaction was rendered enantioselective in 2012 (Scheme 44) [88]. For the enantioselective reaction, styrenes and 1,1-disubstituted styrenes and 1-substituted dienes were the best substrates while internal alkenes proved unreactive. These reactions occurred with excellent enantioselectivity and significant preference

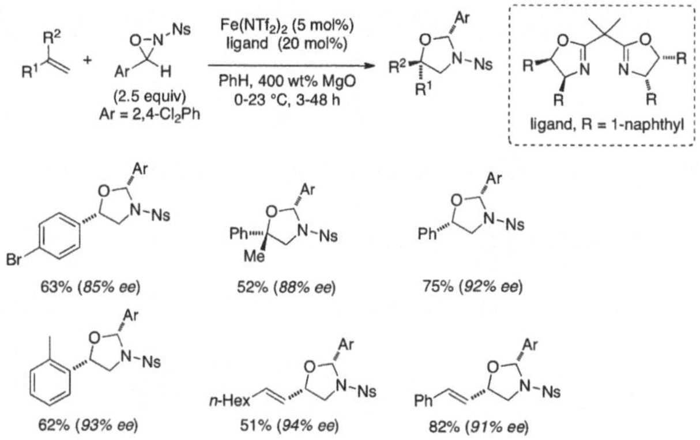

Scheme 44 Enantioselective iron-catalyzed aminoalkoxylation [88]

for the *cis* aminal diastereomer was observed. A mechanism for the oxyamination reaction has not yet been proposed.

3 Alkene Dialkoxylation

Metal-catalyzed alkene dialkoxylation has been used in the synthesis of tetrahydrofurans, lactones, tetrahydropyrans, dioxanes, and morpholines. Some of these methods have been reviewed previously [89, 90].

3.1 Palladium and Copper-Catalyzed Alkene Dialkoxylation Reactions

The palladium(II)-catalyzed intramolecular 1,4-alkoxyacetoxylation of dienes for the synthesis of spirocyclic tetrahydrofurans and tetrahydropyrans was first reported in 1991 (Scheme 45) [91, 92]. The mechanism is thought to involve *trans*-oxypalladation to form a π-allyl intermediate. In the absence of excess nucleophiles, the intermediate palladium(II) acetoxy complex is thought to undergo a *cis* migration (reductive elimination) to yield the major 1,4-*trans* diastereomer.

The analogous synthesis of fused-ring tetrahydropyrans was reported in 2004 [93]. Both the 1,4-*trans* and 1,4-*cis* diastereomers can be obtained selectively in several cases (Scheme 46). While the 1,4-*trans* diastereomer forms in the absence of external nucleophile, in the presence of LiOAc and catalytic LiCl, the 1,4-*cis* diastereomer is favored, presumably due to S_N2-type attack of the π-allylpalladium intermediate with acetate ion.

The synthesis of chiral tetrahydrofurans and a tetrahydropyran via an enantioselective palladium(II)-catalyzed intramolecular alkene dioxygenation

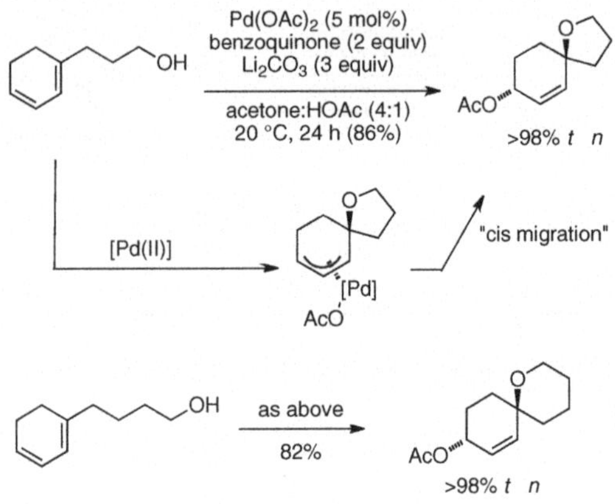

Scheme 45 Palladium-catalyzed 1,4-alkoxyacetoxylation of dienes [91, 92]

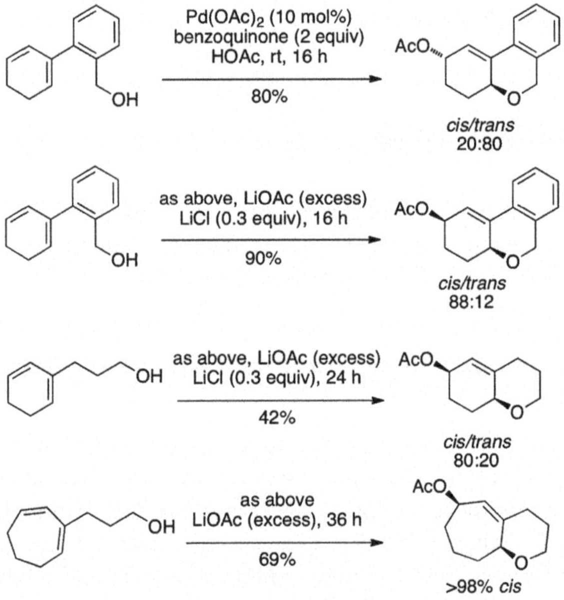

Scheme 46 *Cis*- and *trans*-selective diene 1,4-alkoxyacetoxylation [93]

using molecular O_2 as the terminal oxidant was reported in 2009 (Scheme 47) [94]. The reaction mechanism is thought to involve in situ formation of an *ortho*-quinone methide followed by subsequent addition of an exogenous nucleophile, e.g., MeOH [95]. In this reaction, CuCl was included as a rate-accelerating additive but is not thought to be the primary catalyst.

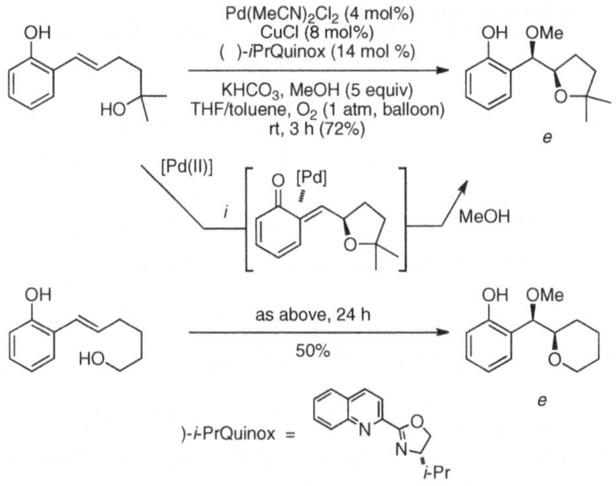

Scheme 47 Enantioselective Pd-catalyzed dialkoxylation [94]

Table 1 Intramolecular *endo*-selective alkene dialkoxylations [96–99]

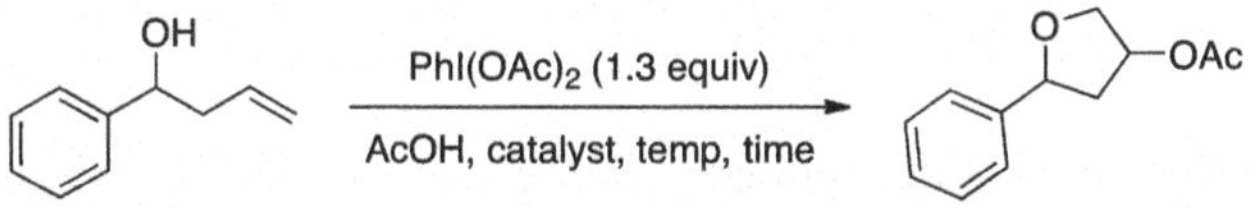

Entry	Catalyst	Conditions	Yield (%)	dr
1	[Pd(dppp)(H_2O)$_2$](OTf)$_2$ (2 mol%)	H_2O, rt, 72 h	78	1.1:1
2	[Bis(NHC)Pd(H_2O)$_2$](OTf)$_2$ (4 mol%)	H_2O, rt, 60 h	60	1.5:1
3	Cu(OTf)$_2$ (10 mol%)	80°C, 16 h	80	1.3:1
4	HOTf (5 mol%)	50°C, 72 h	72	1.2:1

In 2008 and 2010, two independent research groups reported Pd(II)-catalyzed intramolecular alkene dialkoxylation reactions for the synthesis of tetrahydrofurans and lactones using PhI(OAc)$_2$ as the terminal oxidant in the presence of HOAc (Table 1, entries 1 and 2) [96, 97]. Both groups hypothesized that the Pd(II)/Pd(IV) catalytic cycle was involved in the reaction mechanism. An analogous, copper(II)-catalyzed intramolecular alkene dialkoxylation using PhI(OAc)$_2$ was also reported in 2010, where a Cu(III) intermediate was invoked in the catalytic mechanism (Table 1, entry 3) [98]. A subsequent report in 2011 indicated that PhI(OAc)$_2$ under acidic conditions can provide similar product distributions (Table 1, entry 4), thereby calling into question the role of the Pd(II) and Cu(II) species in reactions that employ PhI(OAc)$_2$ under acidic conditions [99].

A copper(II) 2-ethylhexanoate-promoted alkene dialkoxylation was subsequently reported to occur under basic conditions (Scheme 48) [78]. The role of the copper species in promoting this reaction is less ambiguous given the absence of

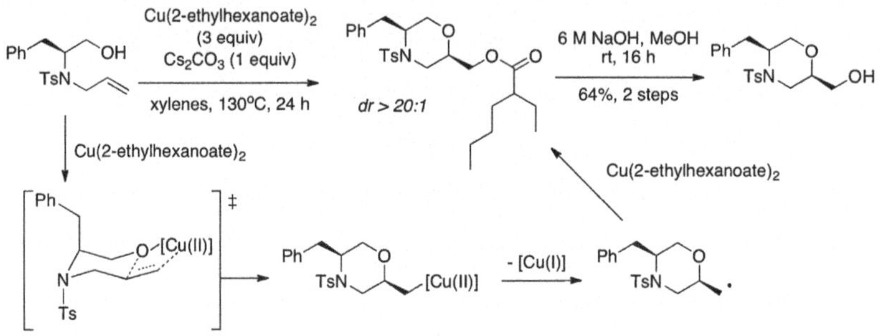

Scheme 48 Copper(II)-promoted alkene dialkoxylation [78]

additional reactive species [78]. The reaction was highly diastereoselective, and alkene addition was proposed to occur via a *cis*-oxycupration mechanism in analogy with an alkene carboetherification reaction proposed to occur via a similar reaction mechanism under related reaction conditions [79].

3.2 Cobalt-Catalyzed Alkene Dialkoxylation Reactions

2,5-*Trans*-disubstituted tetrahydrofurans can be synthesized from 4-pentenols using catalytic amounts of Co(II) complexes in the presence of *t*-butyl peroxide in the presence of O_2 (1 atm) [100, 101]. The method has been optimized for ease of catalyst/ligand removal by use of the water soluble *N*-methylpiperazine (3,5,5-dimethyl-1-(4-methylpiperazine-1-yl)hexane-1,2,4-trione (nmp) ligand (Scheme 49) [101]. The reaction mechanism is thought to involve a carbon radical intermediate which adds to O_2 to form the final C–O bond [102].

3.3 Osmium-Catalyzed Alkene and Diene Dialkoxylations

Variants of osmium-catalyzed alkene dihydroxylation [103] have been reported for the synthesis of lactones [104–106]. Examples of lactone synthesis proceeding via enantioselective alkene dihydroxylation and in situ lactonization are shown below (Scheme 50) [105, 106].

Osmium has also been used to catalyze the stereoselective synthesis of 2,5-*cis*-tetrahydrofurans from 1,5-dienes (Scheme 51) [107]. Initial intermolecular alkene dihydroxylation then facilitates a tethered, intramolecular dihydroxylation process [63].

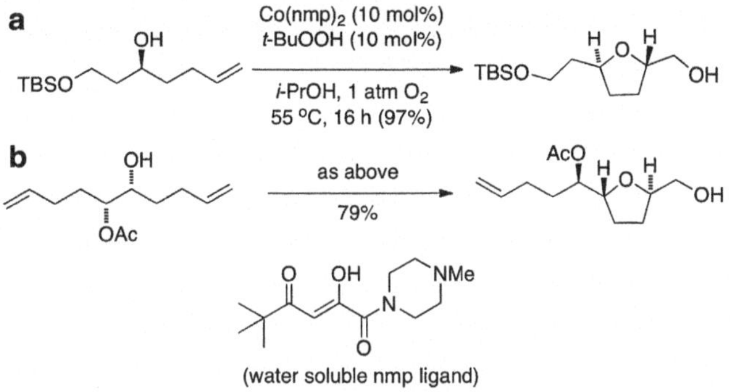

Scheme 49 Mukaiyama aerobic oxidative cyclization [101]

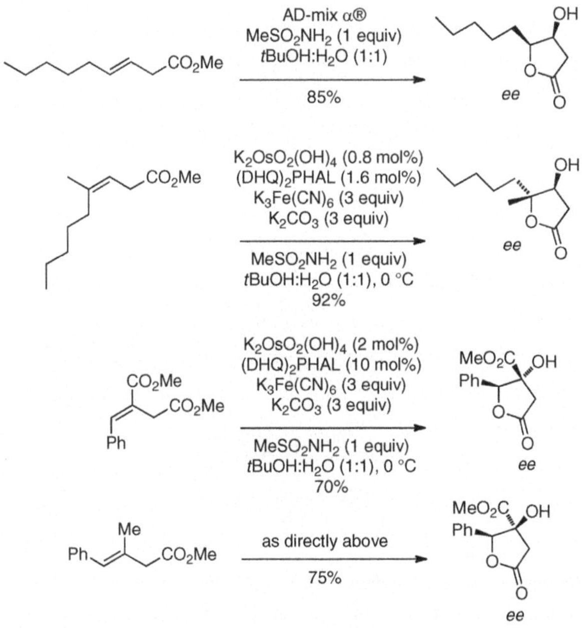

Scheme 50 Lactones via enantioselective Os-catalyzed dihydroxylation [105, 106]

3.4 *Ruthenium-Catalyzed Diene Dialkoxylations*

RuO_4 (derived from $RuCl_3$) has been used to promote and catalyze the oxidation of 1,5-, 1,6-, and 1,7-dienes to the corresponding tetrahydrofurans, tetrahydropyrans, and oxepanes [108]. Some examples are shown below (Scheme 52).

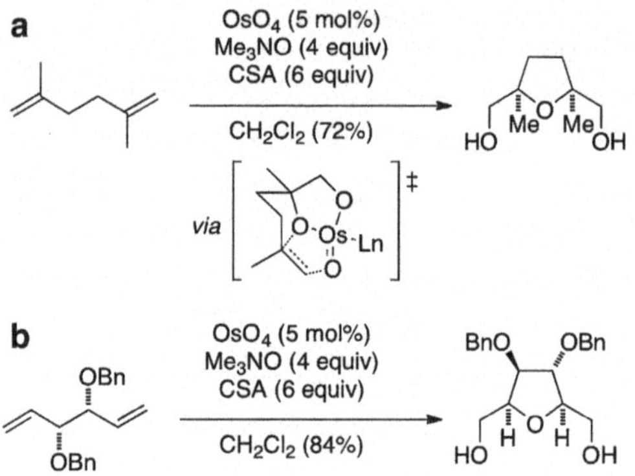

Scheme 51 Osmium-catalyzed dialkoxylation of alkenes [107]

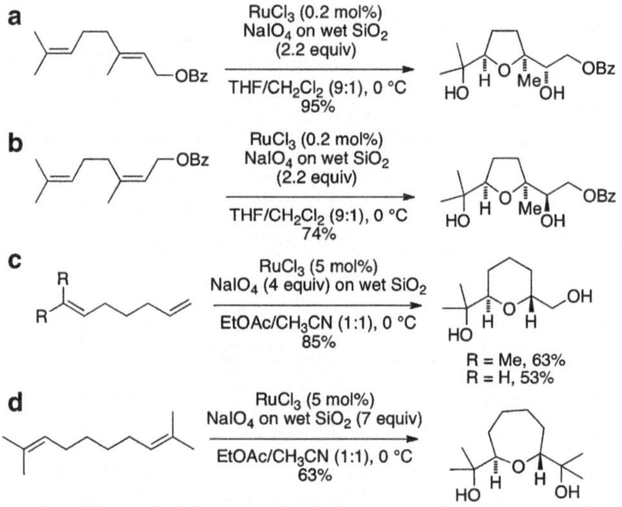

Scheme 52 Ruthenium-catalyzed oxidation of dienes [108]

References

1. Cardona F, Goti A (2009) Metal-catalyzed 1,2-diamination reactions. Nat Chem 1:269–275
2. De JS, Nosal DG, Wardrop DJ (2012) Methods for direct alkene diamination, new & old. Tetrahedron 68:4067–4105
3. Jacques B, Muñiz K (2011) In: Yudin (ed) Catalyzed carbon-heteroatom bond formation. Wiley-VCH Verlag GmbH & Co KGaA, Weinheim, pp 119–135
4. Muniz K, Hovelmann CH, Streuff J, Campos-Gomez E (2008) First palladium- and nickel-catalyzed oxidative diamination of alkenes: cyclic urea, sulfamide, and guanidine building blocks. Pure Appl Chem 80:1089–1096

5. Booker-Milburn KI, Guly DJ, Cox B, Procopiou PA (2003) Ritter-type reactions of N-chlorosaccharin: a method for the electrophilic diamination of alkenes. Org Lett 5:3313–3315
6. Li H, Widenhoefer RA (2010) Intramolecular diamination and alkoxyamination of alkenes with N-sulfonyl ureas employing N-iodosuccinimide. Tetrahedron 66:4827–4831
7. Kong A, Blakey SB (2012) Intramolecular olefin diamination for the stereoselective synthesis of 3-aminopiperidines. Synthesis 44:1190–1198
8. Kim JW, Cho SH, Chang S (2012) Intramolecular oxidative diamination and aminohydroxylation of olefins under metal-free conditions. Org Lett 14:1424–1427
9. Muller CH, Frohlich R, Daniliuc CG, Hennecke U (2012) 2,2′-Bipyrrolidines by an anti-selective alkene diamination. Org Lett 14:5944–5947
10. Chavez P, Kirsch J, Hovelmann CH, Streuff J, Martinez-Belmonte M, Escudero-Adan EC, Martin E, Muniz K (2012) Metal-free diamination of alkenes employing bromide catalysis. Chem Sci 3:2375–2382
11. Schroif-Gregoire C, Travert N, Zaparucha A, Al-Mourabit A (2006) Direct access to marine pyrrole-2-aminoimidazoles, oroidin, and derivatives, via new acyl-1,2-dihydropyridin intermediates. Org Lett 8:2961–2964
12. Hewlett NM, Tepe JJ (2011) Total synthesis of the natural product (±)-dibromophakellin and analogues. Org Lett 13:4550–4553
13. Zöllinger M, Mayer P, Lindel T (2007) Enantioselective total synthesis of (−)-dibromophakellstatin. Synlett 2007:2756–2758
14. Bar GLJ, Lloyd-Jones GC, Booker-Milburn KI (2005) Pd(II)-catalyzed intermolecular 1,2-diamination of conjugated dienes. J Am Chem Soc 127:7308–7309
15. Streuff J, Hövelmann CH, Nieger M, Muñiz K (2005) Palladium(II)-catalyzed intramolecular diamination of unfunctionalized alkenes. J Am Chem Soc 127:14586–14587
16. Muniz K, Hovelmann CH, Streuff J (2008) Oxidative diamination of alkenes with ureas as nitrogen sources: mechanistic pathways in the presence of a high oxidation state palladium catalyst. J Am Chem Soc 130:763–773
17. Yu H, Fu Y, Guo Q, Lin Z (2009) Theoretical investigations on mechanisms of Pd(OAc)2-catalyzed intramolecular diaminations in the presence of bases and oxidants. Organometallics 28:4507–4512
18. Muñiz K (2007) Advancing palladium-catalyzed C–N bond formation: bisindoline construction from successive amide transfer to internal alkenes. J Am Chem Soc 129:14542–14543
19. Sibbald PA, Michael FE (2009) Palladium-catalyzed diamination of unactivated alkenes using N-fluorobenzenesulfonimide as source of electrophilic nitrogen. Org Lett 11:1147–1149
20. Sibbald PA, Rosewall CF, Swartz RD, Michael FE (2009) Mechanism of N-fluorobenzenesulfonimide promoted diamination and carboamination reactions: divergent reactivity of a Pd(IV) species. J Am Chem Soc 131:15945–15951
21. Muñiz K, Streuff J, Chávez P, Hövelmann CH (2008) Synthesis of diamino carboxylic esters by palladium-catalyzed oxidative intramolecular diamination of acrylates. Chemistry 3:1248–1255
22. Chavez P, Kirsch J, Streuff J, Muniz K (2012) Palladium-catalyzed intramolecular diamination of acrylic esters using sulfamates as nitrogen source. J Org Chem 77:1922–1930
23. Muñiz K, Hövelmann CH, Campos-Gómez E, Barluenga J, González JM, Streuff J, Nieger M (2008) Intramolecular diamination of alkenes with palladium(II)/copper(II) bromide and IPy_2BF_4: the role of halogenated intermediates. Chemistry 3:776–788
24. Du H, Yuan W, Zhao B, Shi Y (2007) Catalytic asymmetric diamination of conjugated dienes and triene. J Am Chem Soc 129:11688–11689
25. Zhao B, Du H, Sunliang C, Shi Y (2010) Synthetic and mechanistic studies on Pd(0)-catalyzed diamination of conjugated dienes. J Am Chem Soc 132:3523–3532
26. Du H, Zhao B, Shi Y (2008) Catalytic asymmetric allylic and homoallylic diamination of terminal olefins via formal C–H activation. Jof the Am Chem Soc 130:8590–8591

27. Zabawa TP, Kasi D, Chemler SR (2005) Copper(II) acetate promoted intramolecular diamination of unactivated olefins. J Am Chem Soc 127:11250–11251
28. Zabawa TP, Chemler SR (2007) Copper(II) carboxylate promoted intramolecular diamination of terminal alkenes: improved reaction conditions and expanded substrate scope. Org Lett 9:2035–2038
29. Sequeira FC, Turnpenny BW, Chemler SR (2010) Copper-promoted and copper-catalyzed intermolecular alkene diamination. Angew Chem Int Ed 49:6365–6368, S6365/6361-S6365/6380
30. Wang Y-F, Zhu X, Chiba S (2012) Copper-catalyzed aerobic [3+2]-annulation of N-alkenyl amidines. J Am Chem Soc 134:3679–3682
31. Yuan W, Du H, Zhao B, Shi Y (2007) A mild Cu(I)-catalyzed regioselective diamination of conjugated dienes. Org Lett 9:2589–2591
32. Zhao B, Yuan W, Du H, Shi Y (2007) Cu(I)-catalyzed intermolecular diamination of activated terminal olefins. Org Lett 9:4943–4945
33. Zhao B, Peng X, Zhu Y, Ramirez TA, Cornwall RG, Shi Y (2011) Cu(I)-catalyzed diamination of conjugated dienes. Complementary regioselectivity from two distinct mechanistic pathways involving Cu(II) and Cu(III) species. J Am Chem Soc 133:20890–20900
34. Zhao B, Du H, Shi Y (2008) Cu(I)-catalyzed cycloguanidination of olefins. Org Lett 10:1087–1090
35. Du H, Zhao B, Yuan W, Shi Y (2008) Cu(I)-catalyzed asymmetric diamination of conjugated dienes. Org Lett 10:4231–4234
36. Zhao B, Du H, Shi Y (2009) Cu(I)-catalyzed diamination of conjugated olefins with tunable anionic counterions. A possible approach to asymmetric diamination. J Org Chem 74:8392–8395
37. Wen Y, Zhao B, Shi Y (2009) Cu(I)-catalyzed diamination of disubstituted terminal olefins: an approach to potent NK1 antagonist. Org Lett 11:2365–2368
38. Zhao B, Peng X, Cui S, Shi Y (2010) Cu(I)-catalyzed regioselective diamination of conjugated dienes via dual mechanistic pathways. J Amn Chem Soc 132:11009–11011
39. Muniz K, Streuff J, Hovelmann CH, Nunez A (2007) Exploring the nickel-catalyzed oxidation of alkenes: a diamination by sulfamide transfer. Angew Chem Int Ed 46:7125–7127
40. Iglesias A, Muñiz K (2009) Oxidative interception of the hydroamination pathway: a gold-catalyzed diamination of alkenes. Chemistry 15:10563–10569
41. de HT, Nevado C (2011) Flexible gold-catalyzed regioselective oxidative difunctionalization of unactivated alkenes. Angew Chem Int Ed 50:906–910
42. Donohoe TJ, Callens CKA, Flores A, Lacy AR, Rathi AH (2011) Recent developments in methodology for the direct oxyamination of olefins. Chemistry 17:58–76
43. Donohoe TJ, Callens CKA, Lacy AR, Winter C (2012) Tethered aminohydroxylation reaction and its application to total synthesis. Eur J Org Chem 655–663
44. Nilov D, Reiser O (2002) The sharpless asymmetric aminohydroxylation – scope and limitation. Adv Synth Catal 344:1169–1173
45. Bergmeier SC (2000) The synthesis of vicinal amino alcohols. Tetrahedron 56:2561–2576
46. Farid U, Wirth T (2012) Highly stereoselective metal-free oxyaminations using chiral hypervalent iodine reagents. Angew Chem Int Ed 51:3462–3465
47. Tellitu I, Urrejola A, Serna S, Moreno I, Herrero MT, Dominguez E, SanMartin R, Correa A (2007) On the phenyliodine(III)-bis(trifluoroacetate)-mediated olefin amidohydroxylation reaction. Eur J Org Chem 437–444
48. Lovick HM, Michael FE (2010) Metal-free highly regioselective aminotrifluoroacetoxylation of alkenes. J Am Chem Soc 132:1249–1251
49. Correa A, Tellitu I, Dominguez E, SanMartin R (2006) A metal-free approach to the synthesis of indoline derivatives by a phenyliodine(III) bis(trifluoroacetate)-mediated amidohydroxylation reactions. J Org Chem 71:8116–8319
50. Cochran BM, Michael FE (2008) Metal-free oxidative cyclization of urea-tethered alkenes with hypervalent iodine. Org Lett 10:5039–5042

51. Schmidt VA, Alexanian EJ (2011) Metal-free oxyamination of alkenes using hydroxyamic acids. J Am Chem Soc 133:11402–11405
52. Moriyama K, Izumisawa Y, Togo H (2012) Bronsted acid-assisted intramolecular aminohydroxylation of N-alkenylsulfonamides under heavy metal-free conditions. J Org Chem 77:9846–9851
53. Donohoe TJ, Johnson PD, Pye RJ (2003) The tethered aminohydroxylation (TA) reaction. Org Biomol Chem 1:2025–2028
54. Reddy KL, Sharpless KB (1998) From styrenes to enantiopure a-arylglycines in two steps. J Am Chem Soc 120:1207–1217
55. Donohoe TJ, Johnson PD, Helliwell M, Keenan M (2001) The regioselective aminohydroxylation of allylic carbamates. Chem Commun 2078–2079
56. Donohoe TJ, Johnson PD, Cowley A, Keenan M (2002) The tethered aminohydroxylation (TA) of cyclic allylic carbamates. J Am Chem Soc 124:12934–12935
57. Donohoe TJ, Chughtai MJ, Klauber DJ, Griffin D, Campbell AD (2006) N-sulfonyloxy carbamates as reoxidants for the tethered aminohydroxylation reaction. J Am Chem Soc 128:2514–2515
58. Donohoe TJ, Johnson PD, Pye RJ, Keenan M (2004) Efficient acyclic stereocontrol using the tethered aminohydroxylation reaction. Org Lett 6:2583–2585
59. Donohoe TJ, Bataille CJR, Gattrell W, Kloesges J, Rossignol E (2007) Tethered aminohydroxylation: dramatic improvements to the process. Org Lett 9:1725–1728
60. Donohoe TJ, Callens CKA, Thompson AL (2009) Tethered aminohydroxylation (TA) reaction of amides. Org Lett 11:2305–2307
61. Donohoe TJ, Churchill GH, Wheelhouse KMP, Glossop PA (2006) Stereoselective synthesis of pyrrolidines: catalytic oxidative cyclizations mediated by osmium. Angew Chem Int Ed 45:8025–8028
62. Donohoe TJ, Lindsay-Scott PJ, Parker JS, Callens CKA (2010) New modes for the osmium-catalyzed oxidative cyclization. Org Lett 12:1060–1063
63. Donohoe TJ, Winship PCM, Walter DS (2009) A Lewis acid promoted oxidative cyclization. J Org Chem 74:6394–6397
64. Alexanian EJ, Lee C, Sorensen EJ (2005) Palladium-catalyzed ring-forming aminoacetoxylation of alkenes. J Am Chem Soc 127:7690–7691
65. Szolcsanyi P, Gracza T (2005) Novel Pd(II)-catalyzed N,O-bicyclisation as an efficient route to the 6-oxa-2-azabicyclo[3.2.1]octane skeleton. Chem Commun 3948–3950
66. Liskin DV, Sibbald PA, Rosewall CF, Michael FE (2010) Palladium-catalyzed alkoxyamination of alkenes with use of N-fluoroenzenesulfonamide as oxidant. J Org Chem 75:6294–6296
67. Desai LV, Sanford MS (2007) Construction of tetrahydrofurans by PdII/PdIV-catalyzed aminooxygenation of alkenes. Angew Chem Int Ed 46:5737–5740
68. Liu G, Stahl SS (2006) Highly regioselective Pd-catalyzed intermolecular aminoacetoxylation of alkenes and evidence for cis-aminopalladation and S_N2 C–O bond formation. J Am Chem Soc 128:7179–7181
69. Noack M, Gottlich R (2002) Copper(I) catalysed cyclisation of unsaturated N-benzoyloxyamines: an aminohydroxylation via radicals. Chem Commun 536–537
70. Fuller PH, Kim J-W, Chemler SR (2008) Copper catalyzed enantioselective intramolecular aminooxygenation of alkenes. J Am Chem Soc 130:17638–17639
71. Paderes MC, Keister JB, Chemler SR (2013) Mechanistic analysis and optimization of the copper-catalyzed enantioselective intramolecular alkene aminooxygenation. J Org Chem 78:506–515
72. Sequeira FC, Bovino MT, Chipre AJ, Chemler SR (2012) Multigram synthesis of a chiral substituted indoline via copper-catalyzed alkene aminooxygenation. Synthesis 44:1481–1484
73. Paderes MC, Chemler SR (2009) Diastereoselective pyrrolidine synthesis via copper promoted intramolecular aminooxygenation of alkenes: formal synthesis of (+)-monomorine. Org Lett 11:1915–1918

74. Paderes MC, Chemler SR (2011) Stereoselective copper-catalyzed intramolecular alkene aminooxygenation: effects of substrate and ligand structure on selectivity. Eur J Org Chem 3679–3684
75. Karyakarte SD, Smith TP, Chemler SR (2012) Stereoselective isoxazolidine synthesis via copper-catalyzed alkene aminooxygenation. J Org Chem 77:7755–7760
76. Sanjaya S, Chua SH, Chiba S (2012) Cu(II)-mediated aminooxygenation of alkenylimines and alkenylamidines with TEMPO. Synlett 23:1657–1661
77. Paderes MC, Belding L, Fanovic B, Dudding T, Keister JB, Chemler SR (2012) Evidence for alkene cis-aminocupration, an aminooxygenation case study: kinetics, EPR spectroscopy, and DFT calculations. Chem Eur J 18:1711–1726
78. Sequeira FC, Chemler SR (2012) Stereoselective synthesis of morpholines via copper-promoted oxyamination of alkenes. Org Lett 14:4482–4485
79. Miller Y, Miao L, Hosseini AS, Chemler SR (2012) Copper-catalyzed intramolecular alkene carboetherificaiton: synthesis of fused-ring and bridged-ring tetrahydrofurans. J Am Chem Soc 134:12149–12156
80. Mancheno DE, Thornton AR, Stoll AH, Kong A, Blakey SB (2010) Copper-catalyzed olefin aminoacetoxylation. Org Lett 12:4110–4113
81. Sanjaya S, Chiba S (2012) Copper-catalyzed aminooxygenation of N-allylamidines with PhI $(OAc)_2$. Org Lett 14:5342–5345
82. Michaelis DJ, Shaffer CJ, Yoon TP (2007) Copper(II)-catalyzed aminohydroxylation of olefins. J Am Chem Soc 129:1866–1867
83. Benkovics T, Du J, Guzei IA, Yoon TP (2009) Anionic halocuprate(II) complexes as catalysts for the oxaziridine-mediated aminohydroxylation of olefins. J Org Chem 74:5545–5552
84. Benkovics T, Guzei IA, Yoon TP (2010) Oxaziridine-mediated oxyamination of indoles: an approach to 3-aminoindoles and enantiomerically enriched 3-aminopyrroloindolines. Angew Chem Int Ed 49:9153–9157
85. DePorter SM, Jacobsen AC, Partridge KM, Williamson KS, Yoon TP (2010) N-nosyl oxaziridines as terminal oxidants in copper(II)-catalyzed olefin oxyaminations. Tetrahedron Lett 51:5223–5225
86. Michaelis DJ, Williamson KS, Yoon TP (2009) Oxaziridine-mediated enantioselective aminohydroxylation of styrenes catalyzed by copper(II) bis(oxazoline) complexes. Tetrahedron 65:5118–5124
87. Williamson KS, Yoon TP (2010) Iron-catalyzed aminohydroxylation of olefins. J Am Chem Soc 132:4570–4571
88. Williamson KS, Yoon TP (2012) Iron catalyzed asymmetric oxyamination of olefins. J Am Chem Soc 134:12370–12373
89. Christie SDR, Warrington AD (2008) Osmium and palladium: complementary metals in alkene activation and oxidation. Synthesis 1325–1341
90. McDonald RI, Liu G, Stahl SS (2011) Palladium(II)-catalyzed alkene functionalization via nucleopalladation: stereochemical pathways and enantioselective catalytic applications. Chem Rev 111:2981–3019
91. Backvall J-E, Andersson PG (1991) Stereocontrolled oxaspirocyclization of conjugated dienes via palladium catalysis. J Org Chem 56:2274–2276
92. Nilsson YIM, Araanyos A, Andersson PG, Backvall J-E, Parrain J-L, Ploteau C, Quintard J-P (1996) Synthesis of theaspirone and vitispirane via palladium(II)-catalyzed oxaspirocyclization. J Org Chem 61:1825–1829
93. Verboom RC, Persson BA, Backvall J-E (2004) Palladium(II)-catalyzed intramolecular 1,4-oxyacyloxylation of conjugated dienes. A stereocontrolled route to fused six-membered lactones and pyrans. J Org Chem 69:3102–3111
94. Jensen KH, Pathak TP, Zhang Y, Sigman MS (2009) Palladium-catalyzed enantioselective addition of two distinct nucleophiles across alkenes capable of quinone methide formation. J Am Chem Soc 131:17074–17075

95. Jensen KH, Webb JD, Sigman MS (2010) Advancing the mechanistic understanding of an enantioselective palladium-catalyzed alkene difunctionalization reaction. J Am Chem Soc 132:17471–17482
96. Li Y, Song D, Dong VM (2008) Palladium-catalyzed olefin dioxygenation. J Am Chem Soc 130:2962–2964
97. Wang W, Wang F, Shi M (2010) Bis(NHC)-palladium(II) complex-catalyzed dioxygenation of alkenes. Organometallics 29:928–933
98. Seayad J, Seayad AM, Chai CLL (2010) Copper-catalyzed diacetoxylation of olefins using $PhI(OAc)_2$ as oxidant. Org Lett 12:1412–1415
99. Kang Y-B, Gade LH (2011) The nature of the catalytically active species in olefin dioxygenation with $PhI(OAc)_2$: metal or proton? J Am Chem Soc 133:3658–3667
100. Inoki S, Mukaiyama T (1990) Chem Lett 1:67–70
101. Palmer C, Morra NA, Stevens AC, Bajtos B, Machin BP, Pagenkopf BL (2009) Increased yields and simplified purification with a second-generation cobalt catalyst for the oxidative formation of trans-THF rings. Org Lett 11:5614–5617
102. Schuch D, Fries P, Donges M, Perez BM, Hartung J (2009) Reductive and brominative termination of alkenol cyclization in aerobic cobalt-catalyzed reactions. J Am Chem Soc 131:12918–12920
103. Kolb HC, VanNieuwenhze MS, Sharpless KB (1994) Catalytic asymmetric dihydroxylation. Chem Rev 94:2483–2547
104. Peed J, Davies IR, Peacock LR, Taylor JE, Kociok-Köhn G, Bull SD (2011) Dihydroxylation-based approach for the asymmetric syntheses of hydroxy-γ-butyrolactones. J Org Chem 77:543–555
105. Braukmuller S, Bruckner R (2006) Enantioselective butenolide preparation for straightforward asymmetric synthesis of γ-lactones – paraconic acids, avenaciolide, and hydroxylated eleutherol. Eur J Org Chem 2110–2118
106. Kapferer T, Bruckner R (2006) Asymmetric dihydroxylation of b,g-unsaturated carboxylic esters with trisubstituted C=C bonds – enantioselective synthesis of trisubstituted γ-butyrolactones. Eur J Org Chem 2119–2133
107. Donohoe TJ, Butterworth S (2003) A general oxidative cyclization of 1,5-dienes using catalytic osmium tetroxide. Angew Chem Int Ed 42:948–951
108. Piccialli V (2007) Oxidative cyclizations of dienes and polyenes mediated by transition-metal-oxo species. Synthesis 2585–2607

Top Heterocycl Chem (2013) 32: 77–108
DOI: 10.1007/7081_2013_102

Published online: 30 March 2013

Synthesis of Heterocycles via Palladium-Catalyzed Wacker-Type Oxidative Cyclization Reactions of Hydroxy- and Amino-Alkenes

Nicholas A. Butt and Wanbin Zhang

Abstract Oxygen and nitrogen containing heterocyclic compounds are some of the most important and prominent structures found in biologically active natural and synthetic products, thus their synthesis is of paramount importance to the chemical community. One particularly important route to the synthesis of these structures is that of Wacker-type oxidative cyclizations. Palladium-catalyzed oxidative cyclizations represent an efficient and simple procedure for the synthesis of a variety of heterocyclic structures. The catalytic system can be fine-tuned to promote different oxidative transformations and to induce asymmetry in to the cyclized products, either via the use of chiral ligands or by manipulating chirality present in the starting substrate.

Keywords Cyclization · Heterocycles · Oxidation · Palladium · Wacker

Contents

N.A. Butt and W. Zhang (✉)
Department of Chemistry and Chemical Engineering, Shanghai Jiao Tong University, 800 Dongchuan Road, ZIP 200240 Shanghai, China
e-mail: n.a.butt@sjtu.edu.cn; wanbin@sjtu.edu.cn

Abbreviations

Ac	Acetyl
Ar	Aryl
Bn	Benzyl
Boxax	Binaphthyl-2,2′-bis(oxazoline)
Bu	Butyl
dba	Dibenzylideneacetone
DIPEA	Diisopropylethylamine
DME	Dimethoxyethane
DMF	Dimethylformamide
DMSO	Dimethyl sulfoxide
equiv	Equivalent
Et	Ethyl
h	Hour(s)
i-Pr	Isopropyl
Me	Methyl
MeCN	Acetonitrile
MeOH	Methanol
MOA	Trimethyl orthoacetate
Mol	Mole
MS	Molecular sieves
NHC	*N*-heterocyclic carbene
Ns	Nosyl
OAc	Acetoxy
Ph	Phenyl
p-Tol	*para*-Tolyl
py	Pyridine
pyrox	Pyridine-oxazoline
quinox	2-(4,5-Dihydro-2-oxazolyl)quinoline
rt	Room temperature
s	Second(s)
sprix	Spiro-bis(isoxazoline)
t-Bu	*tert*-Butyl
TFA	Trifluoroacetate
THF	Tetrahydrofuran
THP	Tetrahydropyran
Tol	Toluene
Ts	Tosyl

1 Introduction

The selective Pd(II)-catalyzed oxidation of molecules in organic chemistry is of significant importance to the chemical community [1–4]. Asymmetric oxidation chemistry utilizing palladium catalysts has had an enormous impact on organic chemistry, simplifying reaction procedures for the synthesis of a range of compounds, particularly for the preparation of heterocycles.

Heterocycles are an integral part of many biologically active natural and synthetic products. A number of methodologies and protocols exist for the synthesis of such systems. There has been considerable interest in aerobic oxidation catalysts to prepare heterocycles via oxidative bond forming chemistry; one of the most prominent protocols being Wacker-type cyclization processes [5, 6]. The traditional Wacker process was developed by Wacker-Chemie for the oxidative coupling of ethylene and water to produce acetaldehyde (Scheme 1). The process typically involves a catalyst (usually Pd(II)), an olefin (ethylene), a nucleophile (H_2O), and an oxidant ($Cu(II)/O_2$) [1].

The Wacker process has been successfully applied to the synthesis of a variety of heterocyclic structures via the use of palladium-catalyzed reactions. Originally, stoichiometric quantities of palladium salts were required for the cyclization of alkenyl nucleophiles. More recently Wacker-type catalytic systems have been developed utilizing direct dioxygen catalysis in the presence/absence of other co-oxidants and ligands [1, 2]. In particular, coordinating nitrogen ligands are often used to promote Wacker-type oxidative cyclizations. Types of ligand range from simple heterocycles such as pyridine and sparteine to more complicated structures such as *N*-heterocyclic carbene (NHC) ligands. Different catalytic systems show great versatility, in that they can be tuned to promote different oxidative transformations (Scheme 2) [1]. In addition, asymmetric Wacker-type reactions have been developed to synthesize otherwise difficult to obtain chiral heterocycles. The cyclization products usually generate a new stereogenic center; the stereochemical outcome of which can be controlled by varying the catalytic system [3].

In this chapter we will focus on the Pd-catalyzed Wacker-type oxidative cyclization reactions of hydroxy- and amino-alkenes for the synthesis of oxygen and nitrogen containing heterocycles. Discussions will include C–O and C–N bond formation for the preparation of achiral and chiral heterocycles via aliphatic and aromatic oxygen and nitrogen nucleophiles. The mechanistic details of these reactions will be discussed in the final section of this chapter.

2 C–O Bond Formation

Cyclic ethers and lactones are prevalent in biologically active and medicinally important compounds. The synthesis of such systems has attracted considerable attention. These heterocyclic rings can be readily prepared via Wacker-type

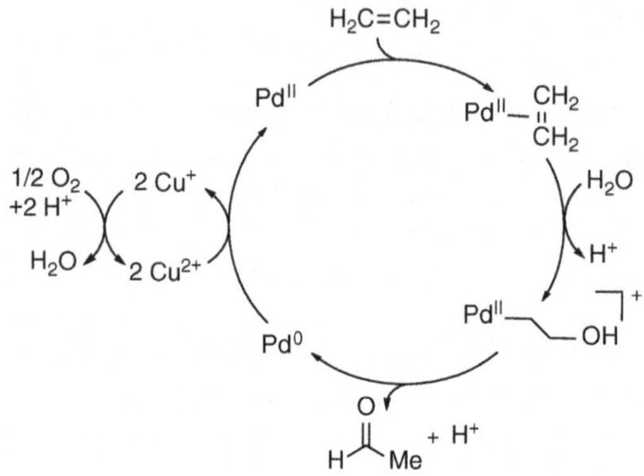

Scheme 1 The original Wacker process developed by Wacker-Chemie [1]

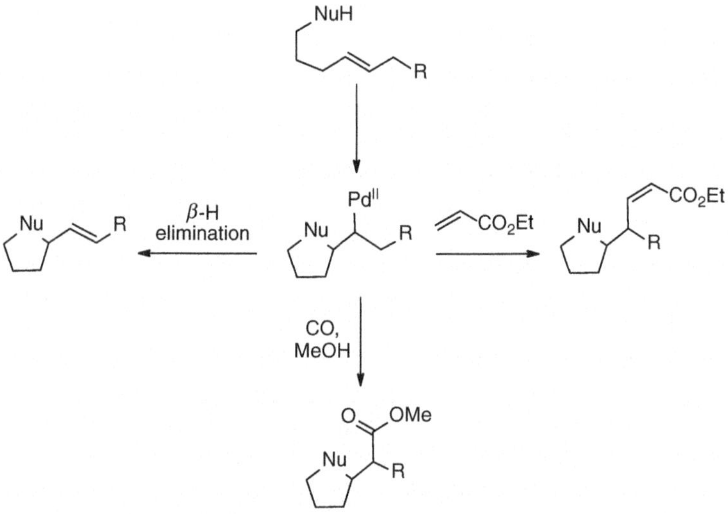

Scheme 2 Typical uses for Pd-catalyzed oxidative cyclizations [1]

cyclization processes and have been utilized in the synthesis of many natural products for 5- and 6-membered ring *O*-heterocycle formation [7–9].

2.1 Achiral Heterocycles

2.1.1 Alcohol Nucleophiles

In 1976, γ,δ-unsaturated alcohols were cyclized to a diastereoisomeric mixture of 2-vinyltetrahydrofurans with a $Pd(OAc)_2$–$Cu(OAc)_2$ catalyst system under an O_2 atmosphere [8]. En route to the synthesis of homoallylic amine **3** from allylic carbamate **1**, Hiemstra et al. utilized a Pd(II)-catalyzed oxidative cyclization to

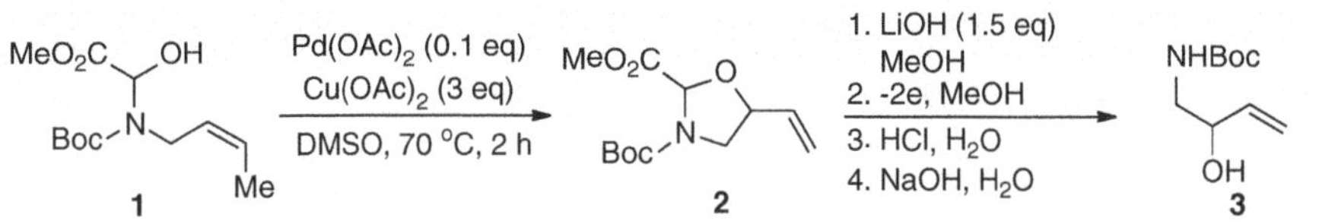

Scheme 3 Hiemstra synthesis of homoallylic amine via a Wacker-type cyclization [10]

generate vinyl substituted oxazolidines **2** [10]. The cyclization has preference for the 5-*exo* cyclization mode with no double bond isomers being formed (Scheme 3). The use of DMSO enhances the regioselectivity of alkene formation. Similar procedures for cyclizations involve replacing stoichiometric $Cu(OAc)_2$ with other oxidative additives and reducing catalyst loadings [11, 12].

Palladium-catalyzed Wacker-type cyclizations can be utilized on a variety of different substrates. α-Alkenyl/α-allyl-β-diketones undergo a $PdCl_2(MeCN)_2$ oxidative alkoxylation (via attack of the enolic oxygen on the Palladium complexed olefin) to form 2,3,5-substituted furans [13]. A stoichiometric amount of oxidant ($CuCl_2$ or quinone) is required for the cyclization to proceed. Less reactive α-allyl-β-ketoesters fail to undergo cyclization.

Wacker-type cyclizations can be applied to the synthesis of carbohydrates and indolizinone-based compounds [14, 15]. C-vinyl furanosides have been prepared from γ,δ-olefinic alcohols via a $Pd(OAc)_2$–NaOAc–O_2/DMSO system [15]. Five-membered rings are preferentially formed using this catalytic system with the β-hydride of the side chain usually being eliminated.

Stoltz's pioneering work on Wacker-type cyclizations has allowed for the methodology to be performed in nonpolar solvents such as toluene [16]. This enabled ligands to be used for direct dioxygen-coupled palladium-catalyzed cyclizations. The conditions are applicable to primary alcohols, phenols, and carboxylic acids (Table 1) [17]. $Pd(TFA)_2$ was used because of its strong counterion effect.

2.1.2 Phenol Nucleophiles

Palladium(II)-induced intramolecular cyclizations of alkenyl phenols have been extensively reported. The first example of the oxidative cyclization of *ortho*-allyl phenols using stoichiometric Pd was reported in 1973 [18]. After 2 years, the first catalytic example of the cyclization of 2-allyphenols into 2-substituted benzofurans via the use of $Pd(OAc)_2$ and Wacker conditions (i.e., using a cuprate salt and molecular oxygen) was reported [19]. The regiochemical outcome of the cyclization (5-membered versus 6-membered ring formation) is affected by the electron density of the palladium species and the type of palladium catalyst [20–22]. For example, 2H-1-benzopyran derivatives have been prepared from *O*-allylic phenols using $Pd(dba)_2$ and $Pd(OAc)_2$ catalysts in DMSO and air as the only reoxidant [22].

The oxidative cyclization of 2-allylphenols has been thoroughly investigated by Bumagin and Stoltz. Bumagin synthesized a number of benzofuran derivatives using a catalytic system of $Pd(OAc)_2$ and $Cu(OAc)_2$ in moist DMF [23]. En route

Table 1 Stoltz's cyclization of alcohol nucleophiles using a catalyst system consisting of 5 mol% $Pd(TFA)_2$, 20 mol% of pyridine, 2 equiv. of Na_2CO_3, 500 mg/mmol of 3 Å MS, 1 atm O_2, in toluene (0.1 M) at 80 °C [17]

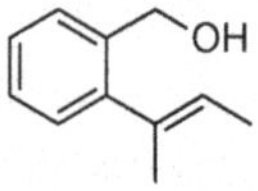	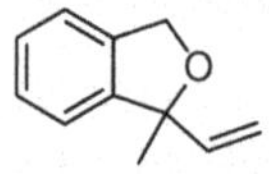	3 h	87%
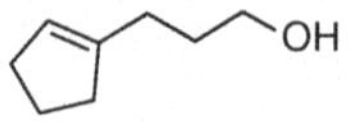	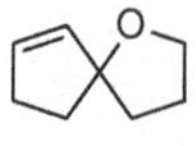	10 h	93%
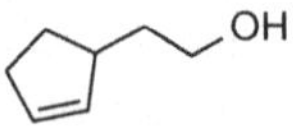	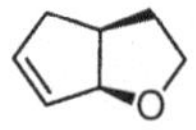	7.5 h	69%
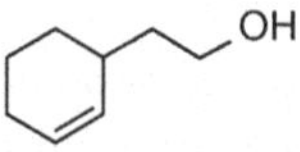	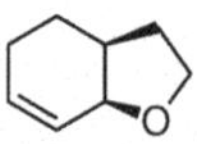	20 h	60%

to investigating the feasibility of enantioselective Wacker cyclizations using phenol derivatives, Stoltz performed an aerobic palladium-catalyzed cyclization of 2-allylphenol derivatives using the nonpolar solvent toluene [16, 17]. The use of DMSO, commonly used in Wacker cyclizations, precludes the use of ligands for enantioselective syntheses because of its highly donating nature as a ligand for palladium. The use of the electron deficient $Pd(TFA)_2$ and pyridine ligand with Na_2CO_3 as a stoichiometric base provided benzofurans in good yields [16, 17]. Electron-rich and deficient phenols are both excellent substrates for this reaction with the former undergoing rapid cyclization under the aforementioned conditions (Table 2).

2.1.3 Carboxylic Acid Nucleophiles

The palladium-catalyzed cyclization of alkenioic acids represents a useful and efficient procedure for the synthesis of lactone-containing molecules. Hayasaka first reported the synthesis of simple alkenyl lactones via the cyclization of alkenoic acids with palladium(II) salts [24]. A catalytic cyclization of alkenoic acids using Pd $(OAc)_2$ and a reoxidant ($Cu(OAc)_2$ or only O_2) was successfully developed by Larock whereby disubstituted alkenes react most rapidly [25]. The reaction conditions are applicable to a range of simple substrates with some cyclized products being obtained in greater than 90% yields (Table 3). This particular methodology is useful for the synthesis of isocoumarin-based products (Table 3, entry 5), of which in previous syntheses using a palladium chloride-based methodology cyclized to a 3:1 mixture of isocoumarin and 3-methylene phthalide [25]. The methodology

Table 2 Stoltz's oxidative cyclization of alkenyl phenols with 5 mol% of $Pd(TFA)_2$, 20 mol% of pyridine, 2 equiv. of Na_2CO_3, 500 mg/mmol of 3 Å MS, 1 atm O_2, 80 °C in toluene (0.1 M) [17]

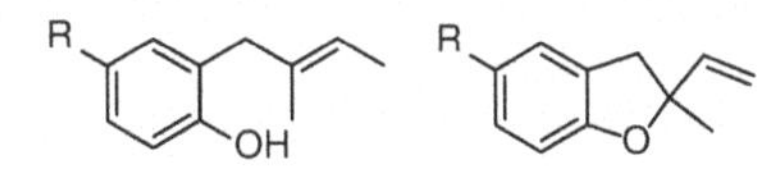

R = H	20 min, 95%
R = Me	20 min, 99%
R = *t*-Bu	25 min, 90%
R = OMe	15 min, 89%
R = COMe	25 h, 93%
R = Br	24 h, 33%

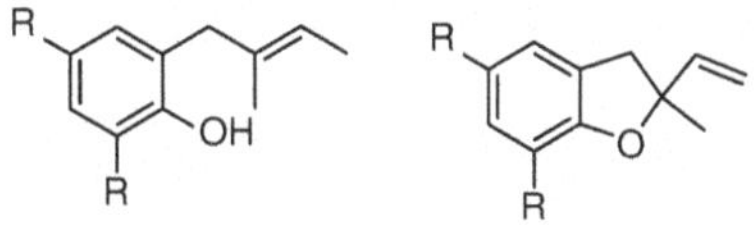

R = Me	20 min, 85%
R = OMe	40 min, 80%

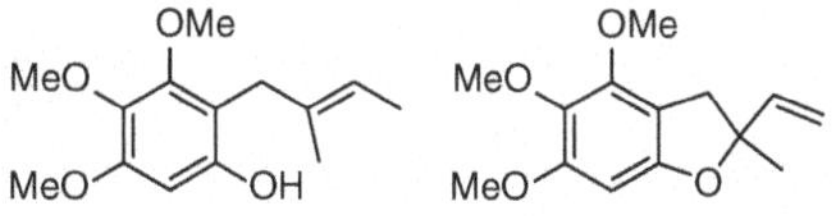

10 min, 86%

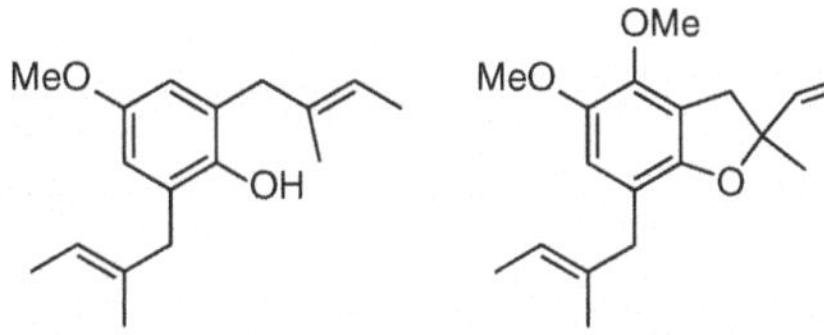

2 h, 93%

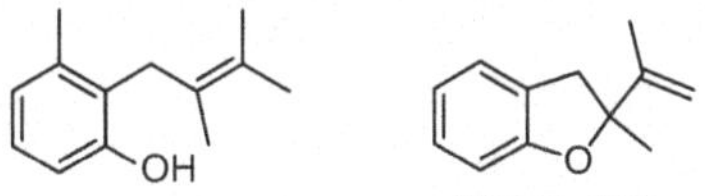

25 min, 80%

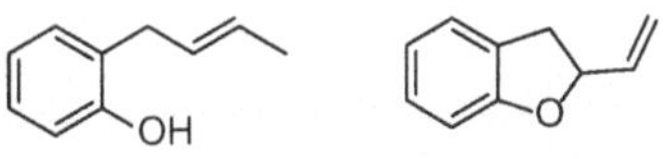

3 h, 74%

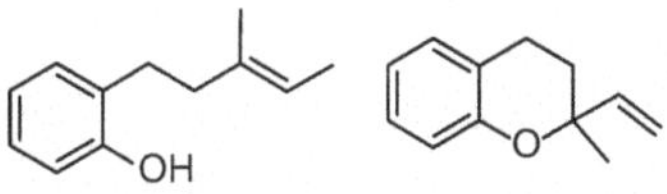

75 min, 85%

developed by Stoltz utilizing a pyridine ligand and $Pd(TFA)_2$ catalyst can reduce reaction times of cyclizations involving carboxylic acids [17].

As mentioned above, carboxylic acid-bearing alkenes can be cyclized to a number of 6-membered ring systems [26–30]. Yudin developed a palladium-catalyzed oxidative activation of arylcyclopropanes to prepare chromene substances [30]. The electronic character of cyclopropane is close to that of olefins with the orbitals of C–H bonds having approximately 33% s-character, and the C–C bonds 17% s-character [30]. This alkene-like character allows for the facile bonding of a Pd species to the cyclopropane ring system. Products such as **7** and **9** can be obtained in moderate yield via ring opening of **4** with a $PdCl_2$ catalyst (Scheme 4). A Wacker-type cyclization using a reoxidant such as $CuCl_2$ in dioxane gives the desired products [30].

Table 3 Larock's conditions for alkenoic acid Wacker-type cyclizations: 0.5 mmol of alkenoic acid, 1.0 mmol of NaOAc, and 5 mol% of $Pd(OAc)_2$ in 10 mL of DMSO under 1 atm of oxygen [25]

	Substrate	Product	time, temp (°C)	yield
1.	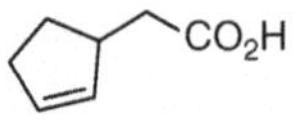	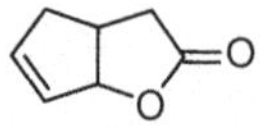	24 h, 25	86%
.	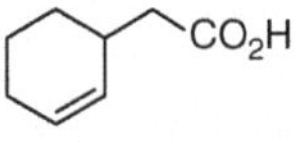	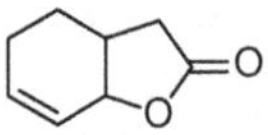	24 h, 25	90%
.	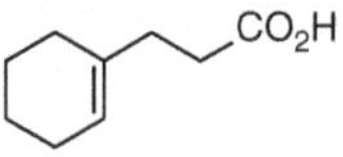	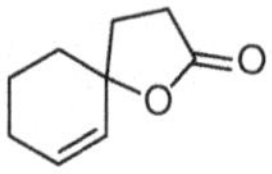	24 h, 80	91%
4.	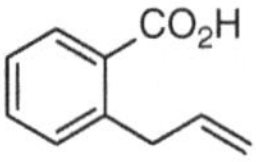	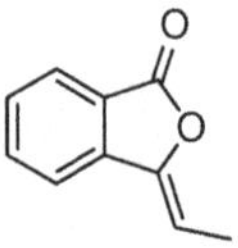	72 h, 80	71%
5.	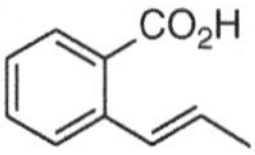	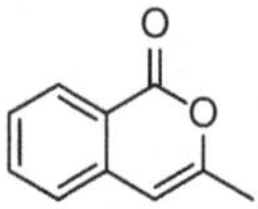	48 h, 80	78%

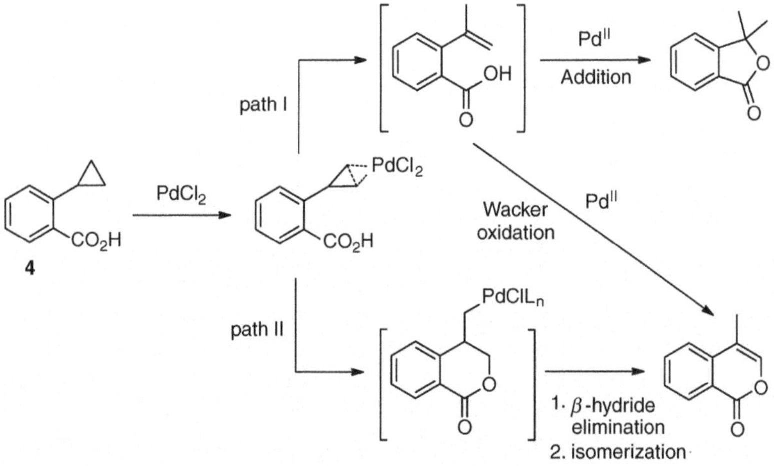

Scheme 4 Cyclopropane ring opening of **4** followed by a Wacker-type cyclization [30]

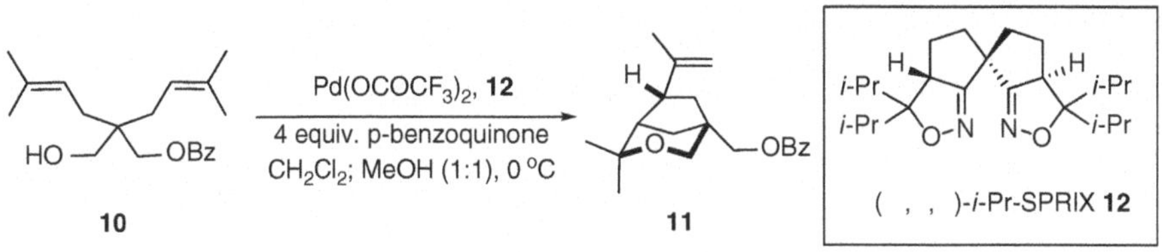

Scheme 5 Oxidative cyclization of alkene alcohol **10** using a SPRIX ligand [31]

2.2 *Chiral Heterocycles*

2.2.1 Alcohol Nucleophiles

A commonly used method to induce asymmetry in a catalytic reaction is to use chiral ligands. Some of the most commonly used ligands for asymmetric Wacker-type reactions are spiro-bis(isoxazoline) (SPRIX) ligands [31–33]. The Sasai group designed and synthesized novel spiro bis(isoxazoline) (SPRIX) ligands possessing a chiral spiro skeleton and two isoxazoline rings for enantioselective Wacker-type cyclizations of alkenyl alcohols [31]. The oxidative cyclization of alkene alcohol **10** using the catalytic system shown in Scheme 5 gave cyclized product **11** in up to 99% ee using the optimized conditions [31], and some minor dihydropyran by-products resulting from β-elimination of the Pd complex. SPRIX ligands have also been applied to the catalytic cyclization of 2-alkenyl-1,3-diketones, with the 6-*endo-trig* cyclization giving rise to chromene derivatives [33].

Gouverneur used substrate chirality to synthesize 2,3-dihydro-4H-pyran-4-one derivatives from chiral β-hydroxyenone with no detectable racemization [34]. Using the optimized conditions shown in Scheme 6 [34], reactions occurred with up to 97% ee. Interestingly, no reaction occurred when a substituent was present on the vinyl carbon adjacent to the carbonyl group. A similar procedure to synthesize a series of multisubstituted chiral dihydropyranones and furanones from *syn*/*anti*-α′,β′-dialkyl-β′-hydroxyenones via a palladium(II) catalyzed diastereoselective synthesis was also developed [35]. The use of a biphasic solvent system (PBS/toluene) can help improve chemical yield and prevent epimerization of the stereocenters [35].

2.2.2 Phenol Nucleophiles

Asymmetric Wacker-type cyclizations of 2-allyl phenols have been well studied. In 1978 Hosokawa and Murahashi demonstrated the first catalytic, asymmetric oxidative cyclization of phenol olefins [36] and used these findings to further probe the catalytic mechanism of oxidative cyclization reactions [37–39]. Z-(But-2-enyl) phenol was converted to optically active 2,3-dihydro-2-vinylbenzofuran with 12% optical yield using a catalytic amount of (−)-β-pinene. An excess of β-pinene inhibited the cyclization because of its ready ability to coordinate with the Pd species, thus preventing coordination of the substrate [37, 38].

Scheme 6 Gouverneur's synthesis of chiral dihydropyranones [34]

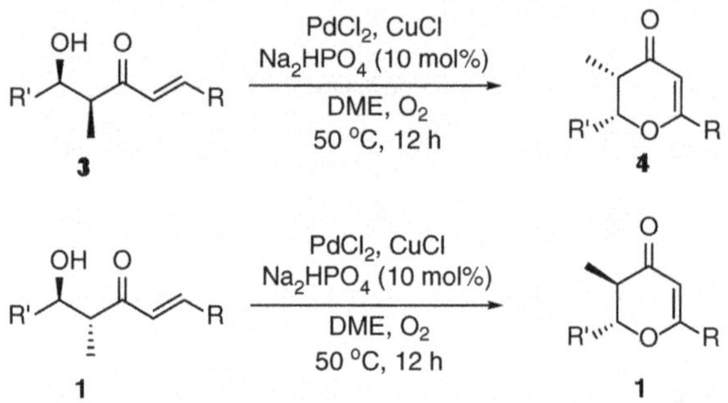

Fig. 1 Commonly used chiral ligands for asymmetric Wacker-type cyclizations [40–47]

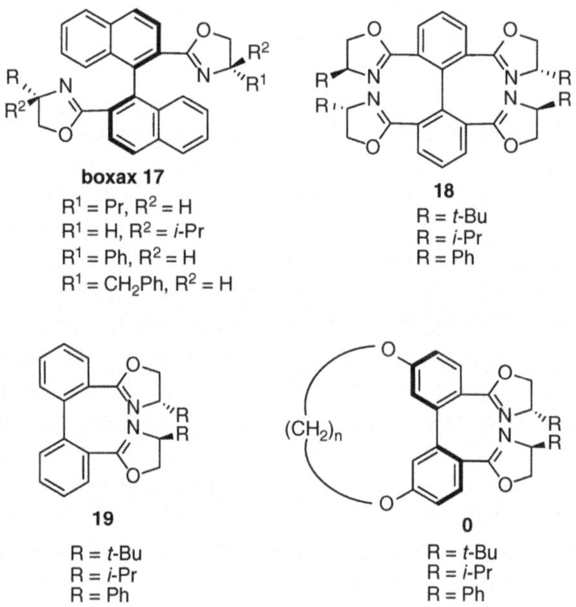

Several chiral ligands (**17–20**) have been developed for the asymmetric oxidative cyclization of phenol alkenes (Fig. 1). Hayashi developed chiral bis(oxazoline) ligands (boxax) **17** consisting of a 1,1′-binaphthyl backbone, which when used in conjunction with palladium bis(triflouroacetate) and *p*-benzoquinone gave dihydrobenzofurans and benzodihydropyrans in up to 97% ee [40–42]. The trifluoroacetate palladium catalyst is required for the reaction to proceed in a facile manner as the trifluoroacetate plays a prominent role in the activation of the coordinated olefin [40]. The catalytic palladium/boxax complex adopts a square-planar configuration with the nitrogen atoms of the oxazoline rings and oxygen atoms of the trifluoroacetate groups being attached to the palladium [41]. A number of boxax ligands have been designed in order to try and increase the reactivity and enantioselectivity [42].

Table 4 Stoltz's aerobic cyclization: reactions performed with 10 mol% of (sp)Pd(TFA)$_2$, 100 mol% of (−)-sparteine, 2 equiv. of $Ca(OH)_2$, 500 mg/mmol of 3 Å MS, 1 atm O_2, in toluene (0.1 M) at 80 °C [17]

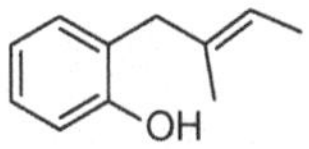	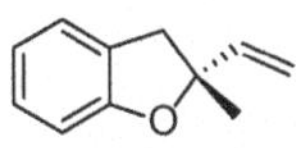	36 h	87%	81%
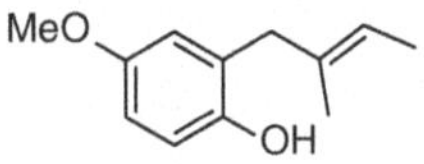	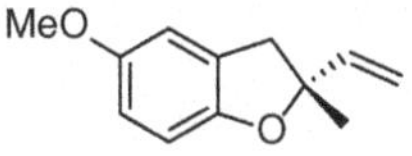	24 h 60 h, 55 °C	64% 57%	88% 90%
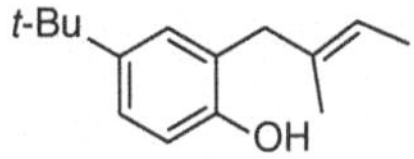	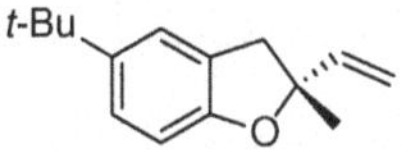	36 h	47%	83%
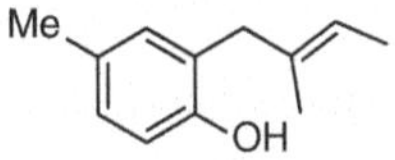	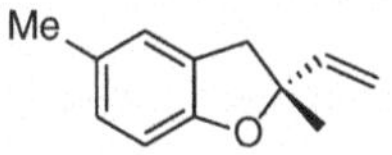	36 h	47%	86%
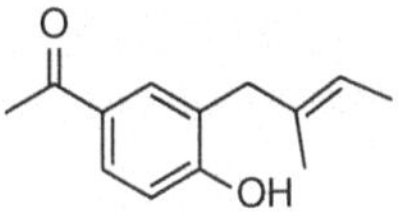	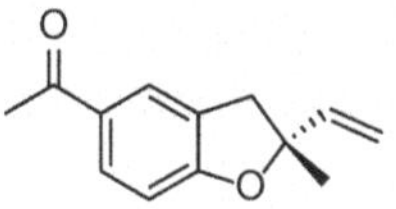	24 h	60%	20%

The Stoltz group has performed a number of reactions to synthesize dihydrobenzofurans using a simple sparteine ligand [17]. Similar enantioselective oxidative cyclizations of alkenyl phenols can be performed using molecular O_2 to reoxidize the palladium species in the absence of other external oxidants. This procedure represented the first extension of a direct dioxygen-coupled racemic reaction to aerobic asymmetric catalysis (Table 4).

Zhang developed axially chiral palladium complexes utilizing tetraoxazoline ligands **18** to give cyclized products **22** from phenol **21** in up to 99% ee with a 1:1 mixture of Pd(TFA)$_2$ and ligand (Scheme 7 above) [43, 44]. Benzodihydropyrans were also obtained in good ee with ligands **18** [45]. Axial chirality in the metal–ligand complex is induced when the ligands coordinate with the palladium metal center. Tropisomeric bisoxazoline ligands **19** and atropisomeric bridged bisoxazoline ligands **20** have also been developed and show similar enantioselectivties to the aforementioned ligands [46, 47].

2.3 Domino Reactions

Wacker-type cyclizations have found use in domino reactions to simplify the process of developing complex molecules. Such reactions are commonly used for the synthesis of chroman derivatives and the furan/pyran containing ionospheres of various antibiotics [48–50]. The domino syntheses often involve a carbonylation reaction with CO insertion proceeding the initial palladium-induced cyclization

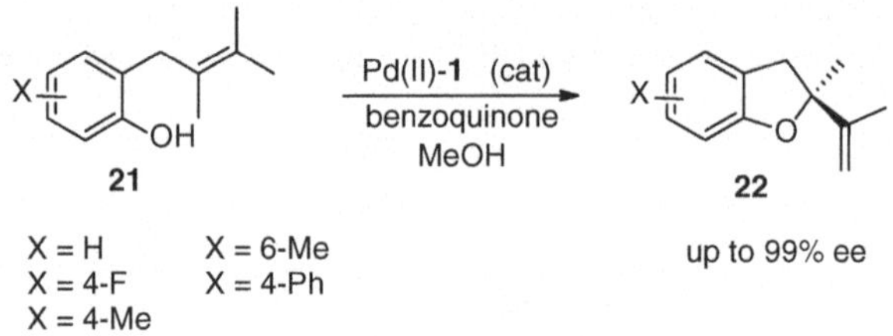

Scheme 7 Cyclization of alkenyl phenol **21–22** [43, 44]

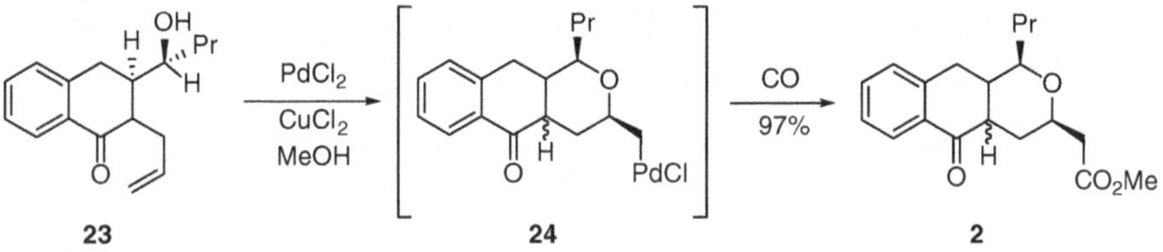

Scheme 8 Oxidative cyclization/CO insertion en route to the synthesis of frenocilin [51]

process. One of the first instances of using Wacker cyclizations in domino reactions was pioneered by Semmelhack et al. who utilized an alkoxycarbonylation sequence to prepare the pyran section **25** of the antibiotic frenocilin, from alkene **23** (Scheme 8) [51].

Wacker-type domino reactions represent an efficient synthetic route towards the synthesis of cyclic lactones via a palladium carbonylation process. Palladium(II)-catalyzed oxidative carbonylation of 3-buten-1-ols and 3-buytn-1-ols can undergo dicarbonylation to give butyrolactones stereospecifically [52]. Of particular use is the applicability of Wacker-type conditions to tandem intramolecular alkoxycarbonylation-lactonizations of 1,3-diols to give bicyclic lactone structures [53, 54]. Unsaturated polyols undergo intramolecular oxycarbonylation reactions with high chemo-, regio-, and diastereoselectivity [55]. In poly alcohol systems (1,2,3-pentenetriols) the diastereoselective course of the cyclization can be controlled by α-*O*-silyl protection [55].

In 2005 Yang utilized a thiourea ligand **26** for a Pd-catalyzed carbonylative annulation of **27** to prepare an important intermediate **28** (with correct stereochemistry) for the synthesis of micrandilactone A (Scheme 9) [56]. The domino Wacker-type process resulted in a 95% yield of product.

Tietze employed boxax ligands for the enantioselective Pd-catalyzed total synthesis of vitamin E [57, 58]. The key Wacker–Heck domino reaction occurs between palladium species **31** and methyl acrylate in the presence of $Pd(TFA)_2$ and a boxax ligand to give the important vitamin E intermediate **33**, in 96% ee (Scheme 10). In addition to synthesizing chroman-derived intermediates [59], chiral bis(oxazoline) ligands have been used in the intramolecular Pd(II)-catalyzed oxycarbonylation of alkene-1,3-diols to give bicyclic lactones, albeit in relatively low yields and ee [60].

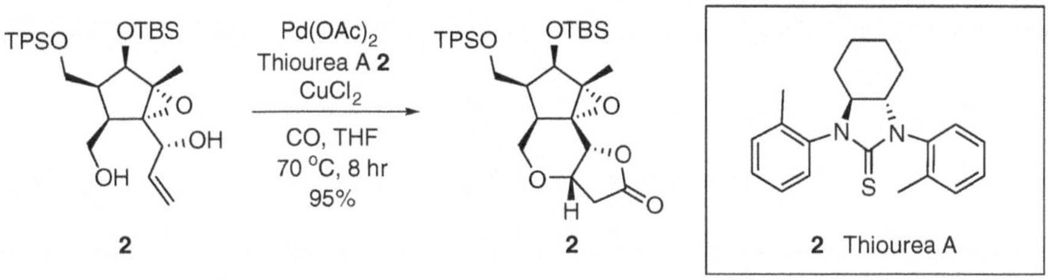

Scheme 9 Synthesis of micrandilactone intermediate **28** via a carbonylative annulation [56]

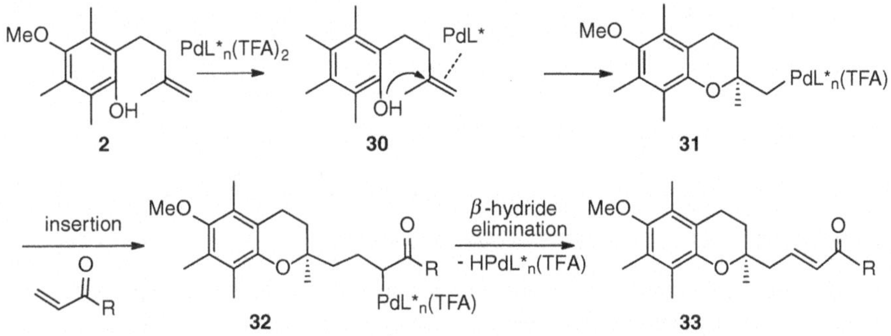

Scheme 10 Synthesis of vitamin E via a Pd(II)-catalyzed domino reaction [57]

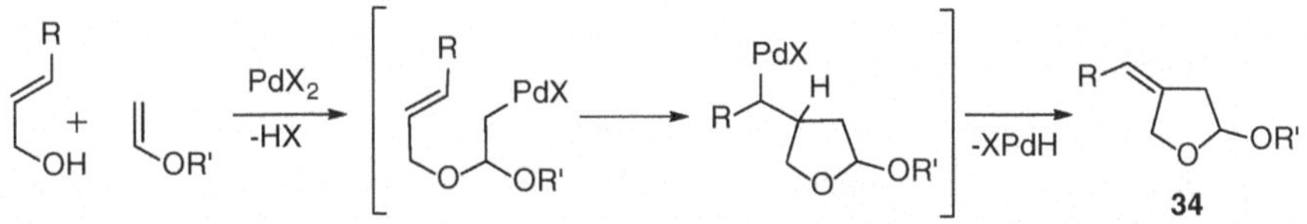

Scheme 11 Intramolecular 5-*exo* cyclization to give 2-alkyloxytetrahydrofurans [62]

Several intermolecular variations on domino Wacker-type processes have also been developed [61–64]. Hosokawa et al. have developed procedures for the synthesis of 2-alkoxytetrahydrofurans from allylic alcohols and vinyl ethers [62, 63]. A catalytic system of $Pd(OAc)_2$, catechol, and $CuCl_2$ in the presence of O_2 results in nucleophilic attack of the alcohol on the vinyl ether followed by an intramolecular 5-exocyclization to give (*Z*)-tetrahydrofuran derivatives **34** (Scheme 11). The reaction appears to be stereospecific with (*E*)- and (*Z*)-allyl alcohols giving the corresponding (*E*)- and (*Z*)-products [63]. It is believed the catechol enhances the stability of the catalyst by constructing a Pd–Cu species bearing catechol as the ligand of Cu and allowing the efficient capture of O_2 and its subsequent activation by the Cu-catechol complex [62]. A Pd-catalyzed stereoselective Oshima–Utimoto reaction has also been developed to synthesis chiral furan derivatives [61].

An intermolecular asymmetric Wacker-type reaction of cinnamyl alcohols and vinyl ethers was also developed utilizing boxax ligands [64]. The catalysis proceeds via a Pd–Cu bimetallic structure and anionic bridging ligands, with the anionic

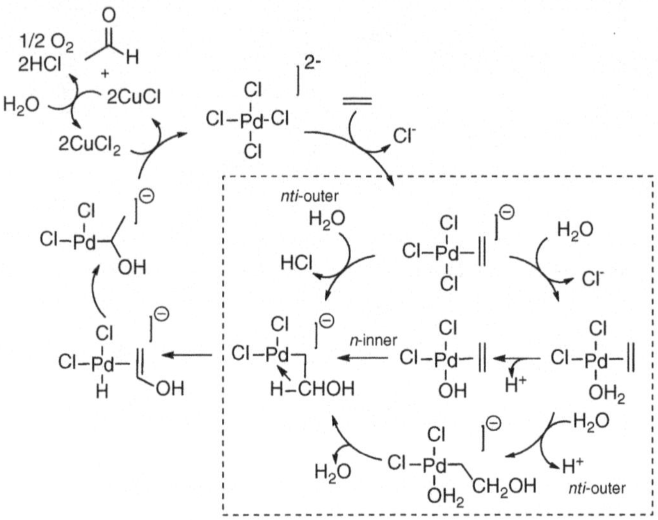

Scheme 12 The catalytic cycle of the Wacker process [67]

ligands being interchangeable between each metal. The enantioselectivity of the reaction is dependent upon the Pd catalyst and chiral ligand, whereas reaction activity depends on the copper salt and catechol.

2.4 Mechanistic Studies

The mechanistic principles underlying the Wacker reaction have been extensively investigated [65–77]. The formation of the olefin palladium complex is understood to be correct (Scheme 12); however, ambiguity in the reaction mechanism arises at the alkoxy/nucleophilic palladation step [67]. Water can attack the palladium-ethene complex either via an inner sphere mechanism, i.e., attack from water already attached to the Pd (*syn*-addition), or via an outer sphere mechanism, i.e., attack from an external water molecule (*anti*-addition).

Henry proposed the rate law shown in Scheme 13, Eq. (1). The original Wacker reaction of ethylene was determined to be first order with respect to ethene and to exhibit first order proton inhibition and a second order chloride inhibition [65, 66]. Using isotopic labeling experiments Henry discarded the original proposal involving the *anti*-addition pathway, giving rise to a new rate term [Scheme 13, Eq. (2)] [72, 73]. Henry proposed the reaction proceeds through a *syn*-addition mechanism at low concentrations of [Cl^-] and through an *anti*-addition mechanism at high [Cl^-] concentrations [74]. Further isotopic labeling experiments performed by the Hayashi group involving the Wacker-type oxidative cyclization of *o*-allylyphenol derivatives appear to corroborate this observation, with the reaction taking place with *syn* stereochemistry in the absence of chloride, and the reaction being *anti* in the presence of chloride [75].

$$\text{rate} = \frac{[\text{PdCl}_4^{2-}][\text{olefin}]}{[\text{H}^+][\text{Cl}^-]^2} \qquad \mathbf{1}$$

$$\text{rate} = \frac{[\text{PdCl}_4^{2-}][\text{olefin}]}{[\text{Cl}^-]^2} \qquad \mathbf{2}$$

Scheme 13 Rate expression for the Wacker oxidation of ethylene with chloride inhibition terms [72]

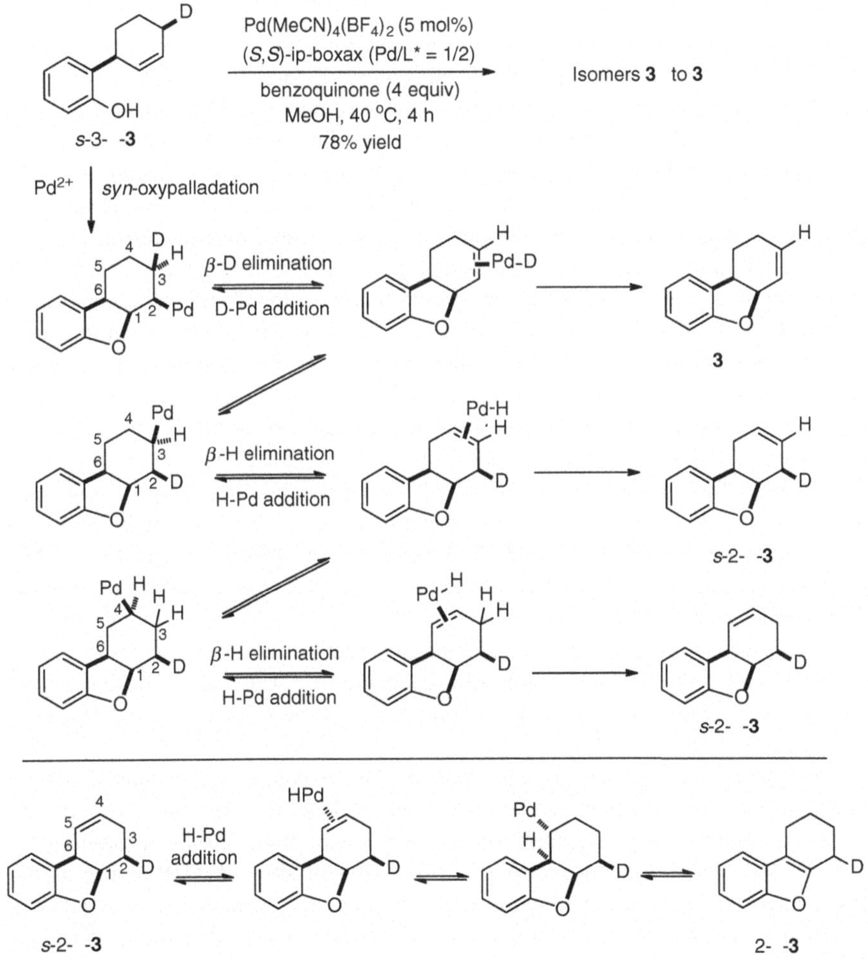

Scheme 14 (**a**) Formation of isomers **36**, *cis*-2-*d*-**37** and *cis*-2-*d*-**38**; (**b**) formation of isomer 2-*d*-**39** [75]

The group of Hayashi probed the stereochemistry at the oxypalladation step of Wacker-type processes using stereospecifically deuterated racemic 6-(2-hydroxyphenyl)-3-deuteriocyclohexene, *cis*-3-*d*-**35** (Scheme 14) [75]. Using reactions with defined stereochemical outcomes the stereochemistry at the oxypalladation step

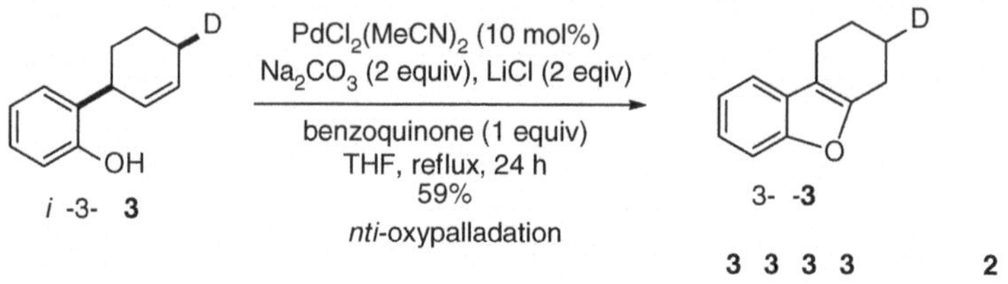

Scheme 15 Predominant formation of *anti*-oxypalladation product 3-*d*-**39** at high Cl^- concentrations [75]

could be determined using the Wacker-type system shown in Scheme 14. Several observations were noted for the reaction: all isomers contained *cis*-fused five- and six-membered ring systems; no deuterium was found in compound **36**; isomers *cis*-2-*d*-**37** and *cis*-2-*d*-**38** had over 95% deuterium incorporation at the 2-position *cis* to the oxygen in isomers **37** and **38**; and finally 2-*d*-**39** contains over 90% deuterium at the 2-position and 5–10% deuterium at the 3- and 4-positions. The stereochemistry of the oxpalladation step was therefore found to be *syn* because of these results [75].

Using the above experiment, Hayashi et al. were able to observe the transfer of deuterium during the course of the reaction (Scheme 14a) [75]. Oxypalladation of *cis*-3-*d*-**35** with phenol oxygen and palladium gives **A**, with Pd being on the same face as the deuterium (*syn* addition). *syn* β-Hydrogen elimination provides intermediate **B**. Dissociation of palladium from the double bond gives isomer **36**. Formation of intermediate **D** via the addition of palladium-deuteride to C-3 (**C**) and subsequent 4-H β-hydrogen elimination gives the *cis*-2-*d*-**37** isomer. In an analogous manner, hydropalladation of **D** gives **E**, and following β-hydrogen elimination isomer *cis*-2-*d*-**38** is obtained. Thermodynamically stable benzofuran **39** is formed via the isomerization of **36**, **37**, and **38** (Scheme 14b, example isomerization of **38**). *syn*-Oxypalladation was further confirmed using similar experiments involving the *trans* isomer of *cis*-3-*d*-**35** [75].

In the presence of chloride ions *anti*-oxypalladation was found to occur giving isomer 3-*d*-**39** as the predominant product (Scheme 15). Deuterium at the 3-position can only be explained by the increased propensity of the substrate to undergo *anti*-oxypalladation [75].

The type of catalysts and ligands used in Wacker-type processes has also been shown to alter the reaction mechanism. Isotopic labeling experiments involving oxidation of simple styrene derivatives using a palladium 2,6-diisopropylphenyl complex in a *tert*-butylhydroperoxide (THP) solvent showed THP acting as the oxygen source for addition into the olefin bond [76]. Sigman discovered that a variation of the Wacker reaction involving the bidentate ligand sparteine complex Pd[(−)-sparteine]Cl_2, and the direct O_2 coupled oxidation of decene, occurs via a mechanism involving a three-water hydrogen-bond bridged chain and subsequent oxypalladation [77]. Stoltz used deuterium labeling experiments to examine the effect monodentate and bidentate ligands had on oxidative cyclizations of primary alkenyl alcohols and alkenyl carboxylic acids [17]. Cyclization of primary alcohols appeared to occur via a *syn*-oxypalladation process – the opposite of the traditional

Wacker reaction, whereas carboxylic acids cyclized in an *anti*-fashion. The use of monondentate and bidentate ligands did not change the stereochemical outcome of the reaction [17].

3 C–N Bond Formation

Nitrogen containing heterocycles are important components of natural products and medicinal compounds, thus methodologies towards their synthesis are constantly evolving. The palladium-catalyzed amination of alkenes represents one significant route toward the synthesis of such systems. As mentioned previously, the conditions required for Wacker-type cyclizations are mild and highly tolerable to various functional groups. The mild conditions can thus be used in the final stages of natural product synthesis whereby many different functional groups are present.

3.1 Achiral Heterocycles

3.1.1 Nonaromatic Nitrogen Nucleophiles

Alkenyl amines are often subjected to Wacker-cyclizations to produce a range of 5- and 6-membered nitrogen heterocycles. In order for cyclic amination to proceed, protection of the amine nitrogen is often required to prevent deactivation of the palladium species via coordination to the metal center. This problem is commonly overcome by using tosyl, acetate, and methyl protected amines; the tosyl variants being particularly useful for their ability to be easily removed. The cyclizations can occur in the presence of reoxidants such as benzoquinone and $CuCl_2$, or in their absence, for example using O_2 as the only reoxidant. External additives and ligands can also have a significant impact on reaction yields and activity.

Hegedus synthesized nonaromatic nitrogen hetereocycles from ω-olefinic tosamides via palladium catalysis [78]. Tosyl amine **40** readily cyclized using a $PdCl_2(MeCN)_2$ catalytic system, benzoquinone as a reoxidant and basic additives to produce enamine product **41** (Scheme 16) [78]. The tosyl group was used to prevent the coordination of the amine to, and henceforth inactivation of, the Pd catalyst. The cyclization is mechanistically related to the Wacker cyclization; the Pd species first coordinates with the olefin, with the Pd-olefin complex subsequently being attacked by an internal nucleophile. Venanzi reported the oxidative cyclization of amino alkenes with a secondary amino group to yield the corresponding cyclic enamines [79]. Similar systems using a $PdCl_2$ catalyst and benzoquinone reoxidant have been used to cyclize unsaturated amines as a key synthetic step in natural product synthesis [80].

Amino alkenes can also be subjected to Wacker-type cyclizations in the absence of any reoxidants such as benzoquinone and $CuCl_2$. Five- and six-membered heterocyclic systems have been obtained using only an O_2 atmosphere and $Pd(OAc)_2$ catalyst in DMSO solvent [12, 81]. Stahl et al. expanded the use of direct

Scheme 16 Oxidative cyclization of an alkenyl tosyl amines utilizing Hegedus' conditions [78]

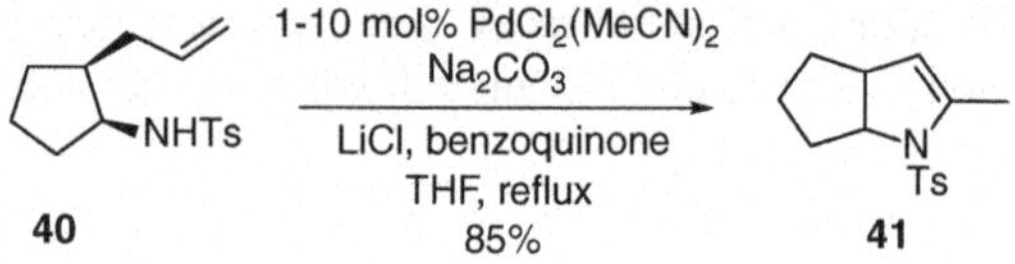

Table 5 Amination of olefin substrates reaction conditions: substrate (0.1 mmol), $[Pd(OAc)_2]$ (5 μmol), pyridine (10 μmol), O_2 (1 atm), xylene (1 mL), 80 °C [82]

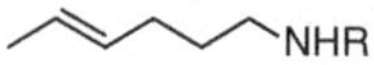 R = Ts R = Ns R = CBz	2 8 48	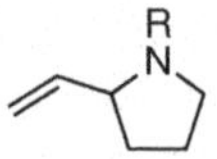	87 87 76
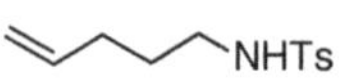	1.5	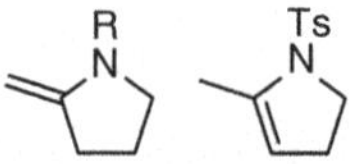	81 (7:3)
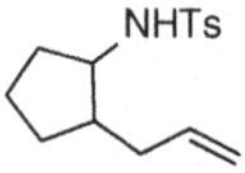	2	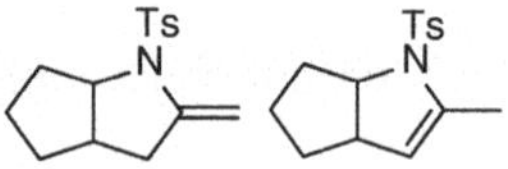	94 (1:1)
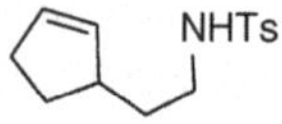	2	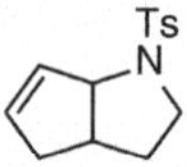	91

dioxygen palladium catalysis for use in nonpolar solvents such as toluene by using pyridine as a ligand [82]. The amination of olefin substrates proceeded with low catalyst loadings (0.2 mol% of $Pd(OAc)_2$) and short reaction times using a $Pd(OAc)_2$/pyridine stoichiometry of 1:2 (Table 5). The use of pyridine as a ligand greatly enhances catalyst activity in nonpolar solvents. The 5-*exo* cyclization gives rise to either tosyl enamides or allylic tosylamides depending on the ability of the molecule to undergo a β-hydride elimination.

A $Pd(OAc)_2$/pyridine catalyst system has also been used to synthesis a series of pyridone derivatives from isoxazolidine-5-spirocyclopropanes [83]. Reduction of the isoxazolidine N–O bond followed by a Wacker-like process gives the pyridone containing compounds [83].

The Stahl group developed a procedure for the regioselective synthesis of six-membered heterocycles (Table 6) [84]. Tethered sulfonamides were oxidatively cyclized using a $Pd(DMSO)_2(TFA)_2$ catalyst and an O_2 atmosphere. The procedure is suitable for the synthesis of a range of six-membered heterocycles such as morpholines, piperazines, piperidines, and piperazinones; all of which can be prepared in good yields.

Table 6 Stahl's synthesis of six-membered heterocycles [84]

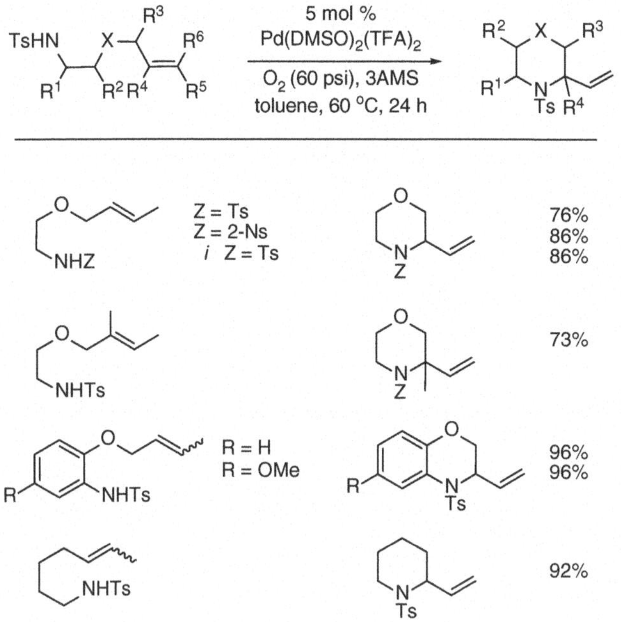

3.1.2 Aromatic Nitrogen Nucleophiles

Indoles and quinolines can be readily prepared from 2-allylaniline derivatives utilizing palladium catalysis. The pioneering work of Hegedus et al. allowed for an efficient and simple synthesis of 2-methlyindoles and quinoline using a $PdCl_2$ catalyst and benzoquinone reoxidant [85, 86]. As expected, amination occurs at the most substituted end of the alkene. The methodology is amenable to a range of electron-rich and poor substituted aniline derivatives and can be applied to carbonylation reactions and domino-type syntheses [86]. The Pd-catalyzed cyclization of 2-allyanilines is particularly useful for the synthesis of substituted indoles such as **44**. *N*-tosyl-3-bromo-2-allylaniline **42** was cyclized to 4-bromo-1-tosylindole **43** using the aforementioned conditions in 77% yield (Scheme 17) [87].

As with other palladium-catalyzed reactions, the reoxidants benzoquinone and $CuCl_2$ can be replaced with O_2/air allowing for easier purification of the products [81, 88, 89]. The pioneering work of Hegedus et al. has been further improved upon by Larock and coworkers who prepared a number of indole and quinoline compounds from 2-allylanilines utilizing a $Pd(OAc)_2/O_2$ catalyst. Olefinic tosylamides could be cyclized to 5- or 6-membered ring products containing an allylic nitrogen system [81]. *O*-allylic *N*-tosylanilides exclusively cyclized to their corresponding 6-membered ring products (up to 86% yield) in contrast to previous syntheses using a $PdCl_2$/benzoquinone catalyst system [81]. Additionally, similar catalytic systems can be used to control the regioselectivity of the palladium(II)-catalyzed cyclization of *N*-allyl-anthranilamides **45** to prepare allylic oxidation

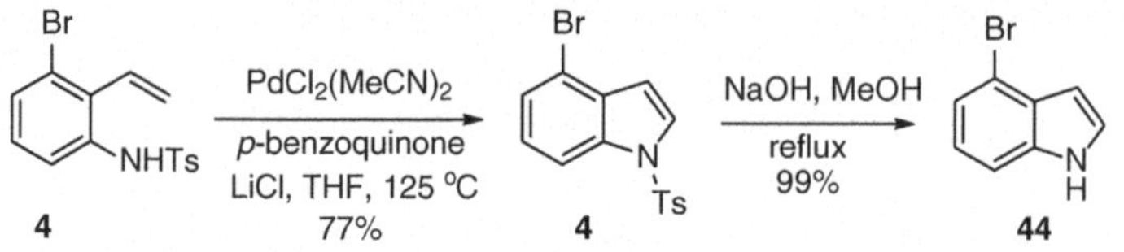

Scheme 17 Oxidative cyclization of *N*-tosyl-3-bromo-2-ethylaniline **42** [87]

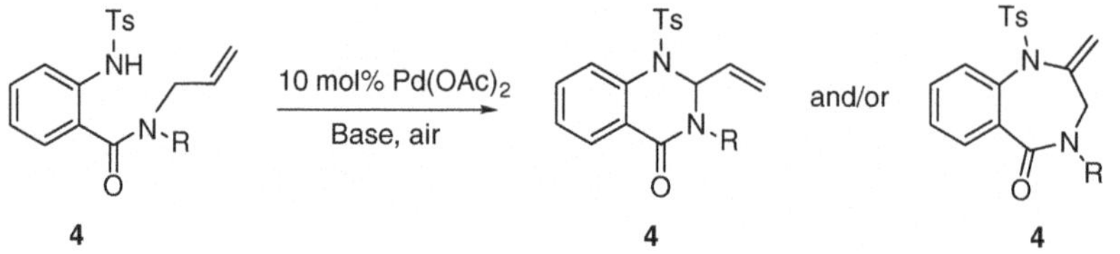

Scheme 18 Oxidative cyclization of *N*-allyl-anthranilamides. R = allyl, Me, Ph [88]

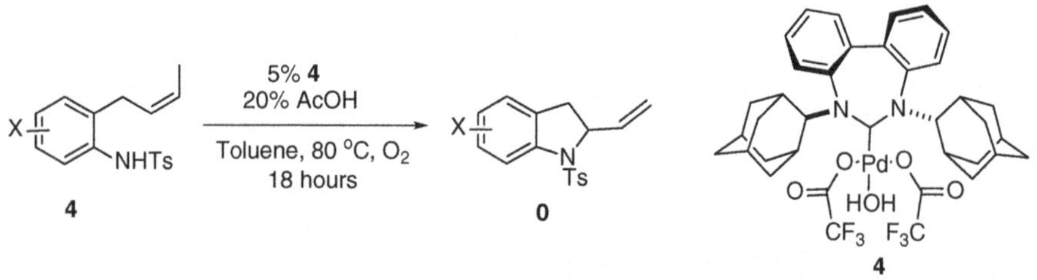

Scheme 19 Oxidative cyclization of *N*-tosylanilide using an NHC coordinated Pd(II) catalyst [90]

products quinazolin-4-ones **46**, and Wacker cyclization products 1,4-benzodiazepin-5-ones **47** (Scheme 18) [88].

Appropriate ligands can be used to increase the efficiency and yield of the intramolecular oxidative amination of alkenes. The group of Wang used a 1,10-phenanthroline ligand with $Pd(OAc)_2$ to synthesize a series of 2-methylquinoline substrates [89]. In 2006 Stahl and coworkers developed novel NHC ligands for use with molecular oxygen as the stoichiometric oxidant [90]. The NHC–palladium catalyst complex $(NHC)Pd(TFA)_2(OH_2)$ **48**, in conjunction with an acid co-catalyst, can facilitate the cyclization of a number of substituted anilines of the type **49** (Scheme 19) [90].

The Zhang group was able to control the regioselectivity of an aza-Wacker cyclization to preferentially synthesize isoindolinones or isoquinolin-1(2H)-ones from the same substrate (Scheme 20) [91]. Substrate **51** was cyclized to isoindoline **52** in good yield when using Phen as a ligand and $Pd(OAc)_2$ in methanol solvent. Isoqiuinolinone **53** was obtained when the ligand and metal source were exchanged for Et_3N and $Pd(MeCN)_2Cl_2/CuCl_2$. The reaction conditions are amenable to a number of substrates; however, attack at the desired vinyl carbon atom is sometimes not possible due to steric reasons (i.e., bulky protecting groups or substituents on the nitrogen atom or on the olefin).

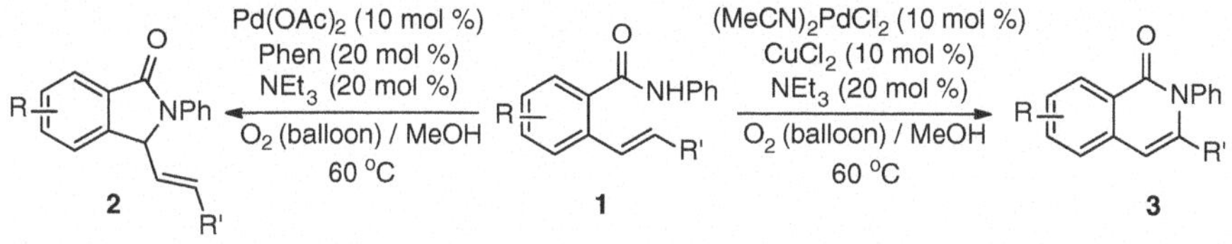

Scheme 20 Selective aza-Wacker cyclization to isoindolinones and isoquinolin-1(2H)-ones [91]

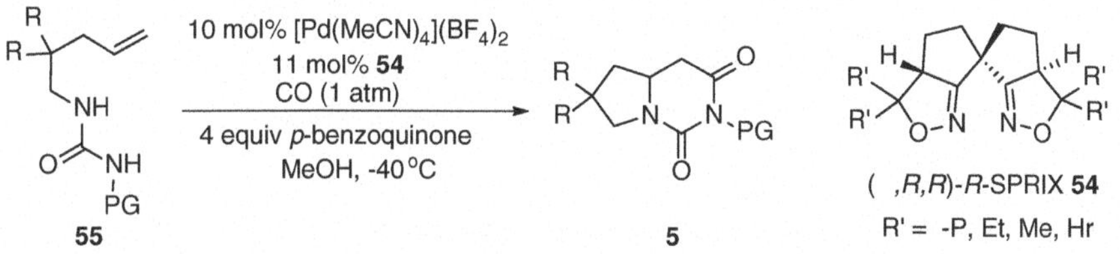

Scheme 21 Oxidative cyclization of alkenyl ureas [92]

3.2 *Chiral Heterocycles*

The asymmetric catalysis of alkenyl amines to form nitrogen heterocycles remains challenging. Although some ligands can induce asymmetry, catalytic asymmetric Wacker-type reactions remain limited in scope.

3.2.1 Nonaromatic Nitrogen Heterocycles

Asymmetric Wacker-type reactions can be performed using chiral ligands or by manipulating chirality in the starting products. Stahl and Sasai have used chiral ligands for the Wacker-type cyclizations of alkenyl amines [92, 93]. Spiro bis (isoxazoline) (SPRIX) ligands **54** were used in the palladium-mediated enantioselective intramolecular oxidative aminocarbonylation of alkenylureas **55**, with enantioselectivities of cyclized products **56** up to 89% ee (Scheme 21) [92]. The substituents R on the substrate contribute significantly to the chemical yield.

The Stahl group developed an asymmetric Wacker-type protocol using pyridine-oxazoline (pyrox) ligands [93] to synthesize chiral pyrrolidine compounds **58**. The greatest enantioselectivity was observed using Bn-quinox ligand **59** (7.5 mol%) with a $Pd(TFA)_2$ catalyst and the conditions shown in Scheme 22, to give pyrrolidine products in up to 98% ee and 68% yield. The reaction conditions are effective with a number of *cis*-alkenes such as ethyl- and benzyl substituted alkenes, as well as gem-dimethyl substituents; all giving rise to products with high ee (>90%) [93].

Stahl also developed a synthesis of enantiopure *cis*-2,5-disubstituted pyrrolidines **61** via *tert*-butanesulfinamide nucleophiles **60** [94]. Chiral γ-aminoalkenes bearing a ^{t}Bu-sulfinyl auxiliary underwent Pd(II)-catalyzed oxidative cyclization to give the desired products with excellent diastereoselectivity (Scheme 23).

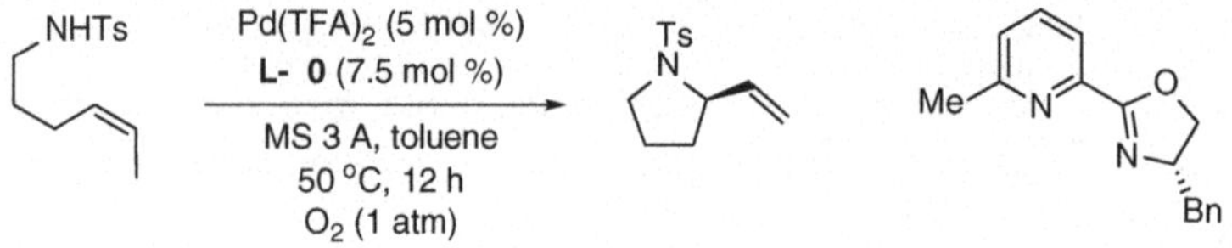

Scheme 22 Enantioselective cyclization of an alkenyl *N*-tosylamide [93]

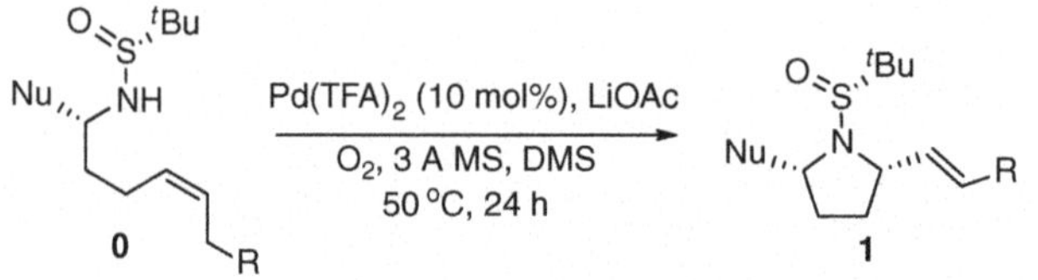

Scheme 23 Oxidative cyclization of alkenes possessing tethered α-substituted ^{t}Bu-sulfinamide nucleophiles [94]

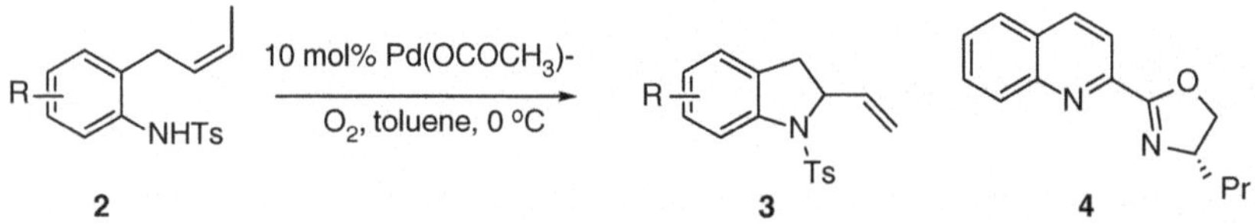

Scheme 24 Zhang's conditions for the cyclization of substrate **62** [96]

3.2.2 Aromatic Nitrogen Heterocycles

A recent method for the synthesis of aromatic nitrogen heterocycles via Wacker-type cyclizations involves the use of NHC ligands [95]. Stahl et al. synthesized seven-membered-ring amidinium salts from enantiomerically pure 2,2′-diamino-6,6′-dimethylbiphenyl, which were subsequently used to prepare chiral NHC–Pd catalysts for intramolecular Pd-catalyzed aminations of substrate **62** (Scheme 24). However enantioselectivity was very poor with only 9% ee [95]. The Zhang group reported the asymmetric aza-Wacker-type cyclization using quinolineoxazoline ligand **64** and Pd(OAc)$_2$ under an O$_2$ atmosphere, to give compound **63** with up to 74% ee (Scheme 24) [96].

Using a ligand closely related to **64**, ^{t}Bu-Pyrox **65**, isoindolinones **67** bearing chiral tetrasubstituted centers could be prepared from substrates **66** in high yields and up to 99% ee (Scheme 25) [97]. A variety of substituted alkenes could be cyclized using the conditions shown, with only a small drop in enantioselectivity as the size of the internal alkene substituent R′ increases.

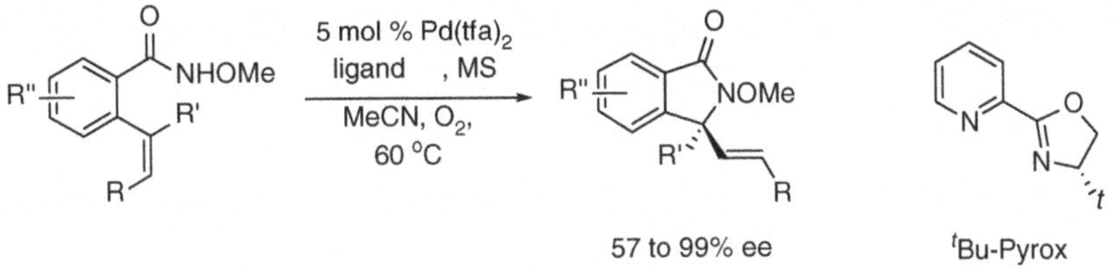

Scheme 25 Enantioselective synthesis of isoindolinones [97]

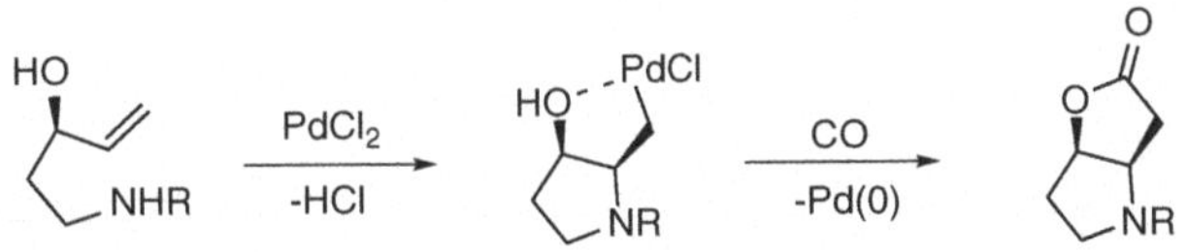

0

Scheme 26 Oxidative cyclization and carbonylation of 3-hydroxy-4-pentenylamines [105]

3.3 *Domino Reactions*

Oxidative palladium(II)-catalyzed cyclizations of alkenyl-amines provide an efficient procedure for synthesizing heterocyclic compounds. Following oxidative cyclization, the resulting palladium complexed intermediate can be exploited to undergo further reactions such as CO insertion and other carbon chain elongation processes. Tandem reactions are particularly useful for natural product synthesis as many new carbon–carbon bonds can be created in one step [98–101].

The Hegedus group was one of the first to pioneer these types of reactions by utilizing *N*-substituted *o*-allyanilines to synthesize dihydroindolacetic acid esters [86]. The conditions must be mild enough to prevent β-hydride elimination and the nitrogen must be protected in order to facilitate a successful reaction and prevent carbonylation of the nucleophilic center. A $PdCl_2/CuCl_2$ catalytic system has been used for the aminocarbonylation of unsaturated amines using urea nucleophiles [102–104] and for the intramolecular aminocarbonylation of 3-hydroxy-4-pentenylamines **68** (Scheme 26) [105]. The conditions were found to be amenable to a range of substrates and have more recently been used in conjunction with chiral ligands to develop enantioselective procedures [105, 106].

The Tamaru group investigated the intramolecular amino carbonylation of *endo*-carbamates and their ability to undergo oxidative cyclizations by subjecting olefin bearing substrates to acidic and buffered Wacker-type conditions (Scheme 27) [102, 103, 107]. *Endo*-carbamates **77** were unable to undergo oxidative cyclization using reactions conditions involving methanol solvent. Slightly modified conditions using trimethyl orthoacetate (MOA) as a solvent were required. MOA can act as a base to generate the conjugate base of **77** (increasing its nucleophilicity) as well as act as a scavenger of HCl, keeping the mixture weakly acidic thus helping the production of the conjugate base of **77**. In **77** the nitrogen lone pair is not

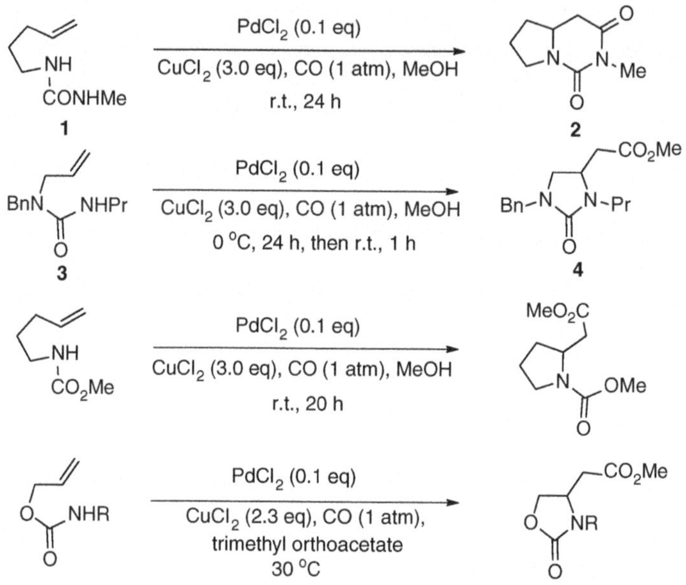

Scheme 27 Oxidative cyclization of carbamates [102, 103, 107]

sufficiently nucleophilic enough to efficiently overlap with the π* orbital of the olefin, thus requiring deprotonation to its conjugate base to increase its nucleophilicity [102, 103, 107].

The group of Yang developed a series of stereoselective aza-Wacker-type tandem reactions [108–110]. The first enantioselective oxidative tandem cyclization using only molecular oxygen as the oxidant was achieved using the catalyst system shown in Table 7 [108]. Using sparteine as a ligand, enantioselectivities of up to 91% were obtained [108]. A ^{t}Bu-QUINOX ligand has since been used to improve the enantioselectivity of oxidative tandem reactions using related substrates [110].

3.4 Mechanistic Studies

The mechanism of aminopalladation reactions have been thoroughly investigated [111–125]. Early studies by Backvall et al. were based on stoichiometric quantities of reagents and showed that oxyamination of alkenes occurred via an aminopalladation-oxidation sequence in overall *cis*-stereochemistry but with a *trans*-aminopalladation step [112, 114]. Taniguchi later reported that the intramolecular aminopalladation step occurs via a *cis*-process [116]. *cis*-Addition of Pd and nitrogen to the propenyl group of enamine **79** would result in the formation of the Pd-σ-complex **80**, which has a *cis*-configuration (Scheme 28). β-Hydride elimination cannot occur because of the *anti* relationship between the Pd and the

Table 7 Wacker-type domino synthesis of indolines [108]

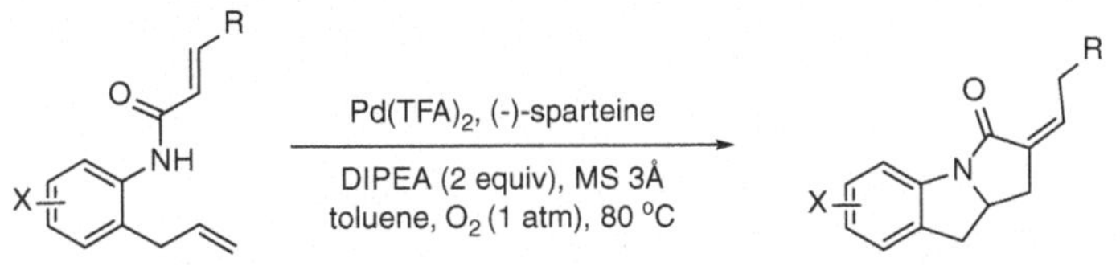

2

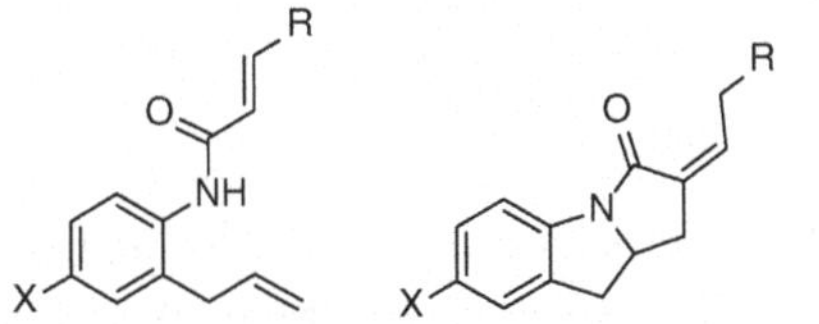 R = H, X = H		20 10 5	24 26 35	75 78 70	83 86 86
R = H, X = Cl		20	48	61	75
R = m-ClPh, X = H		20	48	60	80
R = Me, X = H	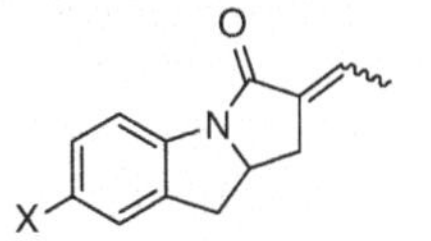	20	48	15	82
	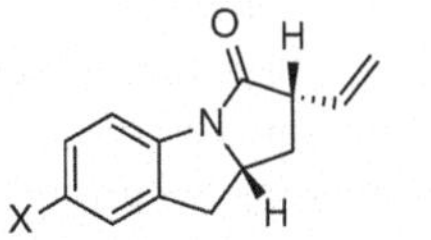	20	48	63 (dr 8:1)	80 (86)
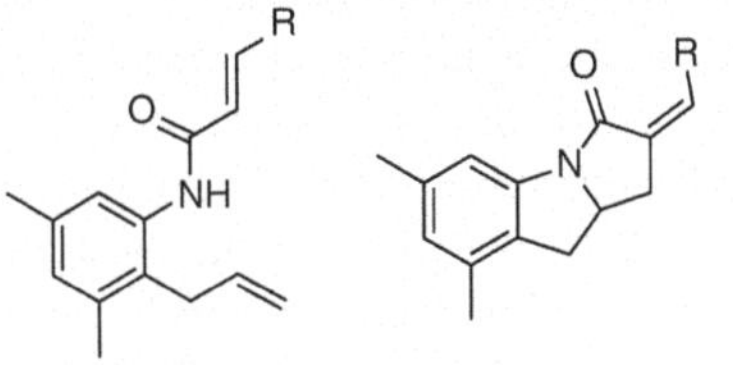 R = H		10	26	63	91
R = m-ClPh		10	26	64	87

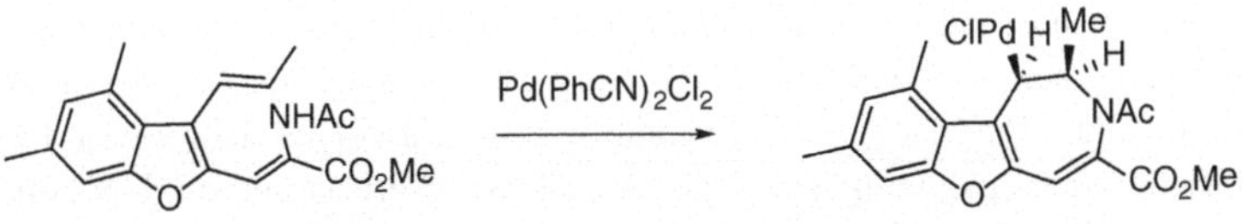

0

Scheme 28 Taniguchi oxidative cyclization of enamine **79** [116]

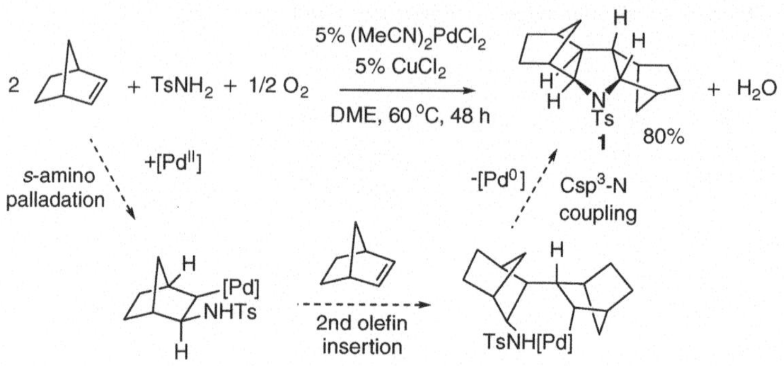

Scheme 29 *cis*-Aminopalladation of norbornene [120]

β-hydrogen atom, allowing compound **80** to be isolated [116]. Following the work of Taniguchi, it is now generally accepted that oxidative amination (Wacker-type) reactions occur predominantly through a *cis*-aminopalladation route [117–119].

Stahl performed intermolecular Wacker-type reactions using a *para*-toluene sulfonamide nucleophile and norbornene to form C_2-symmetric pyrrolidine product **81** (Scheme 29). The norbornene undergoes *cis*-difunctionalization on the *exo* face of the alkene [120]. It is thought this process occurs via *cis*-aminopalladation of norbornene, alkene insertion into the Pd–C bond, and C–N bond formation by reductive elimination [120]. Mechanistic insights into the intermolecular Wacker-type reactions were further carried out by Hartwig [121]. The mechanistic pathway of the intermolecular insertion of ethylene and octene into a palladium diarylamido complex N–C bond was investigated. Complexes of electron-rich amides were found to react more rapidly than electron poor amides with the insertion occurring even at low temperature (−40 °C). Enamine products resulted from *cis*-aminopalladation [121].

Stahl et al. performed a number of mechanistic studies using a series of different palladium(II) catalyst systems and investigated the stereochemistry of aminopalladation using isotopically labeled alkenyl amine substrates [122]. Products obtained via *cis*-aminopalladation were formed with all the catalyst systems apart from the system involving an NHC ligand and benzoic acid additive, in which a mixture of *trans*- and *cis*-products were isolated. Exchanging benzoic acid with basic additives such as Na_2CO_3 allowed for the exclusive formation of the *cis*-aminopalladation product. In addition, Stahl also discovered that the type of nucleophile plays an important role in determining the stereochemical outcome of the reaction, with more acidic nucleophiles favoring the *trans*-aminopalladation process [122]. Stahl also investigated the mechanism of the Wacker-type intramolecular oxidative amination of alkenes using aerobic conditions [123, 124]. The catalytic mechanism of oxidative cyclization using a $Pd(OAc)_2$/pyridine catalyst system involves steady-state formation of a Pd(II)–amidate–alkene intermediate (**A**), alkene insertion into a Pd–N bond (via a pyridine–palladium dissociated pathway or a pyridine–palladium ligated pathway (**B**)), reversible β-hydride elimination (**C**), irreversible AcOH reductive elimination (**D**), and aerobic oxidation of Pd(0) to regenerate the active catalyst (**E**) (Scheme 30) [124].

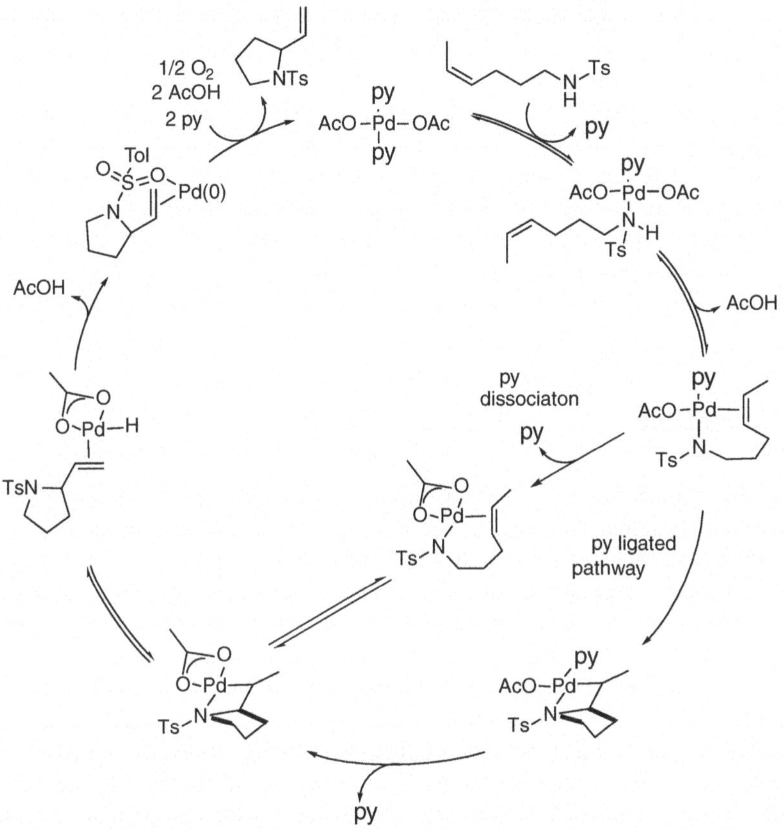

Scheme 30 Catalytic mechanism of $Pd(OAc)_2$/py-catalyzed aerobic oxidative intramolecular amination. *py* pyridine [124]

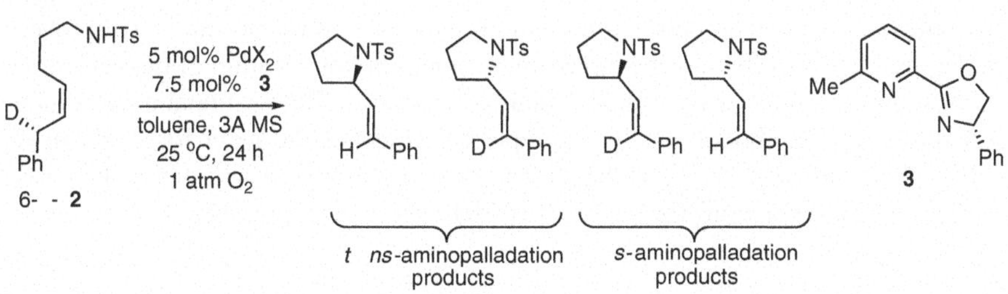

Scheme 31 Oxidative cylization of 6-*d*-**82** to give *trans*- and *cis*-aminopalladation products [125]

In continuing mechanistic studies, the Stahl group prepared deuterated substrate 6-*d*-**82** to probe the effect neutral-donor ligands and anionic ligands had on the aminopalladation step (Scheme 31) [125]. Using the reaction conditions shown in Scheme 31 (PdX_2 = $Pd(TFA)_2$), the major product resulting from cyclization was found to be product **A**, implying the *trans*-aminopalladation pathway was favored over the *cis* route. Analysis of the major enantiomers confirmed the *trans*-pathway was favored (*trans*/*cis* = 91:9). However when the palladium source was replaced with $Pd(OAc)_2$, *cis*-aminopalladation was preferred (*trans*/*cis* = 10:90), with a 9:1:51:39 product ratio for **A**/**B**/**C**/**D** (in overall low enantioselectivity and yield). Analysis of the *trans*-aminopalladation minor products **A** and **B** showed the *trans*-pathway to be relatively enantioselective (e.r. = 9:1), whereas the *cis* pathway exhibits low enantioselectivity. Removing the ligand **83** from the reaction mixture for both palladium sources resulted in the formation of predominantly *cis*-aminopalladation products, thus implying a neutral ligand and the type of anionic ligand can play an important role in determining the most likely reaction pathway [125].

4 Conclusions

Wacker-type reactions represent an important methodology for the synthesis of heterocyclic systems. The past decade has seen great advances in the use of this methodology because of its ease of use, mild reaction conditions, and its applicability to a large number of substrates. The mechanistic principles underlying Wacker-type reactions have also been extensively investigated, with new insights into the understanding of the process being continuously developed.

Of increasing interest is the use of chiral ligands for the synthesis of enantiopure oxygen and nitrogen heterocycles. Several steps have already been taken in order to achieve this goal, with a number of ligands (SPRIX, BOXAX, PYROX, etc.) showing promising enantioselectivtiy for particular substrates. At present, the enantioselective synthesis of nitrogen heterocycles with chiral ligands remains limited in scope; however, this is expected to change with increasing mechanistic understanding of the Wacker-type process.

Wacker-type reactions have also shown their use in the synthesis of complex molecules and domino processes, greatly simplifying reaction procedures. The process can be modified to allow for the relatively simple functionalization of molecules via CO insertion and other carbon chain extension reactions. The continual development of Wacker-type methodology, for example by reducing the need for co-catalysts ($CuCl_2$, benzoquinone, etc.) and developing new chiral ligands will widen the scope for these types of reactions for use by the wider chemical community.

References

1. McDonald RI, Liu G, Stahl SS (2011) Chem Rev 111:2981
2. Beccalli EM, Broggini G, Martinelli M, Sottocornola S (2007) Chem Rev 107:5318
3. Stahl SS (2004) Angew Chem Int Ed 43:3400
4. Backvall JE (1983) Acc Chem Res 16:335
5. Takacs JM, Jiang X (2003) Curr Org Chem 7:369
6. Kotov V, Scarborough CC, Stahl SS (2007) Inorg Chem 46:1910
7. Sohn J-H, Waizumi N, Zhong HM, Rawal VH (2005) J Am Chem Soc 127:7290

8. Hosokawa T, Hirata M, Murahashi SI, Sonoda A (1976) Tetrahedron Lett 17:1821
9. Uenishi J, Ohmi M (2005) Angew Chem Int Ed 44:2756
10. Van Benthem RATM, Hiemstra H, Speckamp WN (1992) J Org Chem 57:22
11. Van Benthem RATM, Hiemstra H, Michels JJ, Speckamp WN (1994) J Chem Soc Chem Commun 357
12. Roenn M, Backvall J, Andersson P (1995) Tetrahedron Lett 36:7749
13. Han X, Widenhoefer RA (2004) J Org Chem 69:1738
14. Mmutlane EM, Harris JM, Padwa A (2005) J Org Chem 70:8055
15. Sharma G, Chander A, Krishnudu K, Krishna P (1998) Tetrahedron Lett 39:6957
16. Trend RM, Ramtohul YK, Ferreira EM, Stoltz BM (2003) Angew Chem Int Ed 42:2892
17. Trend RM, Ramtohul YK, Stoltz BM (2005) J Am Chem Soc 127:17778
18. Hosokawa T, Maeda K, Koga K, Moritani I (1973) Tetrahedron Lett 10:739
19. Hosokawa T, Ohkata H, Moritani I (1975) Bull Chem Soc Jpn 48:1533
20. Hosokawa T, Yamashita S, Murahashi S-I, Sonoda A (1976) Bull Chem Soc Jpn 49:3662
21. Muñiz K (2004) Adv Synth Catal 346:1425
22. Larock R, Wei L, Hightower TR (1998) Synlett 522
23. Roshchin AI, Kel SM, Bumagin NA (1998) J Organomet Chem 560:163
24. Kasahara A, Izumi T, Sato K, Maemura M, Hayasaka T (1977) Bull Chem Soc Jpn 50:1899
25. Larock RC, Hightower TR (1993) J Org Chem 58:5298
26. Minami T, Nishimoto A, Nakamura Y, Hanaoka M (1994) Bull Chem Soc Jpn 42:1700
27. Jabre-Truffert S, Waegell B (1997) Tetrahedron Lett 38:835
28. Jacobi PA, Li Y (2003) Org Lett 5:701
29. Shoji M, Uno T, Kakeya H, Onose R, Shiina I, Osada H, Hayashi Y (2005) J Org Chem 70: 9905
30. He Z, Yudin AK (2006) Org Lett 8:5829
31. Arai MA, Kuraishi M, Arai T, Sasai H (2001) J Am Chem Soc 123:2907
32. Koranne PS, Tsujihara T, Arai MA, Bajracharya GB, Suzuki T, Onitsuka K, Sasai H (2007) Tetrahedron Asymmetry 18:919
33. Takenaka K, Mohanta SC, Patil ML, Rao CVL, Takizawa S, Suzuki T, Sasai H (2010) Org Lett 12:3480
34. Reiter M, Ropp S, Gouverneur V (2004) Org Lett 6:91
35. Reiter M, Turner H, Mills-Webb R, Gouverneur V (2005) J Org Chem 70:8478
36. Hosokawa T, Miyagi S, Murahashi S-I, Sonoda A (1978) J Chem Soc Chem Commun 687
37. Hosokawa T, Uno T, Murahashi S-I (1979) J Chem Soc Chem Commun 4:475
38. Hosokawa T, Uno T, Inui S, Murahashi S-I (1981) J Am Chem Soc 103:2318
39. Hosokawa T, Okuda C, Murahashi S-I (1985) J Org Chem 50:1282
40. Uozumi Y, Kato K, Hayashi T (1997) J Am Chem Soc 119:5063
41. Uozumi Y, Kato K, Hayashi T (1998) J Org Chem 63:5071
42. Uozumi Y, Kyota H, Kato K, Ogasawara M, Hayashi T (1999) J Org Chem 64:1620
43. Zhang YJ, Wang F, Zhang W (2007) J Org Chem 72:9208
44. Wang F, Yang G, Zhang YJ, Zhang W (2008) Tetrahedron 64:9413
45. Liu Q, Wen K, Zhang Z, Wu Z, Zhang YJ, Zhang W (2012) Tetrahedron 68:5209
46. Wang F, Zhang YJ, Wei H, Zhang J, Zhang W (2007) Tetrahedron Lett 48:4083
47. Wang F, Zhang YJ, Yang G, Zhang W (2007) Tetrahedron Lett 48:4179
48. Semmelhack MF, Epa WR, Cheung W-H, Gu Y, Kim C, Zhang N, Lew W (1994) J Am Chem Soc 116:7455
49. Holmes CP, Bartlett PA (1989) J Org Chem 54:98
50. Semmelhack MF, Bozell JJ, Sato T, Wulff W, Spiess E, Zask A (1982) J Am Chem Soc 104: 5850
51. Semmelhack MF, Zask A (1983) J Am Chem Soc 105:2034
52. Tamaru Y (1991) J Org Chem 56:1099
53. Boukouvalas J, Fortier G, Radu L-L (1998) J Org Chem 63:916
54. Semmelhack MF, Hooley RJ, Kraml CM (2006) Org Lett 8:5203

55. Babjak M, Zálupský P, Gracza T (2005) Arkivoc 5:45
56. Tang Y, Zhang Y, Dai M, Luo T, Deng L, Chen J, Yang Z (2005) Org Lett 7:885
57. Tietze LF, Sommer KM, Zinngrebe J, Stecker F (2005) Angew Chem Int Ed 44:257
58. Tietze LF, Stecker F, Zinngrebe J, Sommer KM (2006) Chemistry 12:8770
59. Tietze LF, Spiegl DA, Stecker F, Major J, Raith C, Grosse C (2008) Chemistry 14:8956
60. Kapitán P, Gracza T (2008) Arkivoc 8:8
61. Evans MA, Morken JP (2005) Org Lett 7:3367
62. Minami K, Kawamura Y, Koga K, Hosokawa T (2005) Org Lett 7:5689
63. Kawamura Y, Imai T, Hosokawa T (2006) Synlett 3110
64. Kawamura Y, Kawano Y, Matsuda T, Ishitobi Y, Hosokawa T (2009) J Org Chem 74:3048
65. Henry PM (1964) J Am Chem Soc 86:3246
66. Keith JA, Henry PM (2009) Angew Chem Int Ed 48:9038
67. Comas-Vives A, Stirling A, Lledós A, Ujaque G (2010) Chemistry 16:8738
68. James DE, Hines LF, Stille JK (1976) J Am Chem Soc 98:1806
69. Majima T, Kurosawa H (1977) J Chem Soc Chem Commun 610
70. Stille JK, Divakaruni R (1978) J Am Chem Soc 100:1303
71. Backvall JE, Akermark B, Ljunggren SO (1979) J Am Chem Soc 101:2411
72. Zaw K, Henry PM (1990) J Org Chem 55:1842
73. Francis JW, Henry PM (1991) Organometallics 10:3498
74. Hamed O, Thompson C, Henry PM (1997) J Org Chem 62:7082
75. Hayashi T, Yamasaki K, Mimura M, Uozumi Y (2004) J Am Chem Soc 126:3036
76. Cornell CN, Sigman MS (2005) J Am Chem Soc 127:2796
77. Anderson BJ, Keith JA, Sigman MS (2010) J Am Chem Soc 132:11872
78. Hegedus LS, McKearin JM (1982) J Am Chem Soc 104:2444
79. Pugin B, Venanzi LM (1983) J Am Chem Soc 105:6877
80. Heathcock CH, Stafford JA, Clark DL (1992) J Org Chem 57:2575
81. Larock RC, Hightower TR, Hasvold LA, Peterson KP (1996) J Org Chem 61:3584
82. Fix SR, Brice JL, Stahl SS (2002) Angew Chem Int Ed 41:164
83. Revuelta J, Cicchi S, Brandi A (2005) J Org Chem 70:5636
84. Lu Z, Stahl SS (2012) Org Lett 14:1234
85. Hegedus LS, Allen GF, Bozell JJ, Waterman EL (1978) J Am Chem Soc 100:5800
86. Hegedus LS, Allen G, Olsen D (1980) J Am Chem Soc 102:3583
87. Harrington PJ, Hegedus LS (1984) J Org Chem 49:2657
88. Beccalli EM, Broggini G, Paladino G, Penoni A, Zoni C (2004) J Org Chem 69:5627
89. Zhang Z, Tan J, Wang Z (2008) Org Lett 10:173
90. Rogers MM, Wendlandt JE, Guzei IA, Stahl SS (2006) Org Lett 8:2257
91. Yang G, Zhang W (2012) Org Lett 14:268
92. Tsujihara T, Shinohara T, Takenaka K, Takizawa S, Onitsuka K, Hatanaka M, Sasai H (2009) J Org Chem 74:9274
93. Mcdonald RI, White PB, Weinstein AB, Tam CP, Stahl SS (2011) Org Lett 13:2830
94. Redford JE, Mcdonald RI, Rigsby ML, Wiensch JD, Stahl SS (2012) Org Lett 14:1242
95. Scarborough CC, Bergant A, Sazama GT, Guzei IA, Spencer LC, Stahl SS (2009) Tetrahedron 65:5084
96. Jiang F, Wu Z, Zhang W (2010) Tetrahedron Lett 51:5124
97. Yang G, Shen C, Zhang W (2012) Angew Chem Int Ed 51:9141
98. Ham W-H, Hoon Jung Y, Kyunghae L, Oh C-Y, Lee K-Y (1997) Tetrahedron Lett 38:3247
99. Oh C-Y, Kim K-S, Ham W-H (1998) Tetrahedron Lett 39:2133
100. Szolcsa P, Gracza T, Koman M, Pronayova N, Liptaj T (2000) Tetrahedron Asymmetry 11: 2579
101. Szolcsányi P, Gracza T, Koman M, Prónayovác N, Liptaj T (2000) Chem Commun 471
102. Tamaru Y, Hojo M, Higashimura H, Yoshida Z (1988) J Am Chem Soc 110:3994
103. Tamaru Y, Hojo M, Yoshida Z (1988) J Org Chem 53:5731

104. Harayama H, Takahashi Y, Kimura M, Fugami K, Tanaka S, Tamaru Y (1996) Tetrahedron Lett 37:7287
105. Koóš P, Špánik I, Gracza T (2009) Tetrahedron Asymmetry 20:2720
106. Shinohara T, Arai MA, Wakita K, Arai T, Sasai H (2003) Tetrahedron Lett 44:711
107. Harayama H, Abe A, Sakado T, Kimura M, Fugami K, Tanaka S, Tamaru Y (1997) J Org Chem 62:2113
108. Yip K-T, Yang M, Law K-L, Zhu N-Y, Yang D (2006) J Am Chem Soc 128:3130
109. Yip K, Zhu N, Yang D (2009) Org Lett 11:1911
110. He W, Yip K-T, Zhu N-Y, Yang D (2009) Org Lett 11:5626
111. Åkermark B, Bäckvall JE, Siiralah K, Sjoberg K, Zetterberg (1974) Tetrahedron Lett 15:1363
112. Backvall J, Bjorkman EE (1980) J Org Chem 45:2893
113. Akermark B, Zetterberg K (1984) J Am Chem Soc 106:5560
114. Backvall J, Bjorkman EE (1984) Acta Chem Scand 8:91
115. Hegedus LS, Akermark B, Zetterberg K, Olsson LF (1984) J Am Chem Soc 106:7122
116. Isomura K, Okada N, Saruwatari M, Yamasaki H, Taniguchi H (1985) Chem Lett 385
117. Desai LV, Sanford MS (2007) Angew Chem Int Ed 46:5737
118. Liu G, Stahl SS (2006) J Am Chem Soc 128:7179
119. Mai DN, Wolfe JP (2010) J Am Chem Soc 132:12157
120. Brice JL, Harang JE, Timokhin VI, Anastasi NR, Stahl SS (2005) J Am Chem Soc 127:2868
121. Hanley PS, Markovic D, Hartwig JF (2010) J Am Chem Soc 132:6302
122. Liu G, Stahl SS (2007) J Am Chem Soc 129:6328
123. White PB, Stahl SS (2011) J Am Chem Soc 133:18594
124. Ye X, Liu G, Popp BV, Stahl SS (2011) J Org Chem 76:1031
125. Weinstein AB, Stahl SS (2012) Angew Chem Int Ed 51:11505

Top Heterocycl Chem (2013) 32: 109–156
DOI: 10.1007/7081_2013_105

Published online: 30 March 2013

Synthesis of Saturated Heterocycles via Metal-Catalyzed Alkene Hydroamination or Hydroalkoxylation Reactions

Lisa D. Julian

Abstract The intramolecular hydrofunctionalization of carbon–carbon multiple bonds has emerged as a powerful way to form cyclic structures. A particularly important class of reactions involves the use of amine or alcohol nucleophiles, and alkenes or allenes as electrophiles, to form pyrrolidine, piperidine, tetrahydrofuran, and tetrahydropyran heterocycles in a highly efficient manner using late transition metal catalysts. Asymmetric methods for hydroamination and hydroalkoxylation reactions have recently emerged, allowing for the enantioselective synthesis of such saturated heterocycles. This review covers recent developments (over the last 5–10 years) in late transition metal-catalyzed hydroamination and hydroalkoxylation reactions that generate saturated heterocycles.

Keywords Asymmetric · Catalysis · Heterocycles · Hydroalkoxylation · Hydroamination

Contents

L.D. Julian (✉)
Department of Chemistry, University of Colorado-Denver, Campus Box 194,
PO Box 173364, Denver, CO 80217-3364, USA
e-mail: lisa.julian@ucdenver.edu

1 Introduction

The synthesis of nitrogen and oxygen heterocycles via hydroamination or hydroalkoxylation, respectively, is perhaps the most conceptually simple approach to form this class of cyclized structures from readily available amino- and hydroxyalkenes and allenes (Scheme 1). Such heterocycles are highly prevalent in biologically active molecules; therefore, much effort has been spent on developing catalysts to effect this transformation. Complexes based on rare-earth, alkaline-earth, and group IV metals are known to catalyze the hydrofunctionalization of unsaturated carbon–carbon bonds; however, their lack of stability and functional group tolerance can render these systems impractical for applications toward academic and industrial synthetic targets [1–3]. On the other hand, late transition metal catalysts appear to have a much broader scope and applicability and recent efforts have focused on the development of new late metal complexes for hydrofunctionalizations [4].

Complexes of late transition metals are more stable toward air and moisture and are more tolerant of polar functional groups than early transition metal and lanthanide catalysts. However, until recently, the substrate scope of reactions catalyzed by late transition metals was relatively narrow. This was likely due to the propensity of the heteroatom to unproductively bind to the metal center and due to the lower reactivity of late metal catalysts to promote cyclization, which required most substrates to contain gem-disubstitution on the alkyl linker to accelerate the reaction via the Thorpe–Ingold effect. A number of highly active late transition metal catalysts have now been developed that overcome these limitations, and will be the focus of this review.

One common limitation for hydrofunctionalization is the need for an activating group on the olefin in the form of conjugation (styrenes, dienes) or inclusion of electron-withdrawing groups (Michael acceptors). As such, hydrofunctionalization of unactivated olefins remains a challenge and much effort is ongoing to identify catalysts for this class of substrates. The development of catalysts to control both regioselectivity and enantioselectivity is also of particular importance.

1.1 General Mechanistic Considerations

The intrinsic barrier toward cyclization of amino- and hydroxyalkenes and allenes in the absence of a catalyst is recognized by the presence of two nucleophiles in the substrate, the heteroatom (N or O) and the π-system (alkene, allene, alkyne).

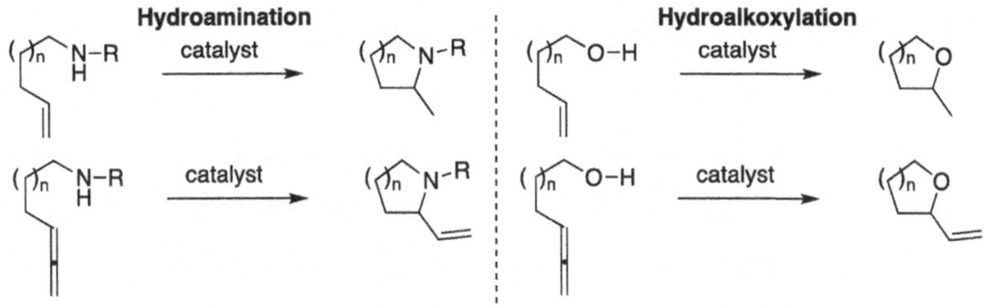

Scheme 1 Hydroamination and hydroalkoxylation of alkenes and allenes to form saturated five- and six-membered heterocycles

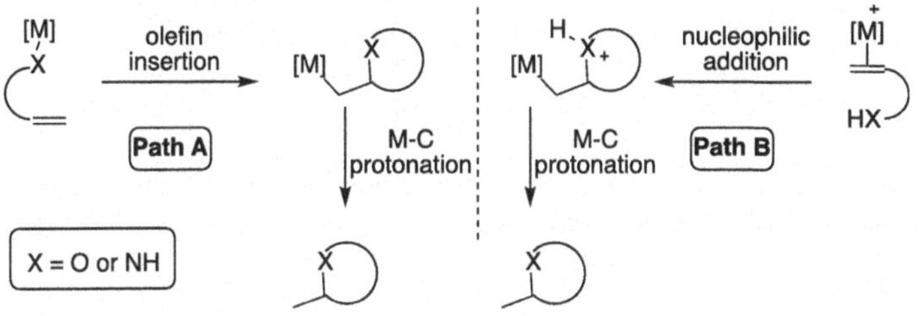

Scheme 2 Proposed limiting mechanisms for late transition metal-catalyzed hydrofunctionalization of unsaturated carbon–carbon bonds

However, the use of a late transition metal catalyst can overcome this barrier, by either activating the heteroatom, allowing for a migratory insertion of the olefin into the M–X bond (Path A, Scheme 2), or by reversing the polarity of the double bond through complexation to form a π-adduct (Path B). Both mechanisms typically favor formation of the Markovnikov product.

Typically, electron-deficient late metal catalysts have been shown to react by outer-sphere nucleophilic attack of the heteroatom onto the coordinated, and therefore activated olefin, allene, or alkyne. Key to the success of this reaction pathway is a favorable equilibrium toward the π-complex (i.e., **II**, Scheme 3) over the heteroatom-bound complex (i.e., **I**). While hydroxyalkenes typically favor the olefin π-complex, amine substrates often favor the heteroatom-bound amine complex, which can deactivate the catalyst. This observation has often led to the use of election deficient N–H donors, such as sulfonamides and carbamates, in late metal-catalyzed hydroamination reactions. The activated olefin undergoes nucleophilic attack by the tethered heteroatom in an *anti* fashion to afford the cyclized adduct **III**, which then undergoes metal–carbon bond cleavage to release the product, upon ligand substitution with aminoalkene [5]. Favorable rates for protonolysis of the M–C bond (**III**→**IV**) over reversion of the cyclized intermediate **III** back to the olefin complex **II** have been shown to be a key factor for achieving efficient catalysis, especially with highly electron-deficient metal complexes [6]. The mechanism for metal–carbon bond cleavage can be envisioned to occur by several different pathways, including direct protonolysis of the M–C bond and C–H reductive

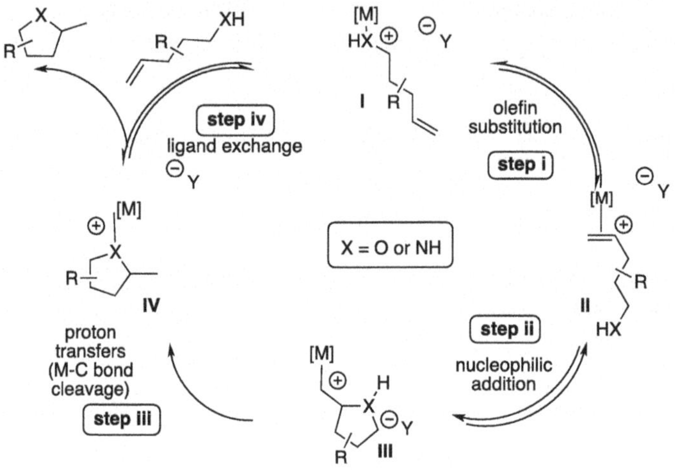

Scheme 3 Catalytic cycle for late metal-catalyzed hydrofunctionalization: nucleophilic attack of heteroatom onto a metal-bound activated olefin

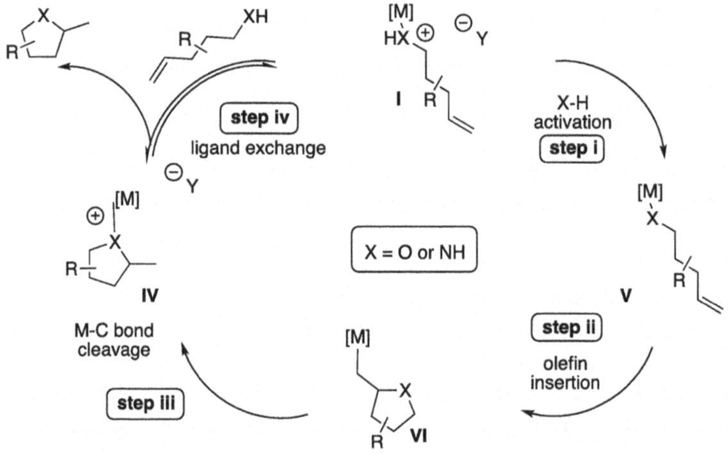

Scheme 4 Catalytic cycle for late metal-catalyzed hydrofunctionalization: olefin insertion into a metal alkoxo- or metal-amido bond

elimination through an intermediate high oxidation state metal–hydride species, although empirical evidence supporting any of the possible mechanisms is lacking.

Although not as common, an olefin insertion mechanism has been proposed for some late transition metal catalysts (Scheme 4). The metal-amido or metal-alkoxo intermediate **V** can be formed either by direct X–H oxidative addition [7], which is more facile for electron-rich metal complexes, or via a base-assisted reaction. Insertion of the olefin into the M–X bond occurs in a *syn* fashion to afford the neutral alkylmetal intermediate **VI**. Finally, metal–carbon bond cleavage, followed

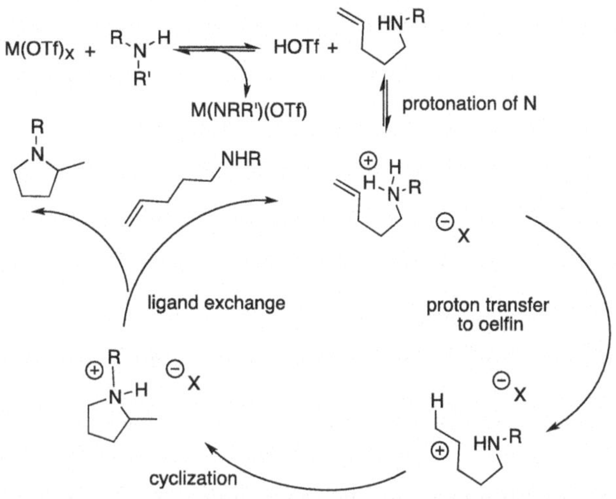

Scheme 5 Bronsted acid-catalyzed hydroamination

by ligand substitution of the product with another equivalent of substrate, completes the catalytic cycle.

There has been some debate as to whether Bronsted acids (i.e., TfOH) are responsible for catalyzing reactions involving some transition metal complexes [8]. Mechanistic studies have provided evidence for a catalytic cycle for hydroamination shown in Scheme 5, whereby TfOH is generated from a reaction of the metal triflate and amine [9, 10]. While the possibility of Bronsted acid catalysis remains ambiguous in some systems, evidence for active participation of the metal has been provided in many other cases. It should be noted that alternative mechanistic pathways have been proposed, especially in cases involving metal–hydride catalyst precursors (Sect. 3.1) [11].

2 Hydroamination of Aminoalkenes

Until recently, the late transition metal-catalyzed addition of N–H groups across *unactivated* alkenes (hydroamination) to form pyrrolidines and piperidines was rare. However, over the last 10 years, researchers have been able to overcome many challenges associated with this process, including low reactivity due to unproductive binding of the amine to the metal center and β-hydride elimination of the intermediate alkylmetal species, through careful ligand design and tuning of the electronic properties of the substrate and metal center. The advances made in developing more active and selective catalysts, along with a current standing of substrate scope, will be reviewed in the following section.

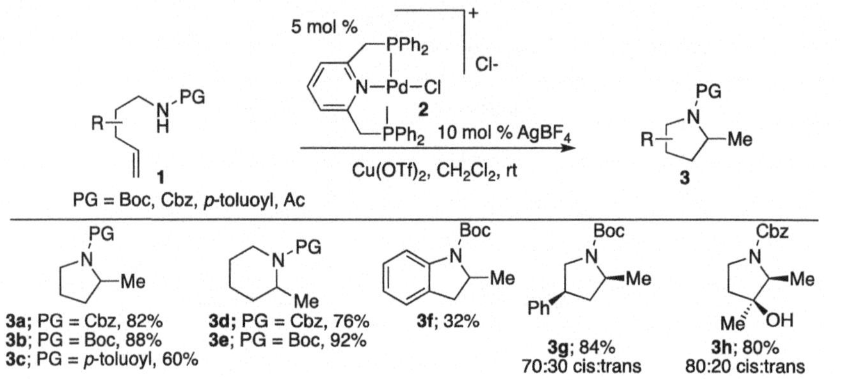

Scheme 6 PNP palladium-catalyzed hydroamination to form pyrrolidines and piperidines

2.1 Substrates Bearing an Electron-Deficient N–H Donor

Amines are often used as ligands for late transition metals as they readily bind to metal centers to form stable complexes. However, in the context of hydroamination, binding of the amine substrate can lead to poisoning of the catalyst, which can inhibit the desired reaction pathway. One strategy that has been successfully employed involves protection of the amine as a sulfonamide, carbamate, or amide group. Cationic palladium and gold complexes have been primarily developed as catalysts for hydroamination of this substrate class. Iron complexes have also been shown to catalyze reactions containing electron-deficient N–H donors, but the scope and limitations are less established [12, 13].

2.1.1 Palladium-Catalyzed Reactions

Olefin complexes of cationic palladium centers and their reactivity toward amine nucleophiles to subsequently form alkylpalladium products have been known since the 1960s [14–16]. However, a palladium catalyst for intramolecular hydroamination of unactivated aminoalkenes was not developed until many years later. In 2006, Michael and coworkers reported that a dicationic palladium pincer complex, generated from the chloride precursor **2** and $AgBF_4$, catalyzed the hydroamination of protected aminoalkenes at room temperature (Scheme 6) [17]. The catalyst was shown to be highly active, as substrates lacking substituents on the alkyl chain that would bias the reaction toward cyclization were readily cyclized to form five- or six-membered rings (**3a–3e**). Various protecting groups on nitrogen were tolerated as well as an unprotected alcohol on the alkyl chain (**3h**), which notably, would not be compatible with lanthanide or group IV catalysts. Substrates containing a stereocenter in the alkyl tether were cyclized with moderate levels of diastereoselectivity. Finally, this methodology was applied to the synthesis of 2,6-disubstituted piperazines **5**, which gave products with high levels of diastereoselectivity (Scheme 7) [18].

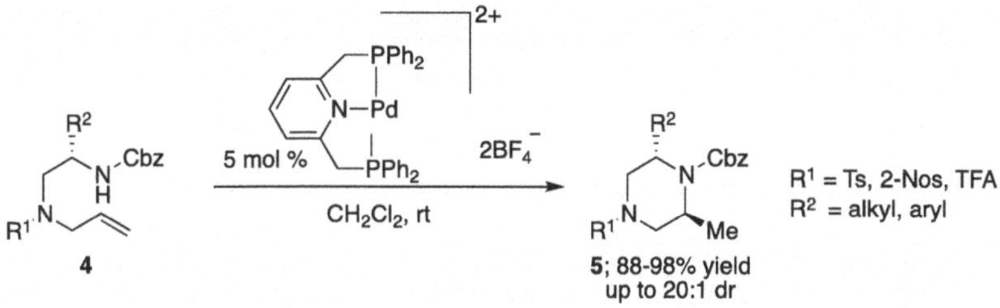

Scheme 7 PNP palladium-catalyzed hydroamination to form piperazines

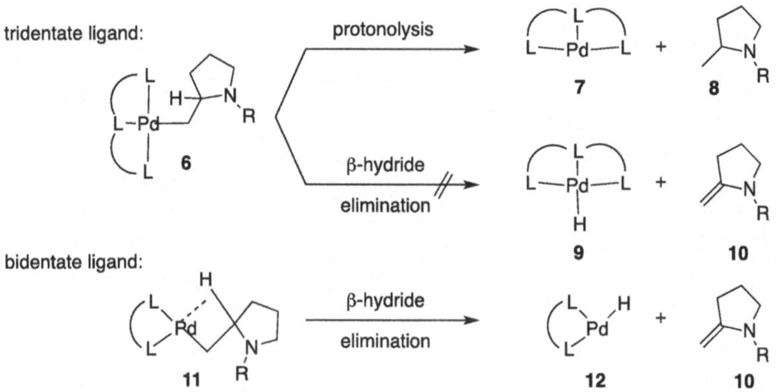

Scheme 8 Inhibition of β-hydride elimination through use of a tridentate pincer ligand

Key to the success of this reaction was the use of a tridentate pincer ligand, which blocks a coordination site on the metal and inhibits β-hydride elimination (Scheme 8). Protonolysis of the alkylpalladium bond of **6** to give **7** is then favored and leads to formation of the desired heterocycle. Detailed mechanistic studies carried out using stoichiometric amount of the dicationic palladium catalyst precursor and substrate provided support for a mechanism involving reversible nucleophilic attack of the amine on the tethered olefin, followed by rate-limiting protonolysis of the palladium–carbon bond [6]. The alkylpalladium complex **6** was shown to be the catalyst resting state, and undesired β-hydride elimination from complex **6** was not observed. In contrast, complexes bearing bidentate ligands (i.e., **11**) would be expected to undergo relatively facile β-hydride elimination.

More recently, Liu and coworkers reported a new catalyst system for intramolecular hydroamination of styrene substrates **13** to form 2-arylpyrrolidines **14** (Scheme 9) [19]. Both electron-rich and electron-poor arenes were tolerated. The authors employed a palladium(II) catalyst precursor in combination with a bulky bipyridine ligand and a stoichiometric amount of *N*-fluorobenzenesulfonamide as a terminal oxidant. The aminoalkene cyclizations are proposed to occur via a mechanism involving olefin insertion into a Pd(II)–H intermediate followed by oxidation to a Pd(IV) species. This catalyst system appears to be limited to styrene-containing substrates as the unactivated alkene *N*-tosyl-*Z*-pentenylamine failed to react under

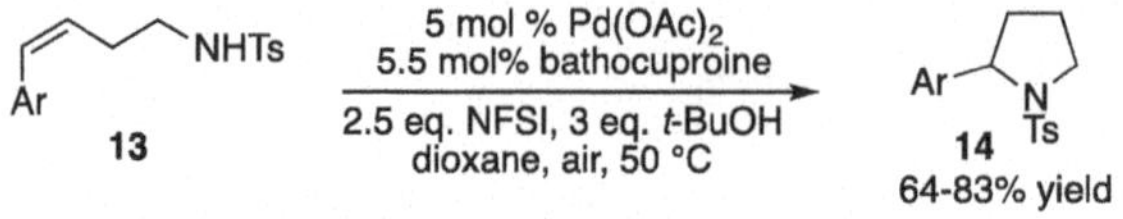

Scheme 9 Palladium-catalyzed intramolecular hydroamination of sulfonamides

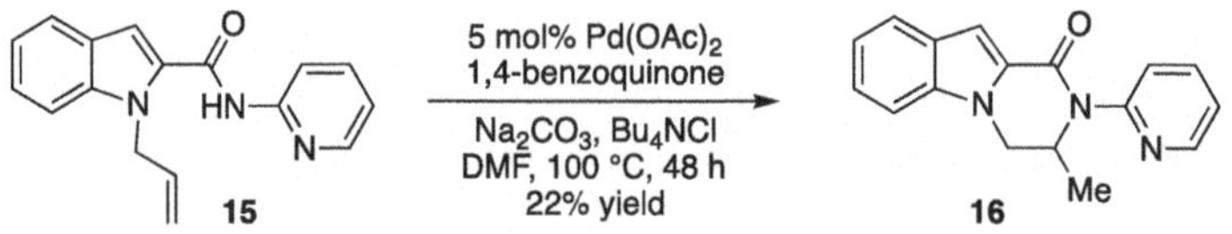

Scheme 10 Palladium-catalyzed hydroamination of *N*-allylindoles

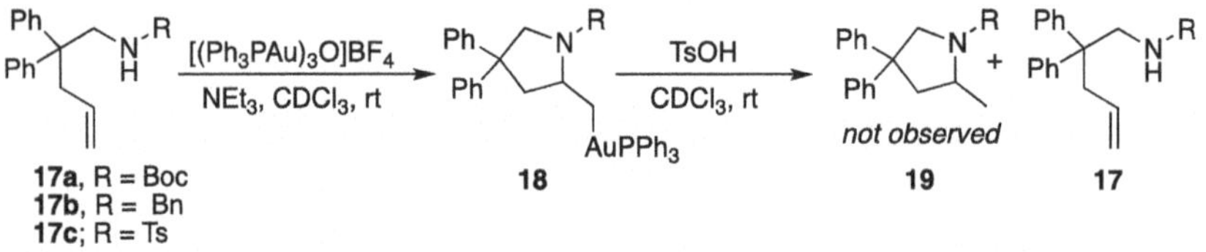

Scheme 11 Mechanistic studies of a gold-catalyzed hydroamination

these conditions. Abbiati and coworkers have also reported a palladium-catalyzed hydroamination reaction of 1-allyl-2-indolecarboxamide **15** to afford polycyclic products **16** (Scheme 10) [20].

2.1.2 Gold-Catalyzed Reactions

Gold complexes have emerged as highly active catalysts for the hydroamination of unsaturated carbon–carbon bonds [21–23]. While reactions involving nucleophilic additions to more reactive alkynes and allenes are well established, only more recently have unactivated alkenes participated in gold-catalyzed hydroamination reactions [24]. An electron-withdrawing group on the nitrogen atom is required for efficient reactivity when cationic gold complexes are employed as catalysts.

In 2010, Toste reported studies on the mechanism of gold-catalyzed hydroaminations of unactivated alkenes [25]. Alkylgold intermediates resulting from *anti*-aminoauration of unactivated olefins in the presence of a base were isolated and characterized, providing direct experimental evidence for the first elementary step of the catalytic cycle (Scheme 11). Alkylgold species are inert toward β-hydride elimination and therefore have a different reactivity compared to the alkylpalladium complexes described above. Interestingly, an aminoalkene containing a basic amine (**17b**, R = Bn) readily cyclized but the resulting alkylgold intermediate (**18**; R = Bn) was highly unstable and not isolable. These results suggest that basic amines fail in hydroamination reactions catalyzed by gold, not due to strong binding of the amine to the gold center, but rather by the inability to

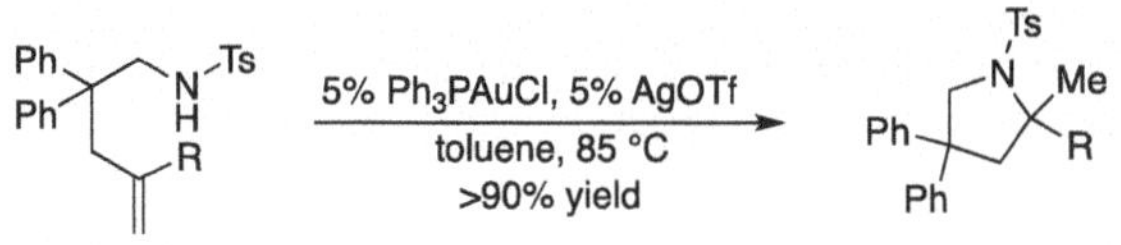

20, R = H or Me **21**

Scheme 12 Gold-catalyzed intramolecular hydroamination of unactivated alkenes

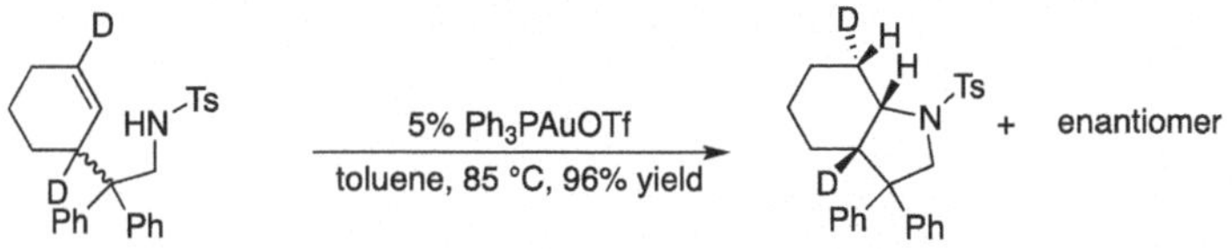

22 **23**

Scheme 13 Deuterium-labeling study

protonate the carbon–gold bond under the reaction conditions. Attempts to protodeaureate the isolable alkylgold complexes (**18**, R = Ts) to form product **19** (R = Ts) by treatment with TsOH resulted only in reversion to aminoalkene **17c**. These results are consistent with a mechanism characterized by reversible nucleophilic attack and rate-limiting protonation of the carbon–gold bond. Computational results corroborate the high barrier for protonolysis. The authors comment that these studies, however, do not rule out a Bronsted acid-catalyzed process, resulting from release of a proton generated in situ.

In 2006, He and coworkers reported the first hydroamination of unactivated aminoalkenes to form pyrrolidines using a cationic gold catalyst PPh_3AuOTf (generated in situ from PPh_3AuCl and AgOTf) at 85°C (Scheme 12) [24]. Only substrates containing a sulfonamide group on the nitrogen were successfully cyclized; those bearing basic amines failed to react. Both terminal and 1,1-disubstituted olefins were cyclized in excellent yields.

The gold-catalyzed cyclization of deuterated substrate **22** was performed to probe reaction mechanism (Scheme 13), and led to exclusive formation of the bicyclic product **23** [24]. This result suggests that the nitrogen atom attacks from the opposite face of a gold(I)-bound olefin to give the *trans*-addition product after protonolysis of the resulting gold(I)–carbon bond.

Widenhoefer has reported the use of a modified cationic gold complex containing a bulky monophosphine ligand (S-Phos) for the hydroamination of aminoalkenes bearing carbamate protecting groups [26]. The high catalytic activity of this complex is attributed to steric rather than electronic factors. A wide range of substrates were cyclized to form five- and six-membered rings, and alcohol and ester functional groups were tolerated. The Widenhoefer group subsequently demonstrated that ammonium salts such as **24** and **26** could also be employed as electron-deficient nitrogen nucleophiles, further exemplifying the utility of this catalyst (Scheme 14) [27]. The biologically relevant tetrahydroisoquinoline heterocycle **27** was synthesized using this method, albeit with a higher catalyst loading and lower yield.

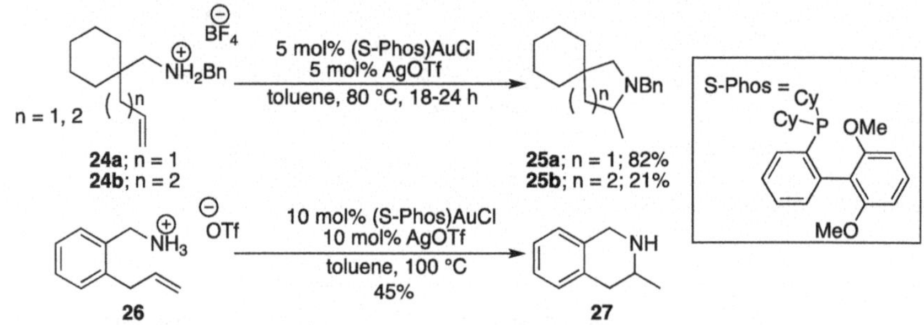

Scheme 14 Hydroamination of ammonium salts

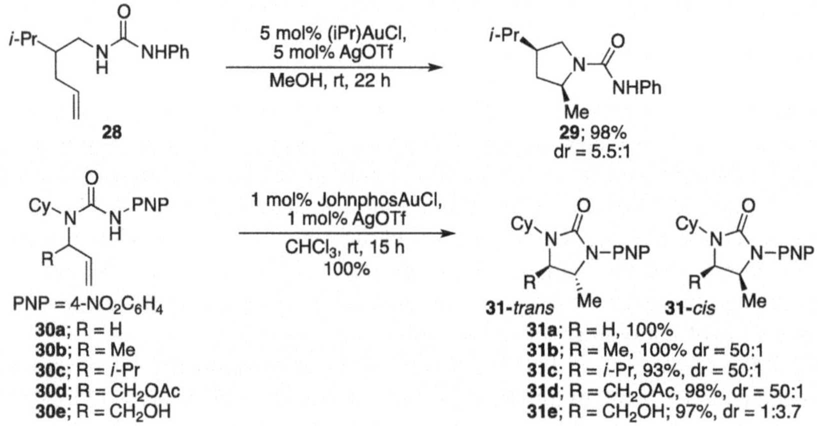

Scheme 15 Hydroamination of urea nucleophiles

Widenhoefer also reported the use of either a bulky monophosphine complex (JohnphosAuOTf) or a cationic *N*-heterocyclic carbene–gold(I) complex [(^{i}Pr)AuOTf] to cyclize unactivated alkenes with a tethered urea nucleophile (Scheme 15) [28]. Pyrrolidines and piperidines could be synthesized from substrates containing an external urea (i.e., **28**) as a protecting group. On the other hand, substrates such as **30** containing the urea group within the tether cyclized to form imidazolidin-2-ones **31**. Diastereoselectivities of >50:1 **31**-*trans*:**31**-*cis* were obtained for substrates bearing an alkyl group on the carbon linker [29]. Interestingly, the unprotected alcohol substrate **30e** gave the *cis* product **31e** as the major product in a 3.7:1 ratio.

The gold-catalyzed intramolecular hydroamination of dienes **32** bearing arylsulfonamide groups was reported by Yeh and coworkers in 2009 to afford the hexahydroindole products **33** (Scheme 16) [30]. Various electron-rich and electron-poor sulfonamides were tolerated, and reactions typically proceeded at 85°C to provide the heterocycles in excellent yields (80–88%). The relative *syn* stereochemistry was observed, which is consistent with a mechanism involving outer-sphere attack of the sulfonamide on the opposite face of the coordinated olefin.

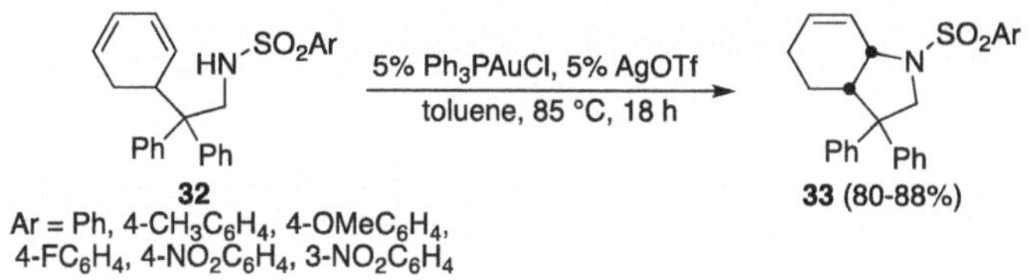

Scheme 16 Hydroamination of dienes to form bicyclic heterocycles

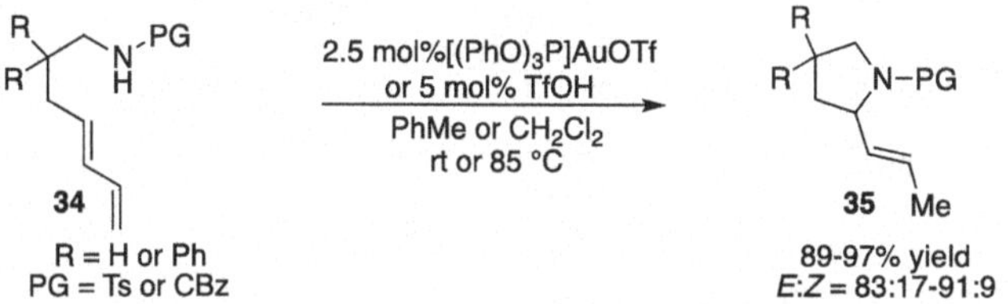

Scheme 17 Gold- and Bronsted acid-catalyzed hydroamination of dienes

In 2011, Najera and Beaza reported the use of either a Lewis or a Bronsted acid catalyst for the cyclization of aminodienes **34** to afford 2-propenyl substituted pyrrolidines **35** with moderate to good *E*/*Z* ratios (Scheme 17) [31, 32]. Notably, substrates lacking substituents at C2 also readily cyclized (**34**; R = H). The gold complex $[(PhO)_3P]AuOTf$ and the Bronsted acid TfOH were found to be the most efficient catalysts for cyclization. However, AgOTf, $FeCl_3{\cdot}6H_2O$, and [(±)-BINAP]$Cu(OTf)_2$ were also found to be competent catalysts. The parallel reactivity of these electron-deficient metal catalysts to TfOH in this report calls into question the role of the metal under these reaction conditions [8].

2.2 *Substrates Bearing a Basic N–H Donor*

Late transition metal-catalyzed intramolecular hydroamination of substrates containing a basic amine was not realized until 2005 [33], and still remains problematic for certain classes of substrates. For example, cyclization of amines onto 1,2-disubstituted amines is rare, as are reactions of aminoalkenes bearing a primary amine nucleophile. The development of new catalysts over the past 7 years has led to significant improvements in reactivity with substrates bearing a basic N–H donor. This section will highlight the discovery of new catalysts that have been designed to overcome challenges associated with basic aminoalkenes.

2.2.1 Platinum-Catalyzed Reactions

Platinum-coordinated olefin complexes have been known to undergo nucleophilic attack by amines for decades [14]. However, a catalytic intramolecular hydroamination process was not realized until relatively recently. Cleavage of the metal–carbon bond to release the product and regenerate the catalyst is the key step

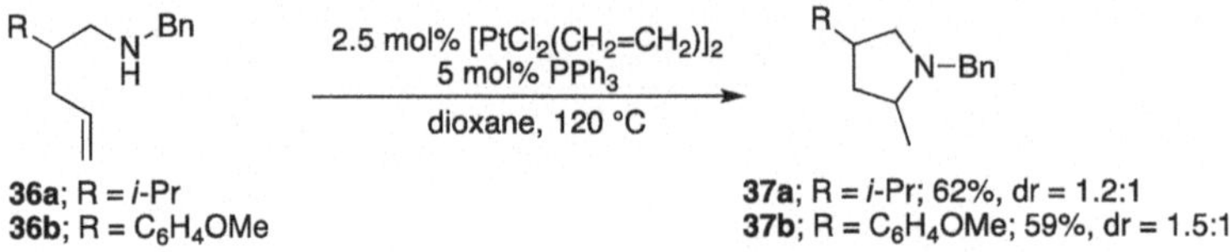

Scheme 18 First platinum-catalyzed hydroamination of basic aminoalkenes

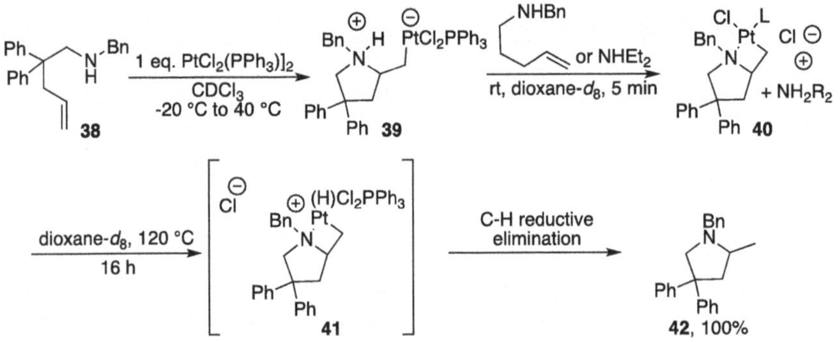

Scheme 19 Proposed mechanism for platinum-catalyzed hydroamination

in the catalytic cycle, and is significantly influenced by the electron density at the metal center. In 2002, further mechanistic studies on nucleophilic additions to isolated dicationic platinum-coordinated olefin complexes showed that although nucleophilic addition is facilitated by an electron-deficient platinum center, protonolysis of the Pt–C bond is disfavored due to the decreased basicity and increased stability of the electron-poor alkylmetal species. These studies provided new insight into the reactivity of platinum–carbon bonds and helped lay the foundation for the logical development of effective hydroamination catalysts [34].

The first late transition metal-catalyzed intramolecular hydroamination of an unactivated alkene with a basic amine was reported in 2005 by Widenhoefer, using a neutral platinum catalyst $[PtCl_2(PPh_3)]_2$ [33]. Aminoalkenes containing secondary alkyl amines with gem-disubstitution on the alkyl linker were readily cyclized to form five- and six-membered rings. The less reactive substrates **36a–b** containing a single substituent on the alkyl chain also reacted to form products **37a–b** in moderate yields and low diastereoselectivity (Scheme 18). Terminal olefins and 1,1-disubstituted olefins were shown to react with amines under these conditions; however, internal olefins were unreactive.

To gain mechanistic information, a stoichiometric reaction of aminoalkene **38** and platinum dimer $[PtCl_2(PPh_3)]_2$ was performed (Scheme 19). The cyclized zwitterionic alkylplatinum intermediate **39** was formed readily and underwent platinum–carbon bond cleavage upon heating to 120°C in the presence of excess amine via the neutral heterobicyclic intermediate **40**. This is proposed to occur by way of intermediate Pt(IV) hydride complex **41** that undergoes C–H bond-forming reductive elimination to release the organic product **42** [33].

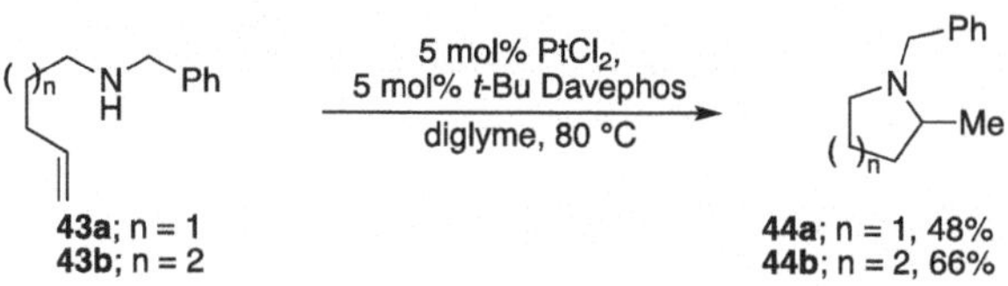

Scheme 20 Platinum-catalyzed hydroamination of *N*-benzyl aminoalkenes

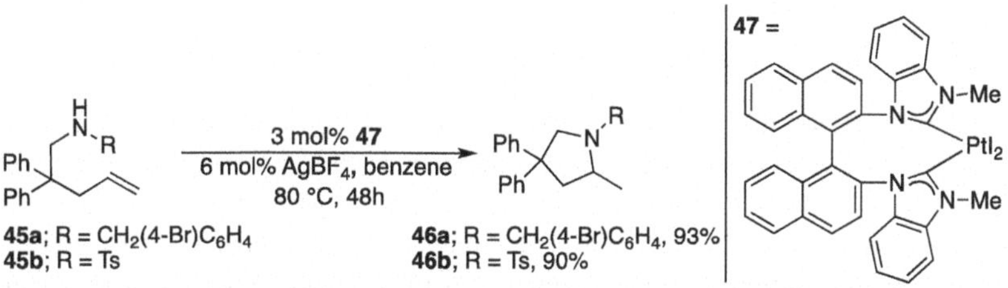

Scheme 21 Pt–NHC-catalyzed hydroamination

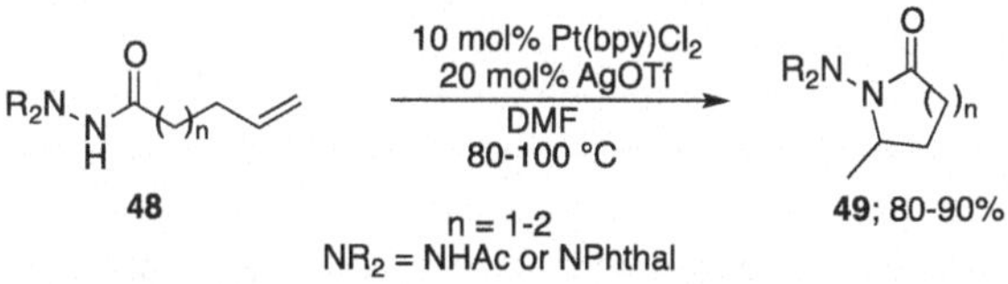

Scheme 22 Platinum-catalyzed hydrohydrazination

This mechanistic hypothesis suggested that a bulky phosphine ligand could facilitate C–H reductive elimination, and use of an *o*-biphenyl phosphine ligand (tBu Davephos, Scheme 20) allowed for reactions to be conducted at lower temperatures (60–80°C versus 120°C) and displayed improved substrate scope [35]. For example, even *N*-benzyl-4-pentenyl-1-amine **43a** and *N*-benzyl-5-hexenyl-1-amine **43b** cyclized to form the corresponding pyrrolidine and piperidine products **44a–b**, which failed using the initially reported Pt/PPh_3-based catalyst system (Scheme 20).

Other ligands, including *N*-heterocyclic carbenes, have since been investigated in platinum-catalyzed hydroaminations [36, 37]. For example, Shi's cationic NHC–Pt(II) complex **47** catalyzed the cyclization of a range of substrates including both basic and electron-deficient *N*-donors (Scheme 21) [36].

Michael and coworkers recently reported a platinum-catalyzed hydrohydrazination of alkenyl hydrazides **48** to form *N*-aminoheterocycles **49** using a cationic platinum catalyst with a bipyridine ligand generated from (bpy)$PtCl_2$ and two equivalents of AgOTf (Scheme 22) [38]. Detailed mechanistic studies support a mechanism involving insertion of the olefin into a platinum-amido bond following initial N–H activation.

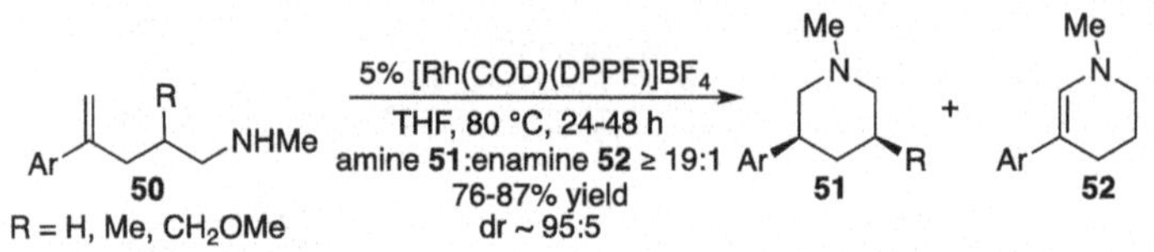

Scheme 23 Anti-Markovnikov hydroamination of styrenes

2.2.2 Rhodium-Catalyzed Reactions

The first Rh-catalyzed hydroamination (of ethylene with secondary amines) was reported by Coulson in 1971 [39]. Since then, ligated rhodium complexes have been developed as catalysts for the intramolecular hydroamination of unactivated aminoalkenes containing a basic amine functionality [40]. In particular, rhodium complexes have been the first class of catalysts to demonstrate reactivity with substrates that contain a primary amine donor [41], and also tolerate a wide range of polar functionality. The most effective catalysts to date are monocationic at the rhodium center and are proposed to react via the nucleophilic addition pathway (Scheme 3). Although primary amines readily bind to rhodium, these substrates still undergo effective catalysis [42].

The intermolecular hydroamination of styrenes to afford anti-Markovnikov products using a rhodium catalyst was reported by Beller in 1997 using the cationic catalyst precursor $[Rh(COD)_2]BF_4$ [43, 44]. In 2006, Hartwig and coworkers reported an intramolecular hydroamination of secondary amines onto styrenes to afford piperidine products resulting from anti-Markovnikov addition via an uncommon 6-*endo*-trig cyclization (Scheme 23) [45–47]. The cationic rhodium complex $[Rh(DPPB)(COD)]BF_4$ was identified as the optimal catalyst precursor for efficient cyclization of aminostyrenes **50** to piperidines **51**, with minimal formation of oxidative amination byproducts **52**, which presumably results from β-hydride elimination of an alkylrhodium intermediate. Substrates with a β-substituent on the alkyl chain (R = Me, OCH_2OMe) afforded products **51** with high *cis* selectivity, which was rationalized by invoking a chair-like transition state in which the aryl and alkyl substituents occupy equatorial positions. Participation of an η^6-Rh–arene complex, formed from coordination of the catalyst with the styrene substrate, may explain the unusual anti-Markovnikov selectivity observed. As such, the development of catalyst systems for the anti-Markovnikov addition of amine to unactivated olefins still remains a significant challenge.

Hartwig and Liu subsequently reported a rhodium catalyst bearing the Davephos ligand that exhibits excellent activity for hydroamination of unactivated alkenes containing both primary and secondary amines (Scheme 24) [41]. Substituted aminoalkenes containing a wide range of functionality, such as aryl nitriles, esters, and alcohols (**53**, Ar = 4-ClC_6H_4, 4-CNC_6H_4, 4-$CO_2MeC_6H_4$), were cyclized at relatively mild temperatures to form pyrrolidines and piperidines selectively. Byproducts resulting from oxidative amination or alkene isomerization were not formed when the Davephos ligand was employed, but were observed when

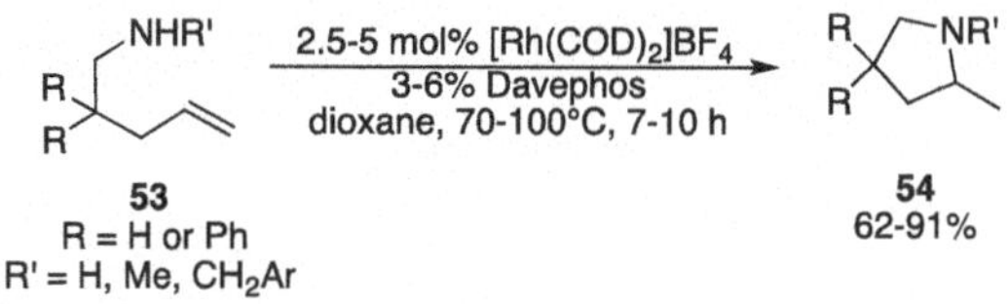

Scheme 24 Rhodium-catalyzed hydroamination of unactivated alkenes using a bidentate η^6-Rh–arene–phosphine complex

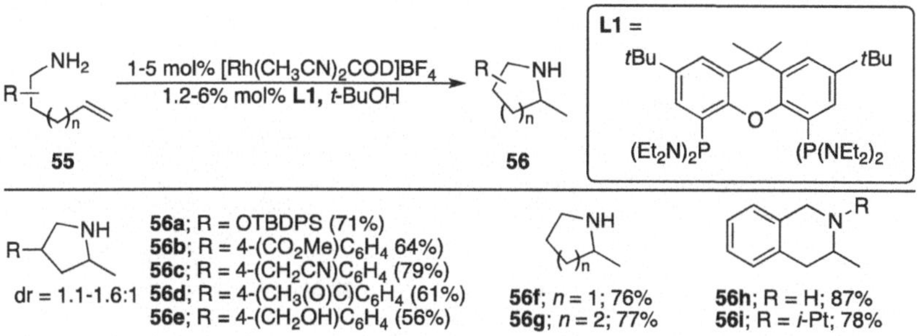

Scheme 25 Rhodium-catalyzed hydroamination of primary aminoalkenes

bidentate phosphine ligands such as DPPB were used. Notably, this was the first late transition metal catalyst to cyclize primary aminoalkenes (**53**; R′ = H); however, higher temperatures and gem-disubstitution (R = Ph) on the alkyl chain were required to facilitate cyclization. Detailed mechanistic studies revealed the active catalyst to be an η^6, κ^1 complex that readily binds olefin to allow for intramolecular nucleophilic attack of the pendant amine [48]. Protonolysis of the Rh–C bond was found to be the rate-limiting step in the catalytic cycle.

In 2010, a rhodium catalyst ligated with an unusual tridentate POP-pincer ligand bearing aminophosphine groups was shown to significantly improve reactivity for primary amine substrates (Scheme 25) [42]. For the first time using a late transition metal catalyst, primary amine substrates that were unbiased toward cyclization and that possessed auxiliary functional groups were cyclized to form five- and six-membered rings. Cyclization of tethered primary amines onto dienes and internal olefins was possible, although these reactions required higher temperatures (100°C) and catalyst loadings, and gave only moderate yields of the cyclized products. In addition, tetrahydroisoquinolines (e.g., **56h–i**), a common structural motif in biologically active molecules, were readily synthesized by hydroamination of 2-allylbenzylamines. Secondary aminoalkenes also readily cyclized under the same conditions. Mechanistic studies indicate that the tridentate "pincer" coordination mode of the ligand is likely involved in inhibiting competing β-hydride elimination reactions that would form imine byproducts. The aminophosphine groups on the ligand are also of key importance, as analogous alkyl- and arylphosphine derivatives lead to inferior results.

The authors provided additional evidence to support a mechanism involving nucleophilic attack of the amine on a coordinated olefin by investigating the

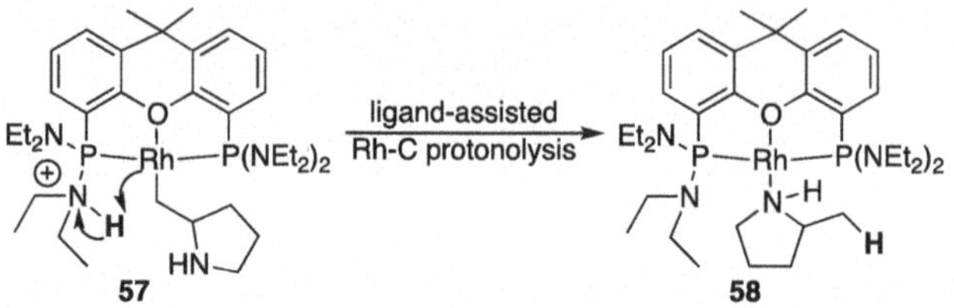

Scheme 26 Possible ligand-assisted Rh–C bond cleavage

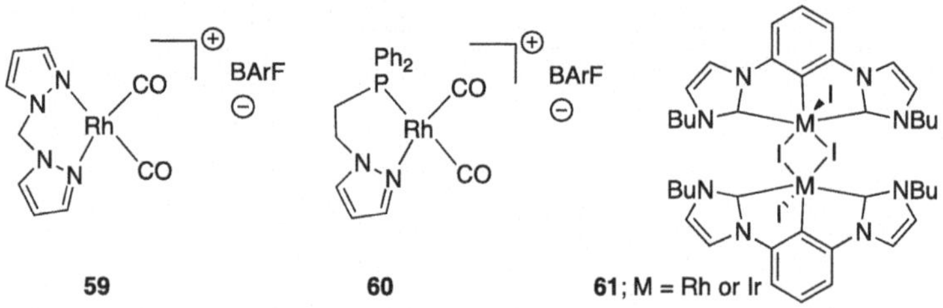

Scheme 27 Rhodium catalysts for hydroamination of aminoalkenes

catalyst resting state and reaction intermediates. However, data imply that these reactions occur with a turnover-limiting step that is different from that of reactions catalyzed by late transition metal complexes of Rh [48], Pd [6], Pt [33], and Ir [49]. This change in the turnover-limiting step and resulting high catalyst activity stem from favorable rates for protonolysis of the Rh–C bond. Probes for the origin of the reactivity of the rhodium complex of **L1** implied that the aminophosphine groups may facilitate proton transfer to the Rh–C bond (e.g., Scheme 26). To date, this cationic P,O,P rhodium catalyst is the most active late transition metal catalyst for the intramolecular cyclization of primary amines.

Rhodium catalysts described in this section demonstrate high reactivity across a broad range of basic aminoalkene substrates and reports of new rhodium catalysts for hydroaminations of unactivated alkenes continue to emerge (Scheme 27) [50–52].

2.2.3 Iridium-Catalyzed Reactions

In 1988, Casalnuovo and coworkers reported the use of an electron-rich iridium complex ($Ir(PEt_3)_2(C_2H_4)Cl$) for the intermolecular hydroamination of aniline and norbornene [53]. This seminal work, along with more recent reports describing Ir-catalyzed hydroamination of aminoalkynes [54], laid the foundation for the development of new iridium catalysts for intramolecular hydroamination of unactivated alkenes [40]. Recently iridium complexes have emerged as highly active catalysts for intramolecular hydroamination of alkenes with either primary or secondary amines. Valuable mechanistic insight has emerged from these recent reports that will guide the development of new catalysts.

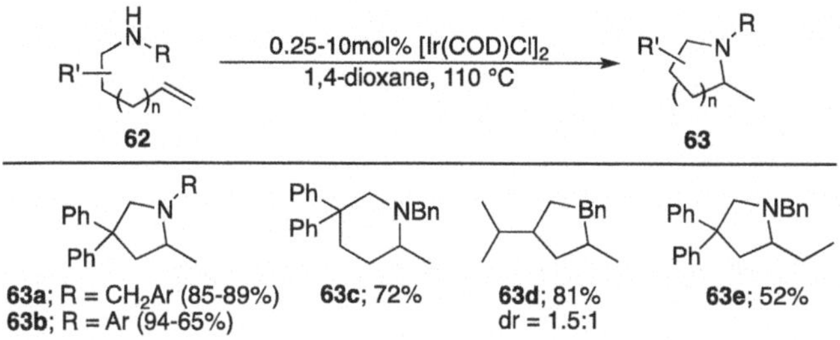

Scheme 28 Intramolecular hydroamination of *N*-alkyl and *N*-aryl aminoalkenes catalyzed by $[Ir(COD)Cl]_2$

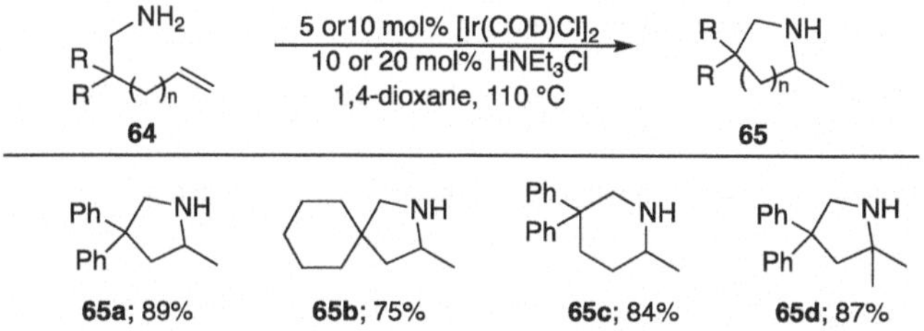

Scheme 29 Intramolecular Ir-catalyzed hydroamination of primary aminoalkenes

In 2009, Stradiotto and coworkers reported that a simple iridium catalyst, $[Ir(COD)Cl]_2$, is effective for the hydroamination secondary alkyl- and arylamines. Interestingly, the addition of added phosphine ligands or salts did not provide any beneficial effect on reactivity [55]. Under typical conditions a variety of substrates **62** were cyclized to afford pyrrolidines and piperidines **63** in excellent yields (Scheme 28). Functional groups such as an aryl chloride, ester, and methoxy ether were tolerated (R = $CH_2(4\text{-Cl})C_6H_4$, $CH_2(4\text{-OMe})C_6H_4$, $CH_2(4\text{-}CO_2Me)C_6H_4$). Unlike the currently available Rh-based catalysts, the $[Ir(COD)Cl]_2$ catalyst is also effective for cyclizing substrates containing *N*-arylamines (**62**; R = Ar). In addition, substrates containing 1,2-disubstituted olefins that are typically more challenging to cyclize underwent hydroaminations, albeit with higher catalyst loadings (5–10%) and longer reaction times (48 h). With the exception of a single example, all substrates required gem-disubstitution on the alkyl chain to bias the substrate toward cyclization.

While $[Ir(COD)Cl]_2$ alone was not effective for substrates bearing primary amines, the addition of a catalytic amount of a proton source (i.e., $HNEt_3Cl$) allowed for successful cyclization of primary aminoalkenes (Scheme 29) [49]. Gem-disubstitution on the alkyl chain was required to bias these substrates toward cyclization.

The authors propose a mechanism involving nucleophilic attack of amine onto a coordinated Ir–olefin, which contrasts with the mechanism of $Ir(PEt_3)_2(C_2H_4)Cl$-catalyzed reactions that operate via the N–H activation pathway [53]. Computational studies revealed that an N–H oxidative addition step with $[Ir(COD)Cl]_2$

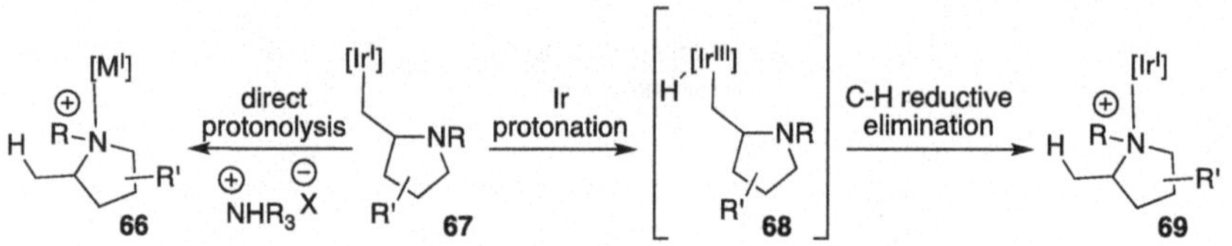

Scheme 30 Proposed mechanisms for iridium–carbon bond cleavage

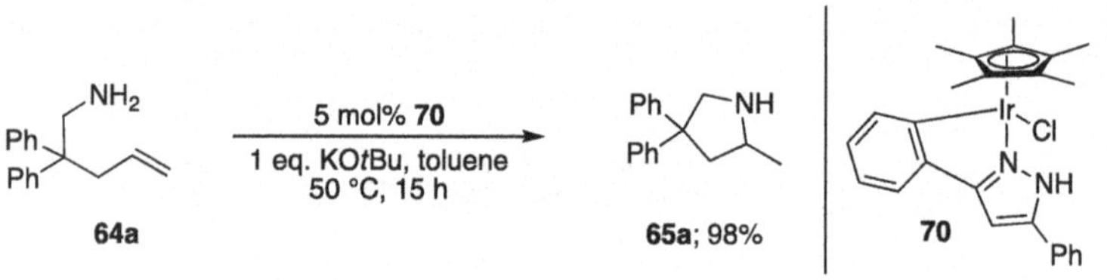

Scheme 31 Iridium-catalyzed hydroamination: metal–ligand cooperation

would be energetically prohibitive in the presence of a lower energy olefin activation pathway. The mechanism for metal–carbon bond cleavage in the [Ir(COD) Cl]$_2$-catalyzed reactions was also interrogated computationally [49], and could occur either through a direct proton transfer from an ammonium salt (**67** → **66**) or via initial formation of an Ir(III)–H species **68** followed by C–H reductive elimination to release the product (Scheme 30). DFT calculations supported the two step process involving initial protonation of the iridium center, followed by a turnover-limiting C–H reductive elimination step from a highly reactive Ir(III)-hydrido intermediate **68**. This is also consistent with the empirically measured primary kinetic isotope effect (KIE = 3.4).

In 2010, Ikariya and coworkers reported the use of a neutral Ir–pyrazolato complex **70** in combination with a strong base (KO*t*-Bu) to catalyze the hydroamination of substrate **64a** [56]. It is proposed that the base serves to deprotonate the pyrazole ligand from the chloride complex to afford an Ir–pyrazolato complex in which the ligand assists nucleophilic attack of amine onto a coordinated olefin (Scheme 31). Following this report, Tobisch published a detailed DFT computational investigation that suggested that a more complex pathway is operative, involving a pyrazolato hydrogen-bonding network that facilitates the Ir–C bond cleavage step rather than nucleophilic addition step [57].

Other iridium complexes, such as the neutral *P*,*O*-phosphino–phenolate complex **71** [58], the cationic *P*,*N*-ferrocenyl iridium complex **72** [59], and the bidentate *N*-heterocyclic carbene complex **73** [60], demonstrated reactivity for intramolecular hydroamination of secondary aminoalkenes, but these more complex catalysts have not yet shown to be advantageous over the simple [Ir(COD)Cl]$_2$ system in initial reports (Scheme 32).

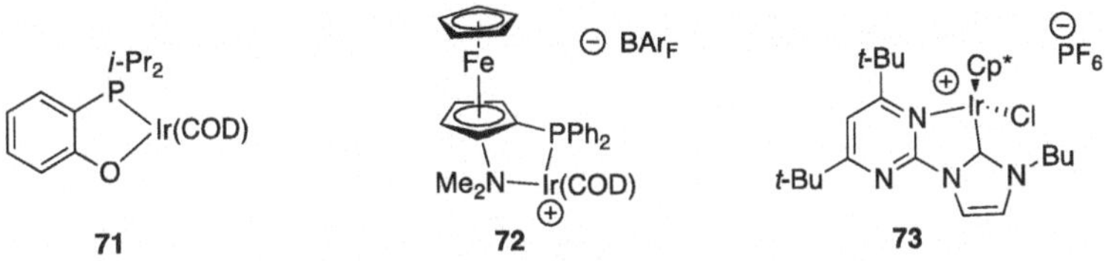

Scheme 32 Bidentate *P,X*-ligated iridium catalysts for hydroamination of secondary aminoalkenes

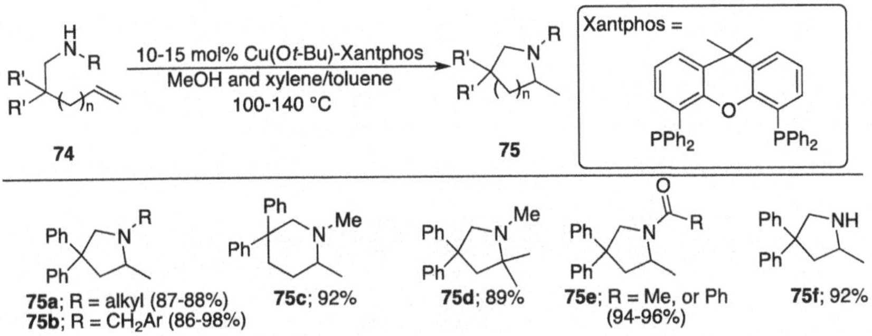

Scheme 33 Cu(O*t*-Bu)–xantphos catalyzed hydroamination

2.2.4 Copper-Catalyzed Reactions

The development of copper complexes as catalysts for hydroamination is desirable due to their significantly lower cost relative to platinum, palladium, rhodium, and iridium catalysts. While reports have begun to emerge over the last 10 years for copper-catalyzed inter- and intramolecular hydroaminations, typically either activated olefins or electron-deficient *N*-protected amines are required for reactions to proceed [61, 62]. Furthermore, Bronsted acid-catalyzed mechanisms have been typically invoked for processes employing Cu(II) salts [9, 63]. To date, there are only two reports of copper-catalyzed intramolecular hydroaminations of basic amines onto tethered unactivated olefins [63, 64].

In 2009, Sawamura and coworkers reported the use of a Cu(O*t*-Bu)–xantphos complex as an efficient catalyst for the hydroamination of both primary and secondary aminoalkene substrates **74** to afford five- and six-membered ring products **75** (Scheme 33) [64]. Interestingly, electron-deficient *N*-donors, such as amide substrates, also reacted to give protected pyrrolidine products in excellent yields. In addition, a variety of *N*-alkyl groups were tolerated, including functionalized benzyl groups containing methoxy, fluoro, cyano, or ester moieties. Unfortunately, gem-disubstitution on the alkyl linker is required to facilitate cyclization, and while substrates containing 1,1-disubstituted olefins participated in the reaction, cyclization of a pendant amine onto a 1,2-disubstituted olefin failed under these conditions.

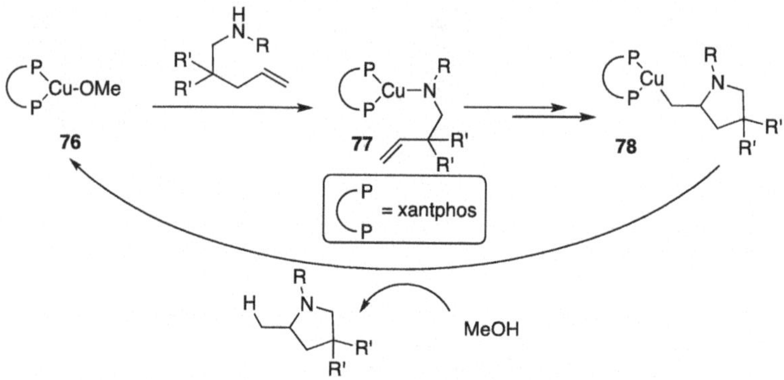

Scheme 34 Proposed mechanism for Cu(I)-catalyzed hydroamination

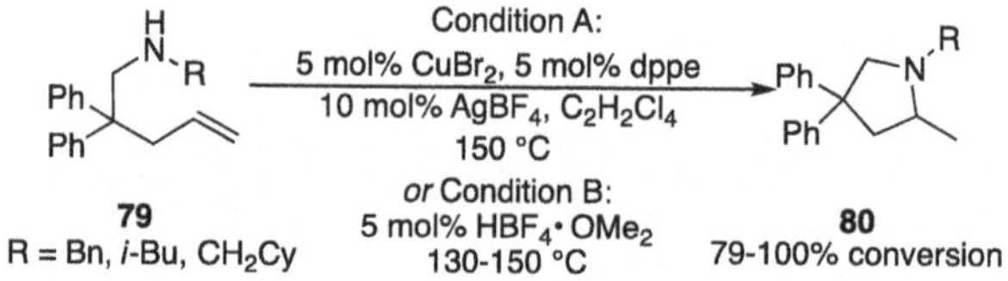

Scheme 35 Cationic copper-catalyzed hydroamination

The authors propose a mechanism involving initial formation of a copper amido complex **77** (Scheme 34). The amido complex **77** undergoes cyclometalation to afford the copper alkyl species **78**, followed by protonolysis to regenerate the copper–alkoxo complex **76** and release the hydroamination product. Enamine products were not observed under these protic conditions, presumably due to the increased rate of protonolysis versus β-hydride elimination in methanol.

A cationic copper complex generated from $CuBr_2$, dppe (diphenylphosphinoethane), and $AgBF_4$ catalyzes the hydroamination of secondary alkylamines **79** onto unactivated olefins (Scheme 35) [63]. The substrate scope is limited to secondary amine substrates containing gem-disubstitution on the alkyl linker. The Bronsted acid (HBF_4–OMe_2) was also found to catalyze hydroaminations under similar conditions (and as such the authors propose a mechanism involving the generation of an equivalent of Bronsted acid upon complexation of the amine substrate to the cationic copper center). The resulting Bronsted acid would serve as the catalyst for cyclohydroamination under these conditions.

2.2.5 Zinc-Catalyzed Reactions

In 2005, Blechert and Roesky reported the first homogeneous zinc catalyst (**83**) for hydroamination of aminoalkynes and aminoalkenes [65]. The zinc(II) center is ligated with an anionic troponiminato group that renders the complex stable toward

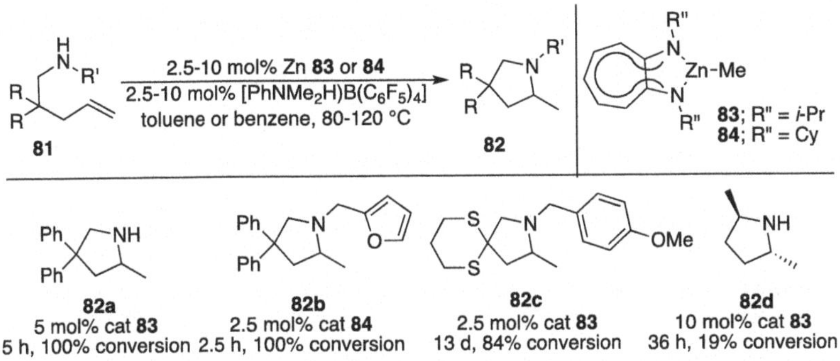

Scheme 36 Zinc-aminotroponiminate-catalyzed hydroamination

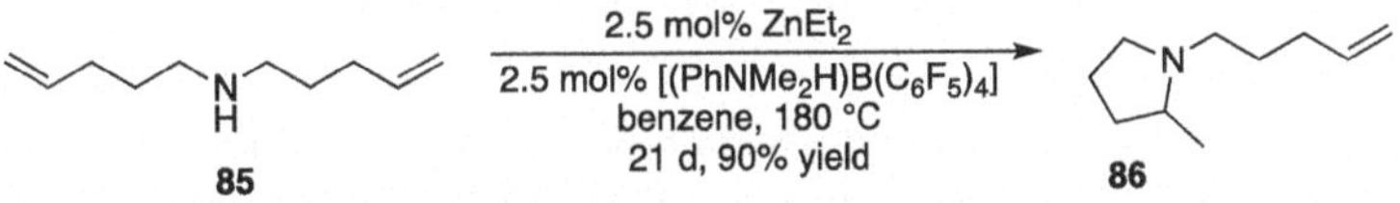

Scheme 37 Et_2Zn-catalyzed hydroamination

air and moisture. In a subsequent report the following year, a modified catalyst (**84**) was found to cyclize secondary aminoalkenes containing various Lewis basic functionalities with improved efficiency (Scheme 36) [66]. Although the catalyst was tolerant of several functional groups, reaction times were often slow and the catalyst showed little reactivity toward substrates lacking backbone substituents. A single example of a substrate without gem-disubstitution on the alkyl chain was reported to cyclize in only 19% conversion after 36 h in the presence of 10 mol% catalyst **83**. Although the mechanism of these reactions is not clear, it is proposed that a cationic zinc complex is the active catalyst, which is generated upon protonation of the alkyl group with a cocatalytic amount of the Bronsted acid $[PhNMe_2H)B(C_6F_5)_4$ [67]. Numerous related aminotroponiminate zinc complexes have also been evaluated in hydroaminations of aminoalkenes [68–72]. However, thus far none have proven superior to **84**, and chiral aminotroponiminate zinc complexes have failed to induce enantioselectivity in hydroamination reactions [72].

In 2009, Roesky and Blechert reported that a combination of Et_2Zn and the Bronsted acid activator $(PhNMe_2H)B(C_6F_5)_4$ catalyzed the hydroamination of unactivated secondary aminoalkenes to form pyrrolidines containing various functional groups [73]. Diethylzinc alone displays higher reactivity than the ligated aminotroponiminate zinc complexes **83** and **84** and was able to cyclize substrate **85** lacking gem-disubstitution on the alkyl chain, albeit with high temperature (180°) and a long reaction time (21 days) to afford pyrrolidine **86** (Scheme 37). All other substrates reported to undergo intramolecular hydroamination in the presence of Et_2Zn had gem-disubstitution on the alkyl chain to facilitate cyclization.

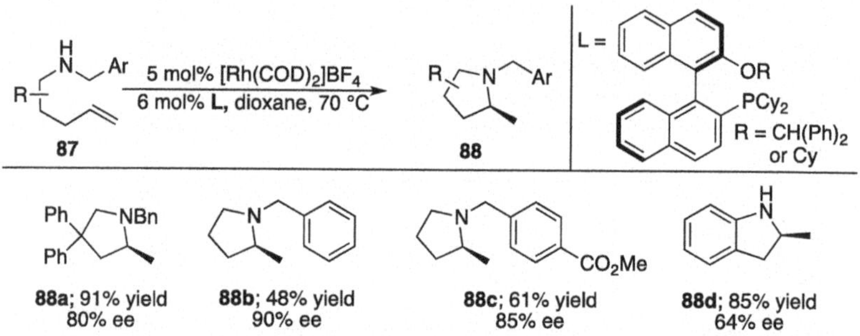

Scheme 38 Rhodium-catalyzed asymmetric hydroamination

2.3 *Asymmetric Hydroaminations of Aminoalkenes*

Although asymmetric hydroaminations to form chiral pyrrolidines and piperidines with lanthanide catalysts have been known since 1992 [74], there are only a few reports describing asymmetric intramolecular hydroaminations to form chiral, non-racemic heterocycles using late transition metal catalysts. This advance is recognized as being especially important for drug discovery programs that require the use of enantiomerically pure heterocyclic compounds for biological studies.

In 2010, Buchwald and Shen reported the use of a chiral rhodium catalyst for the asymmetric hydroamination of secondary aminoalkenes [75]. This work represents the first thorough investigation of chiral ligands and substrates for asymmetric hydroamination of unactivated olefins. They discovered that $[Rh(COD)_2]BF_4$, in combination with an axially chiral MOP ligand (L), gave benzylated pyrrolidines **88** in good yields and enantioselectivities (Scheme 38). It was found that a bulky dialkylphosphine group (i.e., PCy_2) was critical for achieving high enantioselectivities. A variety of secondary aminoalkenes were cyclized. Notably, substrates unbiased by the Thorpe–Ingold effect also reacted to give enantioenriched 2-methylpyrrolidines, albeit in modest yields. Although primary aminoalkene substrates exhibited low reactivity with this catalyst system, one example, 2-allylaniline reacted to give 2-methylindoline in good yield, but with only moderate enantioselectivity (64% ee). This represents the first and only reported example of an asymmetric intramolecular hydroamination of a primary amine substrate using a late transition metal catalyst.

More recently in 2012, Chemler and coworkers reported a copper-catalyzed enantioselective hydroamination to form chiral 2-methylindolines [76]. The ability to form hydroamination products using this catalyst system was initially observed during studies on alkene carboamination reactions [77]. Following elegant mechanistic work, the authors proposed a mechanism that proceeds via a radical pathway involving initial *cis*-aminocupration of the substrate **89** to afford **91**, followed by Cu–C bond homolysis to afford a carbon radical **92** that undergoes either H-atom abstraction to form a hydroamination product **93** or intramolecular coupling with the arylsulfonyl group to form the carboamination product **94** (Scheme 39).

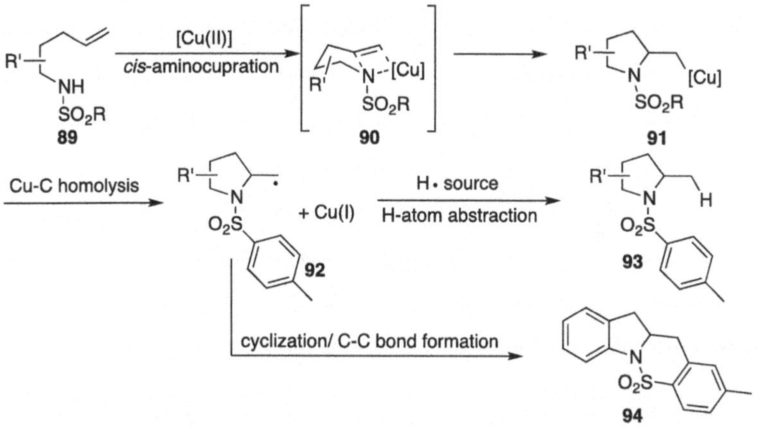

Scheme 39 Proposed mechanism for Cu(II)-catalyzed hydroamination and carboamination

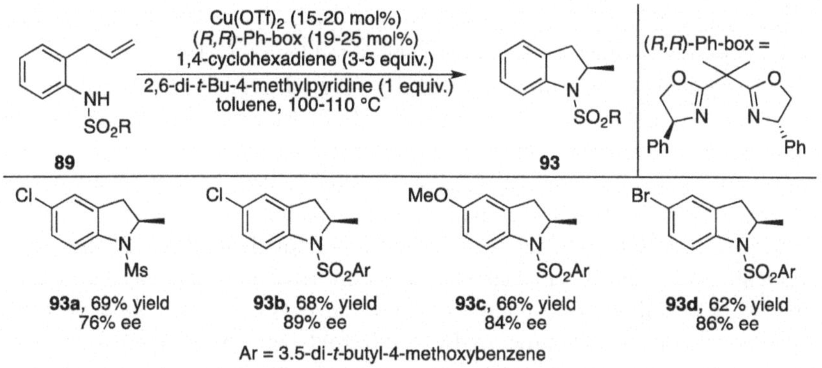

Scheme 40 Asymmetric hydroamination using a chiral copper catalyst

In order to favor hydroamination over carboamination, an H-atom donor, 1,4-cylohexadiene, was added to the reaction mixture (Scheme 40). In addition, use of a mesylate or 3,5-di-*t*-butyl-4-methoxybenzenesulfonate protecting group eliminated or reduced formation of the carboamination product. Asymmetric induction was achieved through the addition of the chiral (*R*,*R*)-Ph-box ligand to afford chiral *N*-sulfonylindolines **93a–d** with good enantioselectivities.

3 Hydroamination of Aminoallenes

Late transition metal-catalyzed additions of amines to allenes have been known for decades [78]. Unsaturated carbon–carbon bonds in the form of allenes and alkynes are generally more reactive toward late transition metal-catalyzed

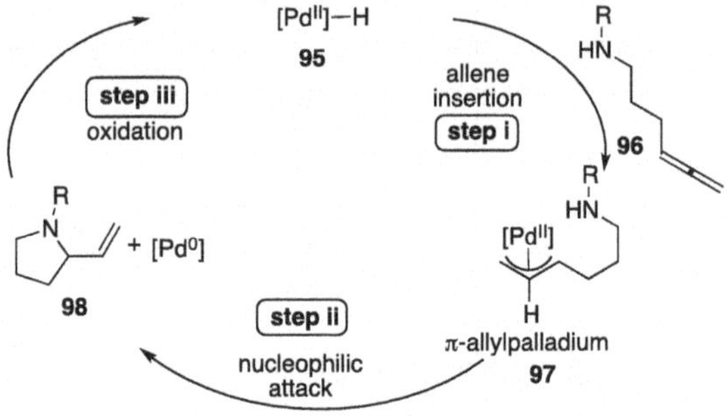

Scheme 41 Palladium–hydride mechanism for allene hydroamination

hydrofunctionalization reactions, in part due to their higher energy π-electrons. The allene C═C π-bond is known to be ~10 kcal/mol less stable than that of an alkene, leading to increased reactivity [79, 80]. Intramolecular cyclizations of aminoalkynes and *endo*-cyclizations of allenes produce unsaturated heterocycles, whereas *exo*-cyclizations onto allenes in the presence of Lewis acidic metal complexes produce saturated 2-alkenyl-substituted heterocycles.

The 5-*exo*-cyclization pathway is usually favored over the 6-*endo*-cyclization pathway for intramolecular hydroaminations of allenes catalyzed by late transition metal complexes. However, selective 5-*endo*-cyclizations are possible if no other pathways for cyclization are available, such as in the case of α-aminoallenes [81, 82]. Various electron-deficient metal complexes, such as metal triflates or cationic species, have been shown to catalyze the addition of nucleophiles to allenes, which are typically proposed to proceed via the olefin activation pathway (Scheme 3). Alternatively metal–hydride complexes have also been shown to be effective catalyst precursors, which proceed by a different mechanism (*vide infra*) [83]. Gold and palladium catalysts are the most studied. However, reports of other late transition metal catalysts for intramolecular allene hydroamination have also been published [84]. Recent advances in the hydroamination of allenes will be highlighted in the following sections.

3.1 Palladium- and Platinum-Catalyzed Reactions

In 1998, Yamamoto reported the first intramolecular hydroamination of allenes with a palladium catalyst [83]. They proposed an alternative mechanism for hydroamination involving insertion of allene **96** into a palladium–hydride bond to give a π-allylpalladium intermediate **97** (Scheme 41). Addition of one equivalent of acetic acid significantly improved reaction rates and efficiency, perhaps facilitating the formation of the Pd–H species [85].

Both five- and six-membered rings could be readily formed using $[(\eta^3\text{-}C_3H_5)PdCl_2]_2$ as the starting complex and dppf as ligand (Scheme 42). Interestingly, both

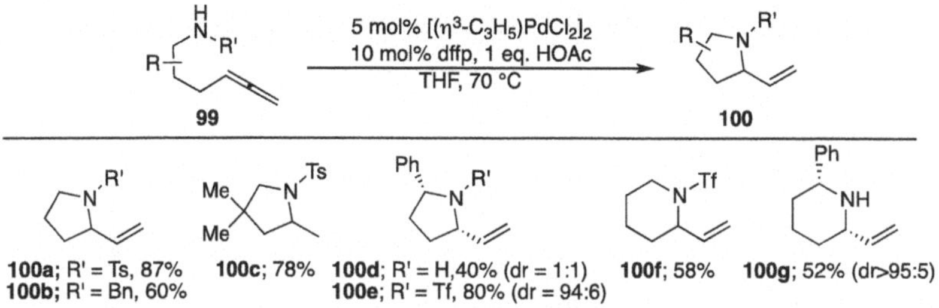

Scheme 42 First palladium-catalyzed intramolecular hydroamination of allenes

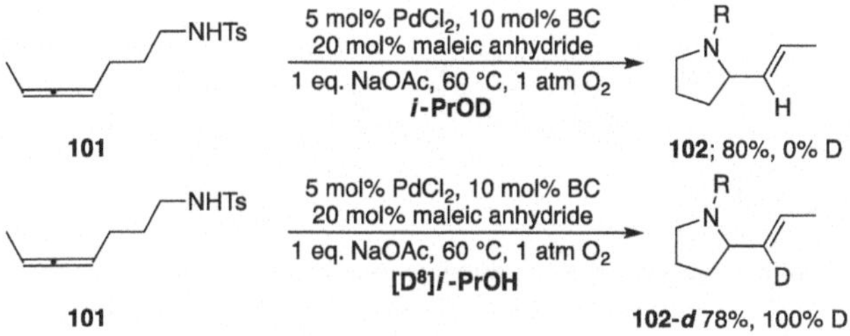

Scheme 43 Palladium-catalyzed allene hydroamination – deuterium-labeling study

electron-deficient *N*-donors, such as amides and tosylamides, and electron-rich primary and secondary amine *N*-donors participated in the cyclization to afford hydroamination products, although electron-deficient amines generally gave better yields. The catalyst was shown to be highly active as substrates lacking the Thorpe–Ingold bias readily cyclized under these conditions.

More than a decade later, Liu and coworkers reported a catalyst system comprised of $PdCl_2$ and the nitrogen-based bathocuprine (BC) ligand, so as to couple the hydroamination process to an aerobic alcohol oxidation process, which would not be compatible with phosphine-based ligands [86]. They also invoked a mechanism involving insertion of the allene into a palladium–hydride bond, followed by nucleophilic attack of the pendant amine on the π-allyl intermediate. Deuterium-labeling studies indicated that the proton that adds across the allenic double bond originates from the α-position of the alcohol co-oxidant (Scheme 43). The substrate scope was expanded to include internal allenes **103**, which produce 2-*trans*-alkenyl and -styrenyl products **104** (Scheme 44). The *trans* olefin geometry was exclusively formed in reactions of 1,3-substituted allenes.

In 1999, Yamamoto discovered that the intramolecular hydroamination of allenes could be accomplished from aminoalkyne starting materials, through in situ formation of the allene **107** (Scheme 45) [87]. Further investigations of this system have since been reported [88], including expansions of scope to generate lactams [89] and tetrahydroisoquinolines [90] (Scheme 45), as well as asymmetric variants discussed in Sect. 3.3. The catalyst system was demonstrated to be

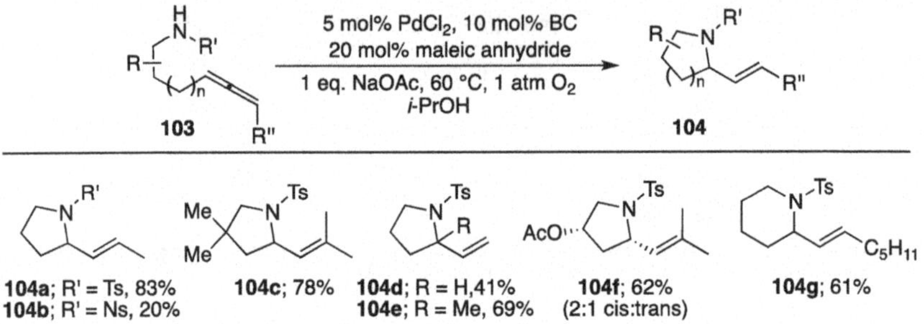

Scheme 44 Palladium-catalyzed hydroamination of allenes under aerobic conditions

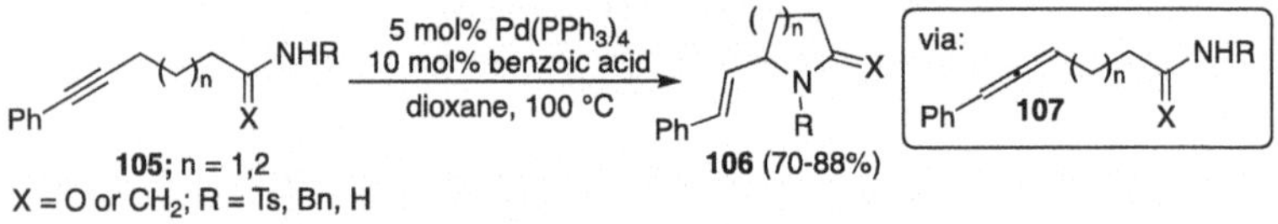

Scheme 45 Palladium-catalyzed isomerization/hydroamination of aminoalkynes

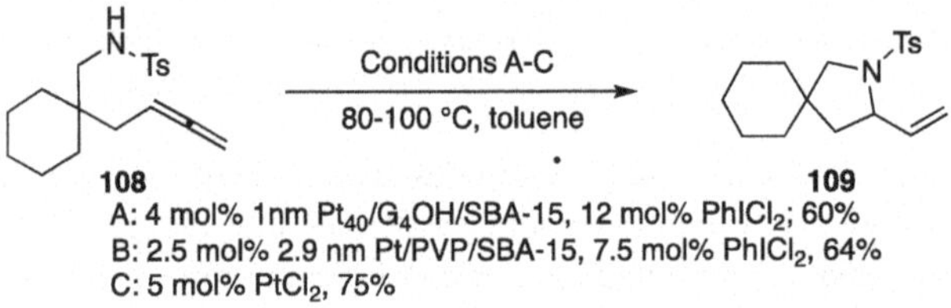

Scheme 46 Platinum nanoparticle-catalyzed hydroamination of aminoallenes

tolerant of a wide range of functional groups (e.g., aryl halides, esters trifluoromethyl), and primary amine nucleophiles were also shown to participate in the hydroamination.

In 2010, Toste demonstrated that platinum nanoparticles could be used as catalysts for the hydroamination of aminoallenes [91]. Although the scope of this reaction has not yet been established, this work represents the first example of a heterogeneous catalyst for hydroamination and serves as a starting point for the development of new heterogeneous catalysts (Scheme 46). Two different types of nanoparticles (Pt_{40}/G4OH and Pt/PVP) were prepared in a range of sizes, and adhered to a solid support SBA-15. Reactions required the presence of the oxidant $PhICl_2$ and under these conditions it was found that the smaller nanoparticles gave superior results, in part due to their increased stability compared to larger nanoparticles. The heterogeneous platinum nanoparticle catalysts were compared to the homogenous catalyst $PtCl_2$ and shown to afford the pyrrolidine product **109** in comparable yields. A π-activation pathway was proposed for this transformation.

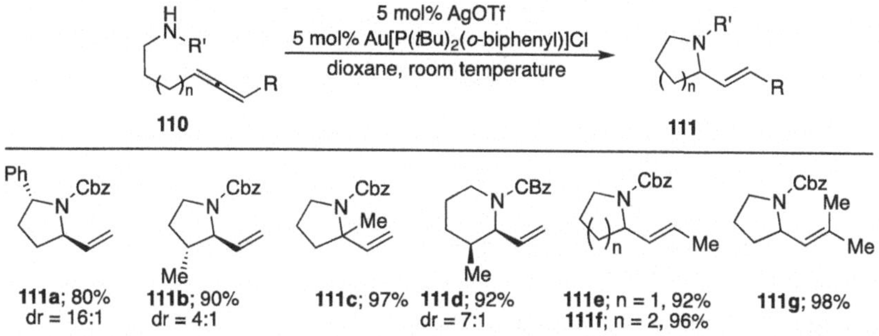

Scheme 47 Hydroamination of aminoallenes with a cationic gold monophosphine complex

3.2 *Gold- and Copper-Catalyzed Reactions*

Electron-deficient gold complexes have emerged as efficient catalysts for the hydroamination reactions of aminoallenes [92]. Cationic gold complexes are soft Lewis acids that tend to be more carbophilic than oxophilic, which results in increased reactivity for π-activation processes and high functional group tolerance. The development of gold complexes as catalysts for intramolecular hydroaminations of aminoallenes with broad substrate scope was established in 2006 by Widenhoefer and coworkers [93]. This group illustrated that the bulky monophosphine gold catalyst Au[P(*t*-Bu)$_2$(*o*-biphenyl)]Cl and a cocatalytic amount of AgOTf were highly active for the hydroamination of aminoallenes containing electron-deficient *N*-donors (i.e., NCbz) (Scheme 47) [93]. The cocatalytic AgOTf serves to generate a cationic gold complex as the active catalyst, which activates the allene moiety allowing for *anti*-nucleophilic attack from the pendant carbamate. A possible Bronsted acid-catalyzed process (via in situ generation of HOTf) was ruled out by conducting control experiments in the presence of HOTf, which showed that aminoallenes failed to cyclize in the absence of a metal catalyst. Reactions were carried out at *room temperature* to afford 2-alkenyl-pyrrolidines and piperidines **111**. The Thorpe–Ingold effect was not required to bias substrates toward cyclization and axially chiral allenes underwent cyclization to form the (*E*)-isomers exclusively. Widenhoefer's catalyst is also active for hydroalkoxylation of allenyl alcohols (see Sect. 4.2) and hydroarylation with carbon nucleophiles.

In 2006, Yamamoto also reported the use of gold(I) or gold(III) salts to catalyze both inter- and intramolecular hydroamination reactions [94, 95]. It was found that AuCl was the optimal catalyst for the synthesis of pyrrolidines and piperidines via cyclization of 1,3-aminoallenes **112** at room temperature in THF (Scheme 48). The products **113** were formed exclusively as the (*E*)-alkene isomer. The gold(III) complex $AuCl_3$ also catalyzed the reaction with similar efficiency, but AuCl is more air stable and thus more practical for use in the laboratory. Nitrogen donors protected with a tosylate or a carbamate group reacted at a faster rate than more basic amines. However, it was still possible to cyclize a secondary benzyl amine

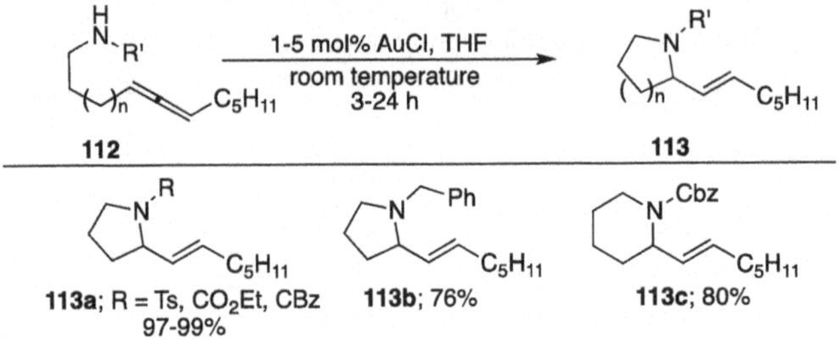

Scheme 48 AuCl-catalyzed hydroamination of aminoallenes

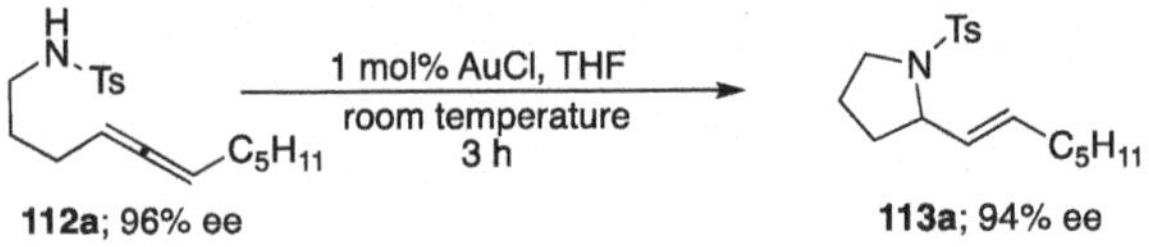

Scheme 49 Chirality transfer from enantioenriched allenes

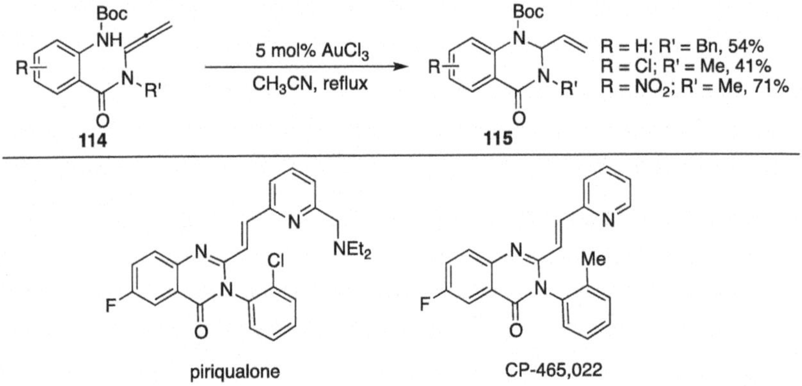

Scheme 50 Synthesis of quinazilones via hydroamination

(**112b**) in 76% yield with a longer reaction time (24 h versus 3 h for electron-deficient amines). Primary amines failed to react under these conditions, likely due to deactivation of the catalyst by the amine. Gem-disubstitution on the alkyl chain was not required to bias the substrate toward cyclization, which highlights the increased reactivity of allenes compared to isolated unactivated alkenes. Finally, it was found that chirality could be transferred from enantioenriched allenes as a method for synthesizing chiral non-racemic pyrrolidines and piperidines (Scheme 49).

Motivated by the biological activity of quinazolin-4-ones, such as piriqualone and CP-465,022, Broggini and coworkers reported the hydroamination of aminoallenes **114** catalyzed by $AuCl_3$ to afford the heterocyclic cores **115** in good yields (41–71%, Scheme 50) [96]. Platinum and palladium complexes also catalyzed the reaction, albeit in low yields. With the recent development of

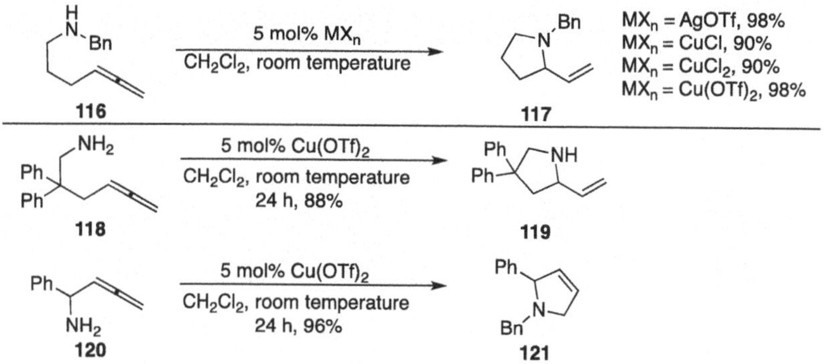

Scheme 51 Copper- and silver-catalyzed hydroaminations of aminoallenes

improved methods, late transition metal-catalyzed hydroamination is beginning to emerge as an option for the synthesis of complex, biologically active molecules.

There are reports of other Lewis acidic metal complexes that catalyze the 5-*exo*-cyclohydroamination of aminoallenes. For example, aminoallenes such as **116** containing a basic amine readily cyclized to form 2-vinyl-pyrrolidines in the presence of copper or silver salts (Scheme 51) [97]. Copper(II) triflate catalyzes the cyclization of primary aminoallenes (e.g., **118**→**119**) as well as 5-*endo*-cyclizations of α-aminoallenes such as **120**. A mechanism involving *anti*-aminometallation of the intermediate π-complex, similar to the mechanism of gold-catalyzed processes, is proposed for electron-deficient Cu and Ag salts.

3.3 Asymmetric Hydroaminations of Aminoallenes

The late transition metal-catalyzed asymmetric intramolecular hydroamination of allenes is more developed than the analogous reactions with unactivated alkenes, in part due to the higher reactivity of allenes in hydroamination reactions [98–100]. One challenge in developing effective chiral gold catalysts stems from the preferred linear geometry of gold(I) complexes, which orients the chiral ligand distant from the substrate. To overcome this limitation, Toste developed a monocationic dinuclear gold(I)–bisphosphine complex using the BINAP ligand which was found to catalyze the asymmetric hydroamination of aminoallenes [101]. The (*R*)-xylyl-BINAP$(AuOPNB)_2$ catalyst is effective for cyclizations of tosyl-protected aminoallenes **122** to chiral vinyl-substituted pyrrolidines **123** (Scheme 52). Chiral piperidines could also be synthesized with good enantioselectivities; however, it was found that a related biphenyl bisphosphine complex (*R*)-ClMeO-BIPHEP $(AuOPNB)_2$ gave superior selectivities. The substrate scope was later expanded to include protected hydrazines and hydroxylamines to form chiral pyrazolidines and isoxazolidines [102].

In 2007, Toste subsequently reported a unique strategy that utilized a chiral anion instead of a chiral ligand for asymmetric hydroamination of aminoallenes and hydroalkoxylation of hydroxyallenes (see Sect. 4.3) [103]. An achiral

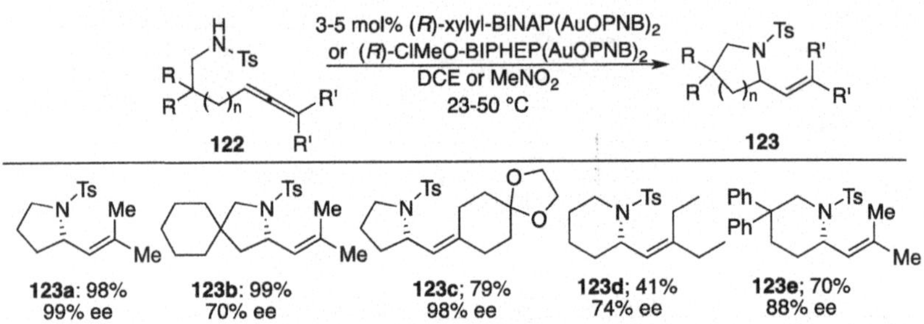

Scheme 52 Dinuclear gold-catalyzed enantioselective hydroamination of aminoallenes

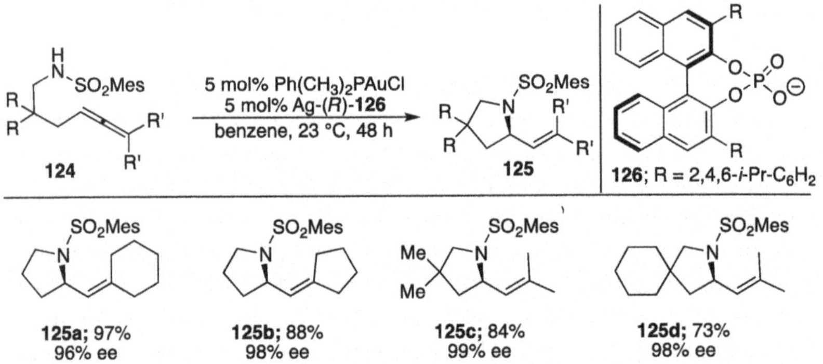

Scheme 53 Asymmetric gold-catalyzed hydroamination mediated by a chiral counterion

gold–phosphine cation [$Ph(CH_3)_2PAu^+$] complexed to a phosphate counterion **126** derived from readily available binaphthol was found to catalyze the hydroamination of sulfonyl-protected aminoallenes **124** to form 2-alkenyl-pyrrolidines **125** in excellent enantioselectivities (Scheme 53).

In 2007, Widenhoefer reported the asymmetric hydroamination of *N*-allenyl carbamates, which had previously failed to react under conditions that were successful for *N*-allenyl sulfonamides [104]. The chiral biaryl complex [(S)-**129**] Au_2Cl_2] in combination with $AgClO_4$ catalyzed the hydroamination of *N*-allenly carbamates and *N*-allenyl carboxamides to form 2-alkenyl-pyrrolidines **128** with fast rates and moderate to excellent enantioselectivities (Scheme 54). There was a pronounced effect on the counterion as the use of AgOTs led to a 1000-fold decrease in reaction rate compared to $AgClO_4$. This enhanced rate was important for achieving high selectivities since reactions could be run at low temperatures. Enantioselectivities were sensitive to the substitution pattern on the alkyl chain linking the nitrogen and allene moieties. For example, aminoallenes containing gem-diphenyl substituents typically reacted with high selectivities; however, cyclohexyl- or unsubstituted derivatives reacted to give lower selectivities. Subsequent work demonstrated that *N*-allenyl ureas were also viable substrates for asymmetric hydroamination employing a similar catalyst system [105].

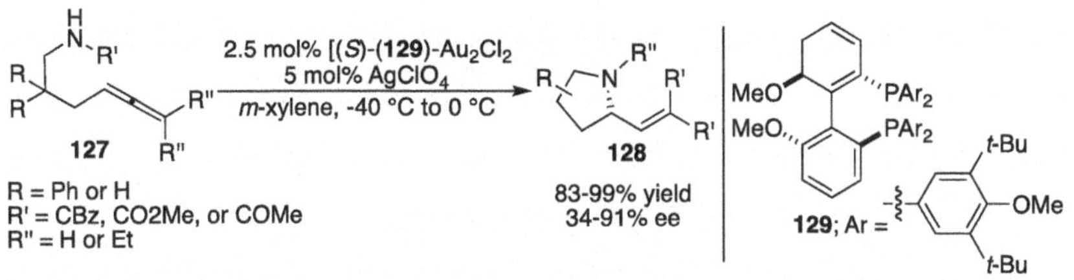

Scheme 54 Asymmetric Au-catalyzed hydroamination of *N*-allenyl carbamates and amides

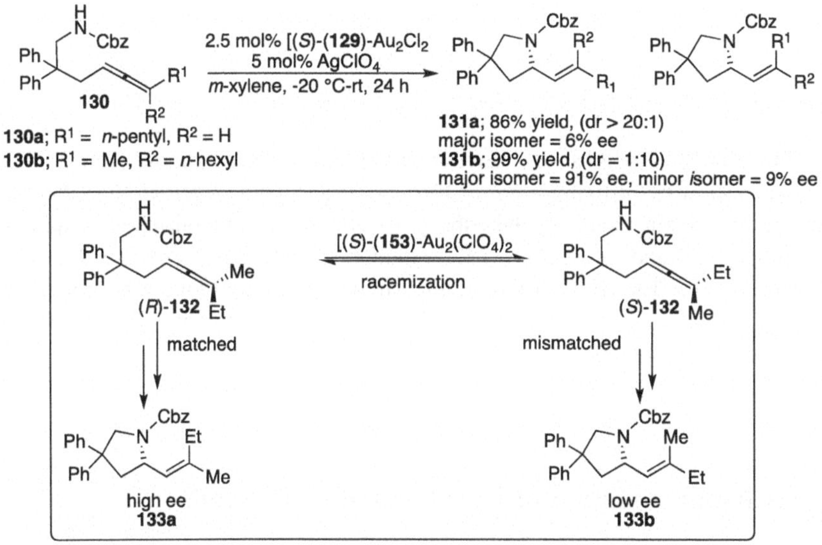

Scheme 55 Dynamic kinetic enantioselective hydroamination of trisubstituted aminoallenes

Use of the chiral gold complex [(S)-**129**]Au_2Cl_2] in reactions of axially chiral 1,3-disubstituted allene substrates such as **130a** led to products **131a** with low enantioselectivity; however, trisubstituted allenes **130b** underwent dynamic kinetic enantioselective hydroamination to afford predominately one of the four possible stereoisomers, due to the ability of such allenes to rapidly racemize in the presence of the gold catalyst (Scheme 55) [106, 107]. These results are in contrast to analogous hydroalkoxylation reactions, where axially chiral hydroxyallenes reacted to give products with high enantioselectivity without prior racemization (see Sect. 4.3). This suggests that the enantiodetermining step is due to an irreversible and selective nucleophilic attack of the amine onto the gold-π-complex, with matched or mismatched reactivity depending on the stereochemistry of the substrate and catalyst, and not as a result of selective formation of the gold allene π complex.

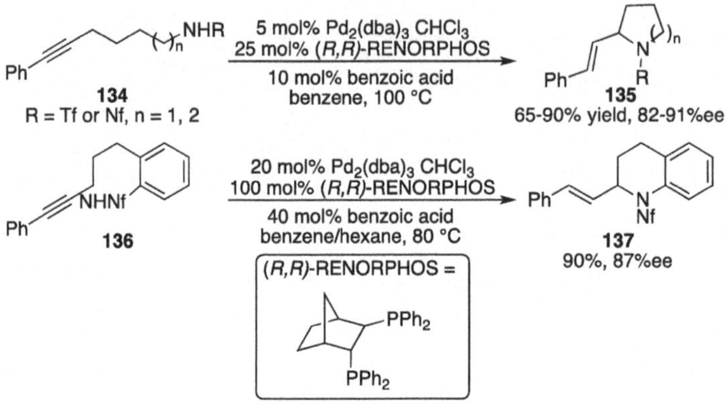

Scheme 56 Asymmetric palladium-catalyzed alkyne isomerization/hydroamination

Yamamoto developed an asymmetric palladium-catalyzed alkyne isomerization/hydroamination by utilizing a chiral bisphosphine ligand (*R*,*R*)-RENORPHOS in combination with catalytic $Pd_2(dba)_3$ to afford enantioenriched 2-alkenyl-pyrrolidines **135** and -piperidines **137** in moderate to excellent enantioselectivities (Scheme 56) [90, 108, 109]. Although the triflate protecting group was suitable, the use of the nonafluorobutanesulfonyl (Nf) group gave better results by allowing for reduced catalyst loadings.

4 Hydroalkoxylation of Unsaturated C–C Bonds

The intramolecular hydroalkoxylation of alkenes is of fundamental importance for the synthesis of oxygen-containing heterocycles [4]. With few exceptions [110], late transition metal-catalyzed hydroalkoxylation reactions proceed through the olefin activation pathway. The reduced nucleophilicity of the oxygen nucleophile compared to amine nucleophiles requires the use of highly Lewis acidic late transition metals, such as metal triflates, tosylates, and halides, as catalysts for hydroalkoxylation. Numerous late transition metals, such as those based on gold, silver, copper, ruthenium, iron, palladium, and platinum, have been developed as catalysts for hydroalkoxylation. It is recognized that the well-studied palladium-catalyzed Wacker oxidation proceeds by initial complexation of an olefin to the electrophilic palladium center followed by nucleophilic attack of an oxygen nucleophile, similar to the first step in hydroalkoxylation [111]. If β-hydride elimination is slowed relative to protonolysis of the metal–carbon bond, then saturated hydroalkoxylation products will be formed instead of Wacker oxidation products. Recent developments in the reactions of hydroxyallenes and hydroxyalkenes for the synthesis of saturated tetrahydrofurans and tetrahydropyrans, as well as asymmetric hydroalkoxylations, will be reviewed in this section.

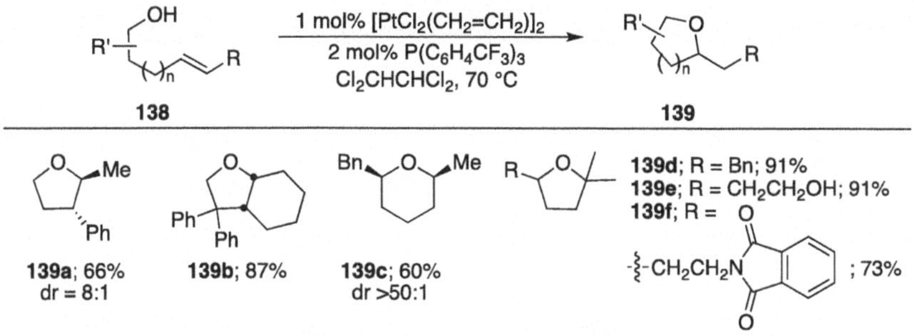

Scheme 57 Platinum-catalyzed hydroalkoxylation of hydroxyalkenes

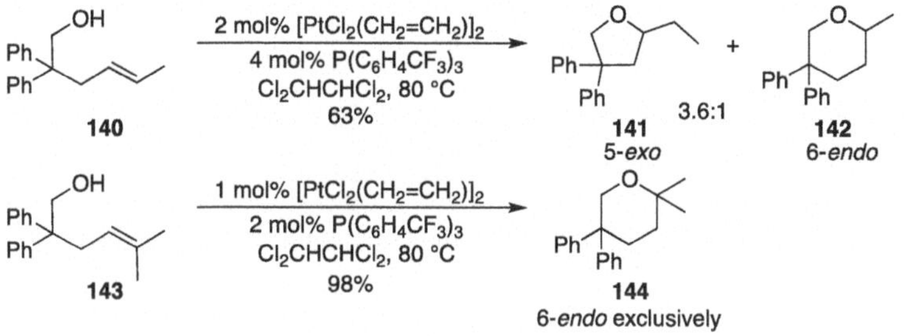

Scheme 58 Competing 5-*exo* and 6-*endo* hydroalkoxylation

4.1 Reactions of Hydroxyalkenes

Similar to hydroamination, hydroalkoxylations of unactivated alkenes are more challenging than analogous reactions with alkynes or allenes and much effort has been spent developing more active and selective catalysts for this class of substrates. In 2004, Widenhoefer and coworkers reported the first late transition metal-catalyzed hydroalkoxylation of unactivated olefins **138** to form five- and six-membered rings **139** [112]. The use of an electron-deficient phosphine (2 mol% $P(C_6H_4CF_3)_3$) in combination with 1 mol% $[PtCl_2(CH_2{=}CH_2)]_2$ allowed for the efficient cyclization of hydroxyalkenes bearing various functional groups (Scheme 57). Notably, the Thorpe–Ingold effect was not required for cyclization.

Hydroalkoxylation of higher substituted olefins is more facile than for corresponding hydroaminations reactions. Substrates with terminal, 1,1-disubstituted, 1,2-disubstituted, and trisubstituted olefins all participated in the hydroalkoxylation reaction. However, for the substrate **140** bearing a *trans*-disubstituted olefin, a 3.6:1 mixture of products was obtained as a result of competing 5-*exo* (**141**) and 6-*endo*-cyclizations (**142**) (Scheme 58). In general, regioselectivity for hydroalkoxylations is strongly governed by the stability of the analogous carbocation that would result from alkene protonation, indicating a late transition state for nucleophilic attack of the heteroatom onto the coordinated olefin. This can often lead to mixtures of 5-*exo* and 6-*endo*-cyclization products, or exclusive

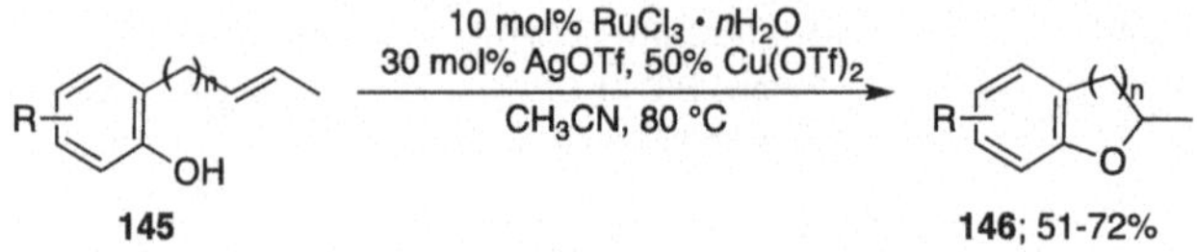

Scheme 59 Ruthenium-catalyzed hydroalkoxylation of unactivated alkenes

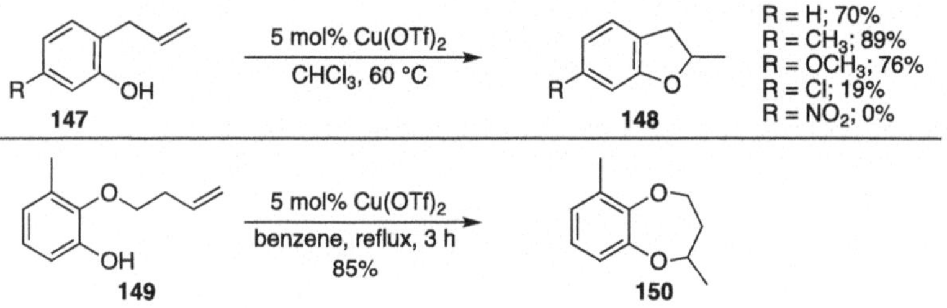

Scheme 60 Copper-catalyzed hydroalkoxylation

formation of 6-*endo* tetrahydropyran products depending on the substitution of the olefin.

In 1998, Furukawa reported the first intramolecular hydroalkoxylation catalyzed by a mixture of $RuCl_3{\cdot}nH_2O$, AgOTf, PPh_3, and $Cu(OTf)_2$ to afford dihydro-2-methyl-benzofuran **146**, without competing β-hydride elimination [113]. Later in 2007, the authors expanded the scope of this reaction (Scheme 59) [114, 115], and also illustrated that $(RuCp^*Cl_2)_2$ in combination with AgOTf was an effective catalyst for the hydroalkoxylation of hydroxyalkenes. The Ru-catalysts gave similar regioselectivities for cyclization as were observed for the Pt-catalysts above. The ruthenium catalyst systems were extremely sensitive to the choice of solvent, silver salt and copper salt. For example, no catalytic activity was observed if $AgBF_4$ was used instead of AgOTf. Although mechanistic data were not obtained, the authors propose the formation of a cationic ruthenium(III) complex that catalyzed hydroalkoxylation via the olefin activation pathway. No explanation for the role of copper triflate was given; however, it alone did not catalyze the reaction in acetonitrile. Hydrocarboxylation was also demonstrated with the $RuCl_3$ catalyst system [114].

Furukawa and Ito later reported that copper(II) triflate alone effectively catalyzed the hydroalkoxylation of unactivated alkenes in nonpolar solvents (Scheme 60) [116]. 2-Allylphenols containing electron-withdrawing groups were poor substrates for hydroalkoxylation, suggesting that the nucleophilicity of the phenol group influences the overall reaction rate. They also demonstrated that efficient cyclization to form a seven-membered ring could be accomplished in good yield (**177–178**; 85%).

In 2009, Hii and coworkers expanded the substrate scope for the $Cu(OTf)_2$-catalyzed hydroalkoxylation (Scheme 61) [117]. They also found that use of triflic acid as catalyst yielded comparable results to the copper-catalyzed process under

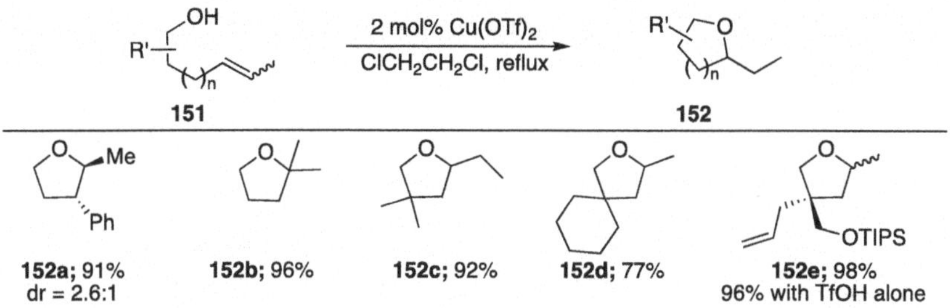

Scheme 61 Copper-catalyzed hydroalkoxylation

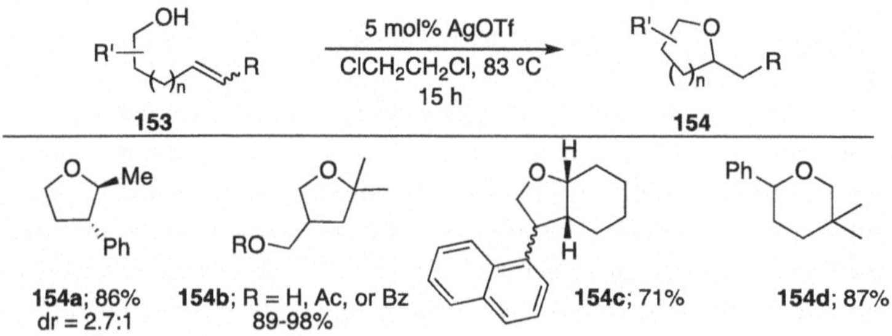

Scheme 62 Silver-catalyzed hydroalkoxylation

otherwise identical conditions (**152e**); however, it was argued that the use of copper is advantageous due to its ease of handling and mild nature.

Simple Lewis acidic silver salts have also been shown to catalyze hydroalkoxylations of unactivated hydroxyalkenes. In 2005, He and coworkers described that the first intramolecular silver-catalyzed hydroalkoxylation of hydroxyalkenes **153** affords tetrahydrofurans and tetrahydropyrans **154** in excellent yields (Scheme 62) [118]. In the initial catalyst screening, the authors found that AgOTf alone gave the best results. Addition of the electron-rich phosphine PPh_3 completely inhibited the reaction. The regioselectivity for *exo*- versus *endo*-cyclizations parallels that of the copper and platinum catalysts described above, suggesting a similar mechanism that likely involves activation of the olefin by the Lewis acid or in situ-generated Bronsted acid followed by nucleophilic attack.

Iron-based Lewis acids have been shown to catalyze hydroalkoxylation of alkenes [119]. For example, the combination of 10 mol% $FeCl_3$ and 30 mol% AgOTf catalyzed the cyclization of hydroxylalkene **155** to tetrahydrofuran product **156** in quantitative yield after 30 min (Scheme 63) [120]. Interestingly, the reaction catalyzed by AgOTf alone afforded the product in only 8% yield after 30 min, suggesting that the iron cocatalyst does indeed accelerate the rate for hydroalkoxylation. Additional control experiments suggest that the iron metal participates in catalysis, and is not merely a source of Bronsted acid.

In 2005 He and coworkers reported the first gold-catalyzed intramolecular hydroalkoxylation of alkenes (Scheme 64) [121]. Heating a mixture of **157** in the

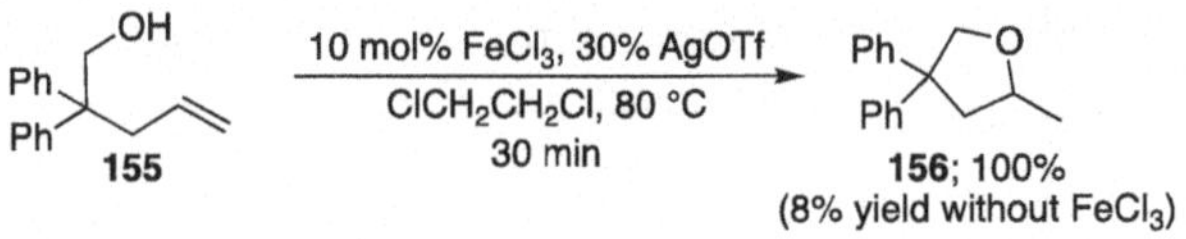

Scheme 63 Iron-catalyzed hydroalkoxylation of an allenyl alcohol

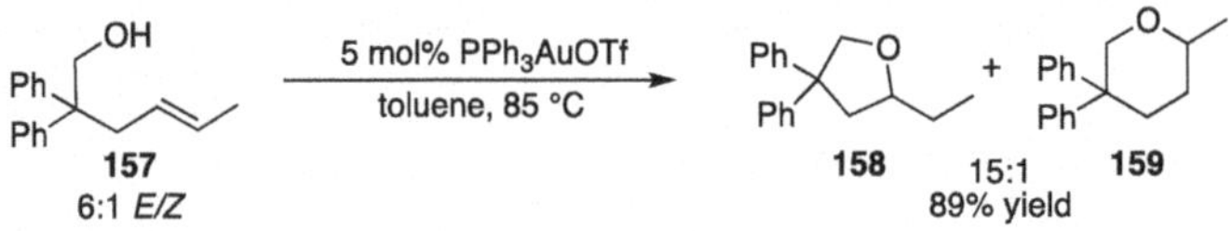

Scheme 64 First gold-catalyzed intramolecular hydroalkoxylation

presence of cationic gold complex PPh_3AuOTf in toluene afforded a 15:1 mixture of products **158** and **159**. It was observed that olefin isomerization could occur under these reaction conditions; therefore, the minor tetrahydropyran product **159** may result from 6-*exo*-cyclization after olefin isomerization to the terminal olefin (versus 6-*endo*-cyclization of substrate **157**). Notably, this catalyst was also highly active for more challenging *intermolecular* reactions of unactivated olefins. In 2012 dienes were also reported to undergo intramolecular hydroalkoxylation in the presence of PPh_3AuOTf [122]. The development of gold nanoclusters stabilized by the hydrophilic polymer poly(*N*-vinyl-2-pyrrolidone) was also reported to catalyze alkene hydroalkoxylation [123].

4.2 *Reactions of Hydroxyallenes*

Since the discovery of the silver-catalyzed hydroalkoxylation of allenes in 1979, numerous late transition metal catalysts have been developed for hydrofunctionalization of allenes [124–126]. Many processes have focused on the synthesis of unsaturated 2,5-dihydrofurans as a result of 5-*endo*-cyclizations from α-hydroxyallenes [127]; however, recent research in the area of gold-catalysis has led to the development of *exo*-hydroalkoxylations of allenes to form saturated 2-alkenyl-tetrahydrofurans and -tetrahydropyrans.

In 2006, Widenhoefer and coworkers reported the use of a Johnphos-gold complex as a catalyst for intramolecular allene hydroalkoxylation which had previously shown efficacy as a catalyst for hydroamination and hydroalkoxylation of alkenes (Scheme 65) [26, 93]. It was found that the regioselectivity for hydroalkoxylation of allene **160** was highly dependent on the counterion. For example, the use of AgOTf in combination with (Johnphos) AuCl afforded a 1.3:1 mixture of 5-*exo* and 6-*exo* products **161** and **162**, but employing AgOTs instead resulted in selective formation of the five-membered ring tetrahydrofuran **161**. Of note, no reaction was observed in the presence of AgOTs alone.

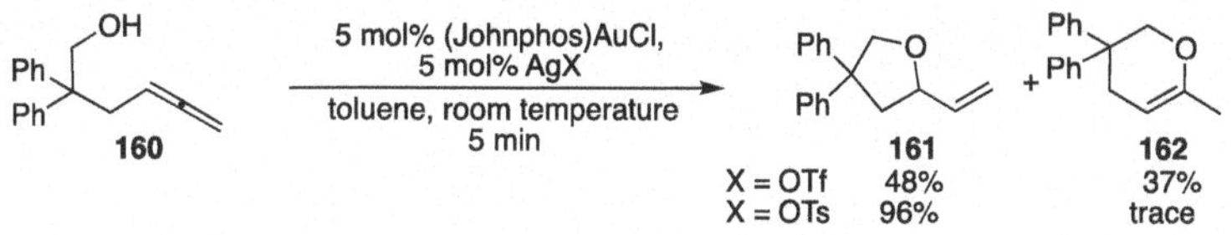

Scheme 65 Counterion effect in gold-catalyzed allene hydroalkoxylation

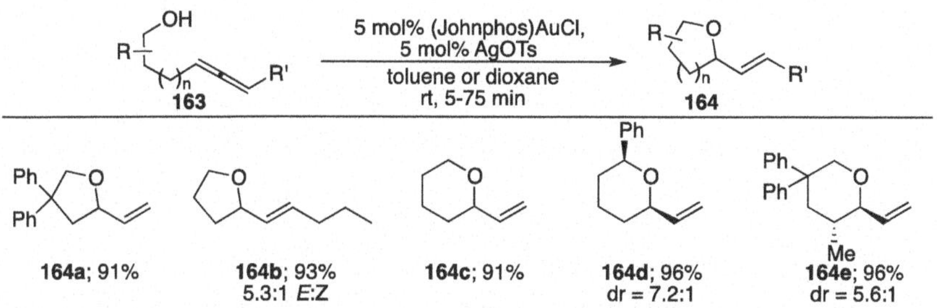

Scheme 66 Gold-catalyzed intramolecular hydroalkoxylation of allenes

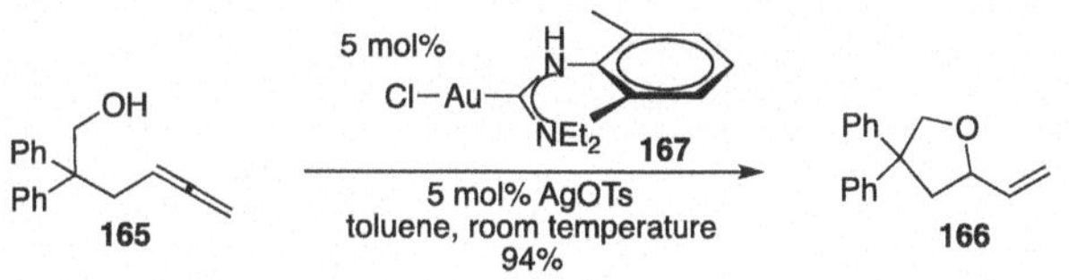

Scheme 67 NHC–gold-catalyzed hydroalkoxylation

This optimized gold catalyst system was evaluated on substrates with various substitution patterns on the alkyl linker and allene (Scheme 66). The catalyst was shown to be highly active as gem-disubstitution on the alkyl linker was not required for cyclization. Both five- and six-membered ring products could be synthesized by this method. Axially chiral allenes containing 1,3-disubstitution cyclized to form *E*-alkenyl products with good selectivity (e.g., **164b**; 5.3:1 *E*:*Z*). Enantioenriched allenes underwent intramolecular hydroalkoxylation with transfer of chirality to afford enantioenriched products.

In 2010, an *N*-heterocyclic carbene-gold complex **167** was shown to catalyze the intramolecular hydroalkoxylation of an allenyl alcohol **165** to 2-vinyltetrahyrofuran **166** (Scheme 67) [128]. Although the scope of the reaction was not investigated, this type of NHC ligand has rarely been used in catalysis. Nonetheless, this complex has proven to be a useful catalyst in other alkene hydrofunctionalizations reactions. A gold complex encapsulated in a supramolecular host has also shown catalytic activity in hydroalkoxylation reactions of allenyl alcohols [129].

The palladium-catalyzed isomerization/hydroamination reaction developed by Yamamoto (described in Sect. 3.1) was also applied to reactions involving oxygen nucleophiles [130, 131]. Hydroxyalkynes **168** were transformed to tetrahydrofurans and tetrahydropyrans **169** via allene **170** in the presence of 10% $Pd(PPh_3)_4$ and 10% benzoic acid in moderate to good yields (Scheme 68). Reactions forming

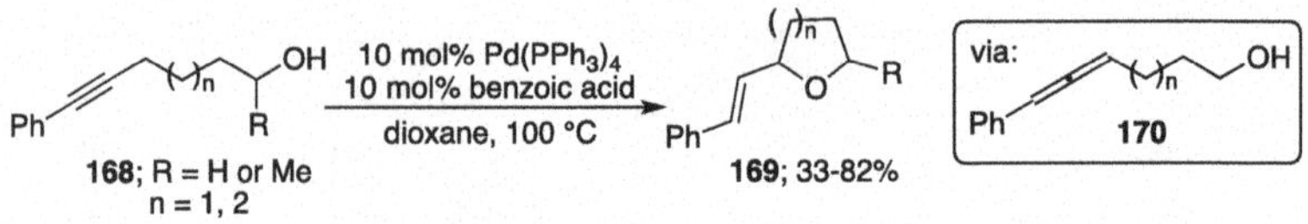

Scheme 68 Palladium-catalyzed isomerization/hydroalkoxylation of hydroxyalkynes

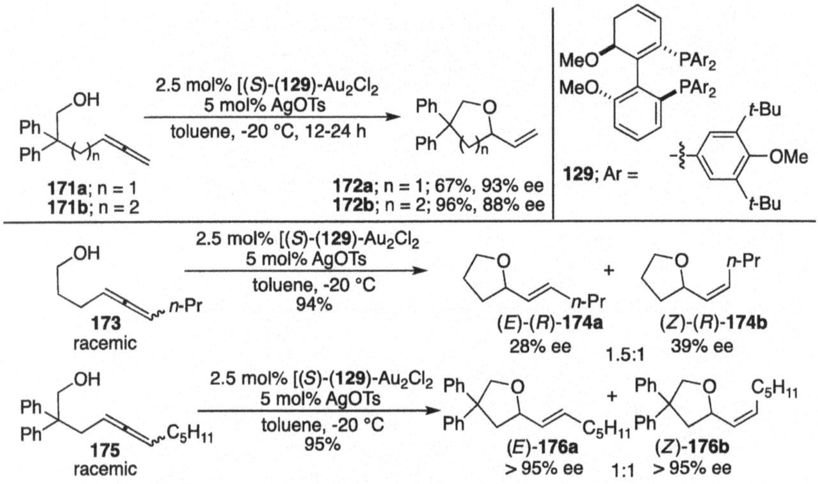

Scheme 69 Asymmetric gold-catalyzed allene hydroalkoxylation

five-membered rings were higher yielding than reactions forming six-membered rings. In the presence of a chiral phosphine, enantioenriched products were formed (see Sect. 4.3).

4.3 Asymmetric Hydroalkoxylation Reactions

The use of a chiral late transition metal catalyst for the enantioselective synthesis of oxygen-containing heterocycles is an attractive approach toward this class of molecules. Although synthetically useful methods for asymmetric intramolecular hydroalkoxylations of unactivated olefins have not yet been reported, hydroxylallenes can be converted to chiral tetrahydrofurans and tetrahydropyrans with good enantioselectivities.

Widenhoefer's chiral dinuclear gold catalyst [(S)-(**129**)-Au_2Cl_2] that was used for asymmetric allene hydroamination was also successfully applied to asymmetric allene hydroalkoxylation to afford both tetrahydrofuran (**172a**) and tetrahydropyran (**172b**) products (Scheme 69) [100, 132]. In order to achieve high enantioselectivities, gem-disubstitution on the alkyl chain was required. For example, hydroxyallene **173** was cyclized to give a 1.5:1 mixture of *E*:*Z* products in excellent yield, but poor enantioselectivity. Notably, both isomeric products **174a-b** were of

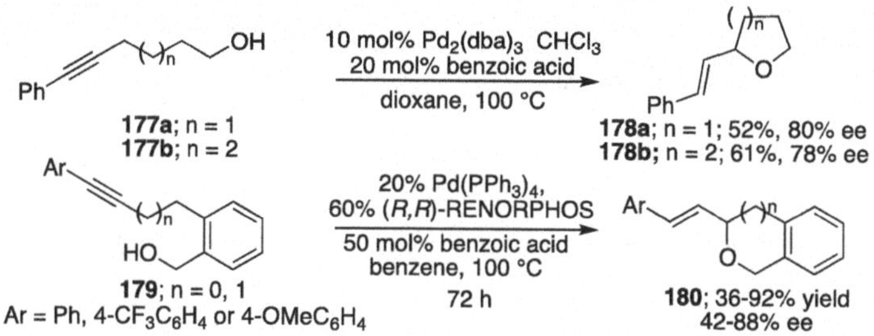

Scheme 70 Palladium-catalyzed asymmetric isomerization/hydroalkoxylation of hydroxyalkynes

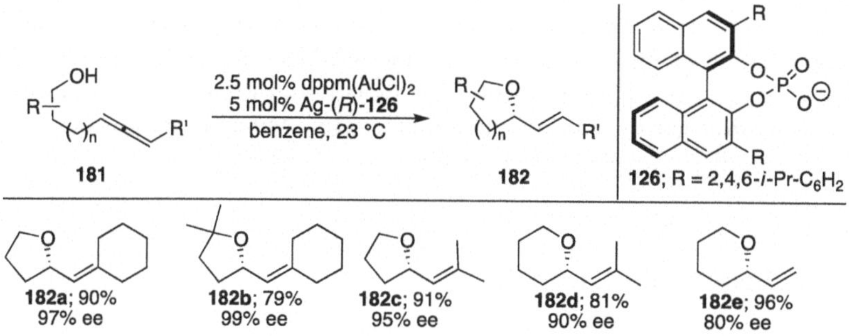

Scheme 71 Chiral counterion-mediated asymmetric hydroalkoxylation

the same absolute configuration. While enantioselectivities were low for the 1,3-disubstituted allene substrate **173** lacking substituents on the alkyl linker, high enantioselectivities (>95% ee) were obtained for substrates of type **175**, bearing an internal allene and gem-disubstitution on the alkyl linker. A 1:1 mixture of *E*:*Z* products (*E*)-**176a** and (*Z*)-**176b** was formed and hydrogenation of this mixture generated the analogous saturated tetrahydrofuran in 90% ee, confirming that the chiral center in both geometric isomers were formed with the same sense on chiral induction.

Yamamoto reported an asymmetric version of the palladium-catalyzed alkyne isomerization/hydroalkoxylation by utilizing a chiral bisphosphine ligand (*R*,*R*)-RENORPHOS in combination with catalytic $Pd_2(dba)_3$ to afford enantioenriched 2-alkenyl heterocycles (Scheme 70) [109].

Toste utilized a chiral counterion strategy to render hydroalkoxylation of allenyl alcohols enantioselective [103]. The conditions employed were similar to those described above in analogous asymmetric allene hydroamination reactions (Scheme 71). However, the authors found that the bidentate ligand dppm (diphenylphosphinomethane) gave higher enantioselectivities for hydroalkoxylation than the monodentate dimethylphenyl phosphine ligand ($PPhMe_2$), which was optimal for hydroamination. Reactions conducted in more polar solvents such as acetone led to significantly reduced enantioselectivities, which highlights the importance for the formation of a tight ion pair in this methodology.

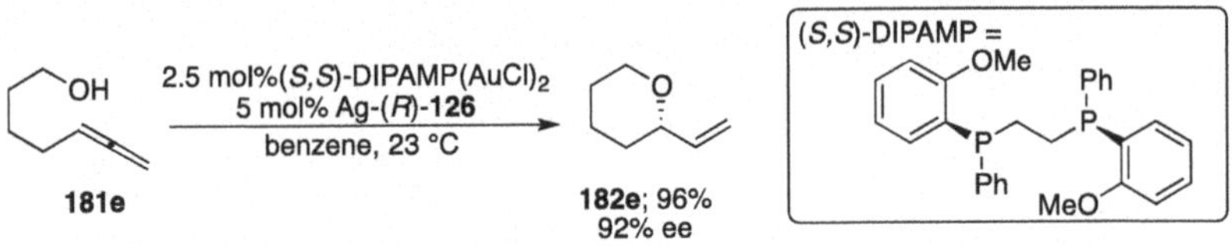

Scheme 72 Combined chiral ligand and chiral counterion strategy for asymmetric hydroalkoxylation

Toste also found that the combination of a chiral phosphine ligand with the chiral counterion could lead to enhanced enantioselectivities. For example, 2-vinyltetrahydropyran **182e** was formed in only 80% ee under the standard conditions employing the achiral dppm ligand. However, the same product could be formed in 92% ee if instead the chiral (*S*,*S*)-DIPAMP ligand was used (Scheme 72). The combination of the (*S*,*S*) ligand with the (*R*)-**126** counterion represented the "matched" case, since pairing of the (*S*,*S*) ligand with the (*S*)-enantiomer of the chiral counterion led to reduced enantioselectivity. Mikami and coworkers also observed a similar synergistic effect [133].

5 Conclusion and Outlook

Over the last decade, significant progress has been made toward the development of chemoselective and stereoselective late transition metal catalysts for hydrofunctionalization of unsaturated C=C bonds to form saturated heterocycles. Advances have been achieved for hydroamination and hydroalkoxylation reactions of challenging substrates bearing unactivated olefins, especially through the use of highly active platinum, rhodium, palladium, and gold catalysts. In addition, investigations into reaction mechanism have revealed valuable insight that has driven the structure-based design of new ligands. The use of late transition metal catalysts has led to methods that tolerate numerous functional groups (e.g., alcohols, esters, nitriles, ketones) that would otherwise not be compatible with early transition metals or lanthanide catalysts. Although much progress has been achieved, many challenges still remain, including limited substrate scope for asymmetric cyclizations with unactivated olefins, and cyclizations of substrates bearing di- and trisubstituted olefins. In addition, the development of new catalysts for enantioselective transformations will undoubtedly be the focus of continued research.

References

1. Muller TE, Hultzsch KC, Yus M, Foubelo F, Tada M (2008) Hydroamination: direct addition of amines to alkenes and alkynes. Chem Rev 108:3795–3892
2. Muller TE, Beller M (1998) Metal-initiated amination of alkenes and alkynes. Chem Rev 98:675–704

3. Hong S, Marks TJ (2004) Organolanthanide-catalyzed hydroamination. Acc Chem Res 37:673–686
4. Patil NT, Kavther RD, Shinde VS (2012) Transition metal-catalyzed addition of C-, N-, and O-nucleophiles to unactivated C-C multiple bonds. Tetrahedron 68:8079–8146
5. Chianese AR, Lee SJ, Gagne MR (2007) Electrophilic activation of alkenes by platinum(II): so much more than a slow version of palladium(II). Angew Chem Int Ed 46:4042–4059
6. Cochran BM, Michael FE (2008) Mechanistic studies of a palladium-catalyzed intramolecular hydroamination of unactivated alkenes: protonolysis of a stable palladium alkyl complex is the turnover-limiting step. J Am Chem Soc 130:2786–2792
7. Ozerov OV, Guo C, Papkov VA, Foxman BM (2004) Facile oxidative addition of N-C and N-H bonds to monovalent rhodium and iridium. J Am Chem Soc 126:4792–4793
8. Schlummer B, Hartwig JF (2002) Bronsted acid-catalyzed intramolecular hydroamination of protected alkenylamines. Synthesis of pyrrolidines and piperidines. Org Lett 4:1471–1474
9. Rosenfeld DC, Shekhar S, Takemiya A, Utsunomiya M, Hartwig JF (2006) Hydroamination and hydroalkoxylation catalyzed by triflic acid. Parallels to reactions initiated with metal triflates. Org Lett 8:4179–4182
10. McBee JL, Bell AT, Tilley TD (2008) Mechanistic studies of the hydroamination of norbornene with electrophilic platinum complexes: the role of proton transfer. J Am Chem Soc 130:16562–16571
11. Sanford MS, Groves JT (2004) Anti-Markovnikov hydrofunctionalization of olefins mediated by rhodium-porphyrin complexes. Angew Chem Int Ed 43:588–590
12. Komeyama K, Morimoto T, Takaki K (2006) A simple and efficient iron-catalyzed intramolecular hydroamination of unactivated olefins. Angew Chem Int Ed 45:2938–2941
13. Dal Zotto C, Michaux J, Zarate-Ruiz A, Gayon E, Virieux D, Campagne JM, Terrasson V, Pieters G, Gaucher A, Prim D (2011) $FeCl_3$-catalyzed addition of nitrogen and 1,3-dicarbonyl nucleophiles to olefins. J Organomet Chem 696:296–304
14. Panunzi A, De Renzi A, Palumbo R, Paiaro G (1969) Addition reactions on coordinated olefinic ligands. II. The reaction between amines and monoolefins coordinated in *cis*-dichloro (olefin)(*t*-phosphine)platinum(II) complexes. J Am Chem Soc 91:3879–3883
15. Hahn C, Vitagliano A, Giordano F, Taube R (1998) Coordination of olefins and N-donor ligands at the fragment [2,6-bis(diphenylphosphino)methyl)pyridine]-palladium(II). Synthesis, structure, and amination of the new dicationic complexes $[Pd(PNP)(CH_2{=}CHR)](BF_4)_2$ (R = H, Ph). Organometallics 17:2060–2066
16. Hahn C, Morvillo P, Vitagliano A (2001) Olefins coordinated at a highly electrophilic site – dicationic palladium(II) complexes and their equilibrium reactions with nucleophiles. Eur J Inorg Chem 419–429
17. Michael FE, Cochran BM (2006) Room temperature palladium-catalyzed intramolecular hydroamination of unactivated alkenes. J Am Chem Soc 128:4246–4247
18. Cochran BM, Michael FE (2008) Synthesis of 2,6-disubstituted piperazines by a diastereoselective palladium-catalyzed hydroamination reaction. Org Lett 10:329–332
19. Xu T, Qiu S, Liu G (2011) Palladium-catalyzed inter- and intramolecular hydroamination of styrenes coupled with alcohol oxidation using *N*-fluorobenzenesulfonamide as the oxidant. J Organomet Chem 696:46–49
20. Abbiati G, Beccalli E, Broggini G, Martinelli M, Paladino G (2006) Pd-catalyzed cyclization of 1-allyl-2-indolecarboxamides by intramolecular amidation of unactivated ethylenic bond. Synlett 1:73–76
21. Liu X-Y, Li C-H, Che C-M (2006) Phosphine gold(I)-catalyzed hydroamination of alkenes under thermal and microwave-assisted conditions. Org Lett 8:2707–2710
22. Kitahara H, Sakurai H (2011) Catalytic activity of gold nanoclusters in intramolecular hydroamination of alkenes and alkynes with toluenesulfonamide under aerobic and basic conditions. J Organomet Chem 696:442–449
23. Ramachary DB, Narayana VV (2011) Sequential combination of ruthenium-, base-, and gold-catalysis – a new approach to the synthesis of medicinally important heterocycles. Eur J Org Chem 3514–3522

24. Zhang Z, Yang C-G, He C (2006) Gold(I)-catalyzed intra- and intermolecular hydroamination of unactivated olefins. J Am Chem Soc 128:1798–1799
25. LaLonde RL, Brenzovich WE, Benitez D, Tkatchouk E, Kelley K, Goddard WA III, Toste FD (2010) Alkylgold complexes by the intramolecular aminoauration of unactivated alkenes. Chem Sci 1:226–233
26. Han X, Widenhoefer RA (2006) Gold(I)-catalyzed intramolecular hydroamination of alkenyl carbamates. Angew Chem Int Ed 45:1747–1749
27. Bender CF, Widenhoefer RA (2008) Gold(I)-catalyzed intramolecular hydroamination of unactivated C=C bonds with alkyl ammonium salts. Chem Commun 2741–2743
28. Bender CF, Widenhoefer RA (2006) Room temperature hydroamination of *N*-alkenyl ureas catalyzed by a gold(I) *N*-heterocyclic carbene complex. Org Lett 8:5303–5305
29. Li H, Song F, Widenhoefer RA (2011) Gold(I)-catalyzed intramolecular hydroamination of *N*-allylic *N*-arylureas to form imidazolidin-2-ones. Adv Synth Catal 353:955–962
30. Yeh M-CP, Pai H-F, Lin Z-J, Lee B-R (2009) Stereoselective synthesis of hexahydroindoles and octahydrohepta[*b*]pyrroles via gold(I)-catalyzed intramolecular 1,4-hydroamination of 1,3-dienes. Tetrahedron 65:4789–4794
31. Baeza A, Najera C (2011) Intramolecular hydroamination of conjugated dienes catalyzed by Lewis and Bronsted acids. Synlett 5:631–634
32. Taylor JG, Whittall N, Kuok Hii K (2006) Copper-catalyzed intermolecular hydroamination of alkenes. Org Lett 8:3561–3564
33. Bender CF, Widenhoefer RA (2005) Platinum-catalyzed intramolecular hydroamination of unactivated olefins with secondary alkylamines. J Am Chem Soc 127:1070–1071
34. Hahn C, Morvillo P, Herdtweck E, Vitagliano A (2002) Coordination of alkenes at a highly electrophilic site. New dicationic platinum(II) complexes: synthesis, structure, and reactions with electrophiles. Organometallics 21:1807–1818
35. Bender CF, Hudson WB, Widenhoefer RA (2008) Sterically hindered mono(phosphines) as supporting ligands for the platinum-catalyzed hydroamination of amino alkenes. Organometallics 27:2356–2358
36. Zhang R, Xu Q, Mei L, Li S, Shi M (2012) A *N*-heterocyclic carbene (NHC) platinum complex as pre-catalyst for the intramolecular hydroamination of olefins with secondary alkylamines and oxidative amination of ω-alkenic amines. Tetrahedron 68:3172–3178
37. Lavery CB, Ferguson MJ, Stradiotto M (2010) Platinum-catalyzed alkene cyclohydroamination: evaluating the utility of bidentate ligation and phosphine-free catalyst systems. Organometallics 29:6125–6128
38. Hoover JM, DiPasquale A, Mayer JM, Michael FE (2010) Platinum-catalyzed intramolecular hydrohydrazination: evidence for alkene insertion into a Pt-N bond. J Am Chem Soc 132:5043–5053
39. Coulson DR (1971) Catalytic addition of secondary amines to ethylene. Tetrahedron Lett 12:429–430
40. Hesp KD, Stradiotto M (2010) Rhodium- and iridium-catalyzed hydroamination of alkenes. ChemCatChem 2:1192–1207
41. Liu Z, Hartwig JF (2008) Mild, rhodium-catalyzed intramolecular hydroamination of unactivated terminal and internal alkenes with primary and secondary amines. J Am Chem Soc 130:1570–1571
42. Julian LD, Hartwig JF (2010) Intramolecular hydroamination of unbiased and functionalized primary aminoalkenes catalyzed by a rhodium aminophosphine complex. J Am Chem Soc 132:13813–13822
43. Beller M, Eichberger M, Trauthwein H (1997) Anti-Markovnikov functionalization of olefins: rhodium-catalyzed oxidative amination of styrenes. Angew Chem Int Ed 36:2225–2227
44. Beller M, Trauthwein H, Eichberger M, Breindl C, Herwig J, Muller TE, Thiel OR (1999) The first rhodium-catalyzed anti-Markovnikov hydroamination: studies on hydroamination and oxidative amination or aromatic olefins. Chem Eur J 5:1306–1319
45. Takemiya A, Hartwig JF (2006) Rhodium-catalyzed intramolecular, anti-Markovnikov hydroamination. Synthesis of 3-arylpiperidines. J Am Chem Soc 128:6042–6043

46. Utsunomiya M, Kuwano R, Kawatsura M, Hartwig JF (2003) Rhodium-catalyzed anti-Markovnikov hydroamination of vinylarenes. J Am Chem Soc 125:5608–5609
47. Utsunomiya M, Hartwig JF (2003) Intermolecular, Markovnikov hydroamination of vinylarenes and alkylamines. J Am Chem Soc 125:14286–14287
48. Liu Z, Yamamichi H, Madrahimov ST, Hartwig JF (2011) Rhodium phosphine-π-arene intermediates in the hydroamination of alkenes. J Am Chem Soc 133:2772–2782
49. Hesp KD, Tobisch S, Stradiotto M (2010) $[Ir(COD)Cl]_2$ as a catalyst precursor for the intramolecular hydroamination of unactivated alkenes with primary amines and secondary alkyl- or arylamines: a combined catalytic, mechanistic, and computational investigation. J Am Chem Soc 132:413–426
50. Hua C, Vuong KQ, Bhadbhade M, Messerle BA (2012) New rhodium(I) and iridium(I) complexes containing mixes pyrazolyl-1,2,3-triazolyl ligands as catalysts for hydroamination. Organometallics 31:1790–1800
51. Nguyen TO, Man BY-W, Hodgson R, Messerle BA (2011) Intramolecular hydroamination of aminoalkenes using rhodium(I) and iridium(I) complexes with N, N- and P, N-donor ligands. Aust J Chem 64:741–746
52. Bauer EB, Andavan GTS, Hollis TK, Rubio RJ, Cho J, Kuchenbeiser GR, Helgert TR, Letko CS, Tham FS (2008) Air- and water-stable catalysts for hydroamination/cyclization. Synthesis and application of CCC-NHC pincer complexes of Rh and Ir. Org Lett 10:1175–1178
53. Casalnuovo AL, Calabrese JC, Milstein D (1988) Rational design in homogeneous catalysis. Ir(I)-catalyzed addition of aniline to norbornylene via N-H activation. J Am Chem Soc 110:6738–6744
54. Burling S, Field LD, Messerle BA, Rumble SL (2007) Late transition metal catalyzed intramolecular hydroamination: the effect of ligand and substrate structure. Organometallics 26:4335–4343
55. Hesp KD, Stradiotto M (2009) Intramolecular hydroamination of unactivated alkenes with secondary alkyl- and arylamines employing $[Ir(COD)Cl]_2$ as a catalyst precursor. Org Lett 11:1449–1452
56. Kashiwame Y, Kuwata S, Ikariya T (2010) Metal-pyrazole bifunction in half-sandwich C-N chelate iridium complexes: pyrazole-pyrazolato interconversion and application to catalytic intramolecular hydroamination of aminoalkene. Chem Eur J 16:766–770
57. Tobisch S (2012) Metal-ligand cooperation in catalytic intramolecular hydroamination: a computational study of iridium-pyrazolato cooperative activation of aminoalkenes. Chem Eur J 18:7248–7262
58. Hesp KD, McDonald R, Stradiotto M (2010) Intramolecular hydroamination of unactivated alkenes with secondary alkylamines catalyzed by iridium phosphine-phenolate complexes. Can J Chem 88:1–9
59. Metallinos C, Zaifman J, Dodge L, Pilkington M (2009) Palladium(II), platinum(II), and iridium(I) complexes of 2-phosphino-1-dimethylaminoferrocenes: a survey of structure and catalysis. Organometallics 28:4534–4543
60. Specht ZG, Cortes-Llamas SA, Tran HN, van Niekerk CJ, Rancudo KT, Golen JA, Moore CE, Rheingold AL, Dwyer TJ, Grotjahn DB (2012) Enabling bifunctionality and hemilability of *N*-heteroaryl NHC complexes. Chem Eur J 17:6606–6609
61. Munro-Leighton C, Delp SA, Alsop NM, Blue ED, Gunnoe R (2008) Anti-Markovnikov hydroamination and hydrothiolation of electron-deficient vinlyarenes catalyzed by well-defined monomeric copper(I) amido and thiolate complexes. Chem Commun 111–113
62. Rao W, Kothandaraman P, Koh CB, Chan PWH (2010) Copper(II) triflate-catalyzed intramolecular hydroamination of homoallylic amino alcohols as an expedient route to *trans*-2.5-dihydro-1*H*-pyrroles and 1,2-dihydroquinolines. Adv Synth Catal 352:2521–2530
63. Michon C, Medina F, Capet F, Roussel P, Agbossou-Niedercorn F (2010) Inter- and intramolecular hydroamination of unactivated alkenes catalyzed by a combination of copper and silver salts: the unveiling of a Bronsted acid catalysis. Adv Synth Catal 352:3293–3305

64. Ohmiya H, Moriya T, Sawamura M (2009) Cu(I)-catalyzed intramolecular hydroamination of unactivated alkenes bearing a primary or secondary amino group in alcoholic solvents. Org Lett 11:2145–2147
65. Zulys A, Dochnahl M, Hollmann D, Lohnwitz K, Herrmann J-S, Roesky PW, Blechert S (2005) Intramolecular hydroamination of functionalized alkenes and alkynes with a homogeneous zinc catalyst. Angew Chem Int Ed 44:7794–7798
66. Donchnahl M, Pissarek J-W, Blechert S, Lohnwitz K, Roesky PW (2006) A new homogeneous zinc complex with increased reactivity for the intramolecular hydroamination of alkenes. Chem Commun 3405–3407
67. Lohnwitz K, Molski MJ, Lulh A, Roesky PW, Dochnahl M, Blechert S (2009) Aminotroponiminate zinc complexes with different leaving groups as catalysts for the intramolecular hydroamination of alkenes. Eur J Inorg Chem 1369–1375
68. Donchnahl M, Lohnwitz K, Lulh A, Pissarek J-W, Biyikal M, Roesky PW, Blechert S (2010) Functionalized aminotroponiminate zinc complexes as catalysts for the intramolecular hydroamination of alkenes. Organometallics 29:2637–2645
69. Meyer N, Lohnwitz K, Zulys A, Roesky PW, Dochnahl M, Blechert S (2006) Aminotroponate zinc complexes as catalysts for the intramolecular hydroamination of alkenes and alkynes. Organometallics 25:3730–3734
70. Duncan CT, Flitsch S, Asefa T (2009) Aminotroponiminate-zinc complex-functionalized mesoporous materials: efficient and recyclable intramolecular hydroamination catalysts. ChemCatChem 1:365–368
71. Jenter J, Lulh A, Roesky PW, Blechert S (2011) Aminotroponiminate zinc complexes as catalysts for the intramolecular hydroamination. J Organomet Chem 696:406–418
72. Meyer N, Roesky PW (2009) Chiral aminotroponiminate zinc complexes. Organometallics 28:306–311
73. Pissarek J-W, Schlesiger D, Roesky PW, Blechert S (2009) Diethylzinc: a simple and efficient catalyst for the swift hydroamination at room temperature. Adv Synth Catal 351:2081
74. Gagne MR, Brard L, Conticello VP, Giardello M, Stern CL, Marks TJ (1992) Stereoselection effects in the catalytic hydroamination/cyclization of aminoolefins at chiral organolanthanide centers. Organometallics 11:2003–2005
75. Shen X, Buchwald SL (2010) Rhodium-catalyzed asymmetric intramolecular hydroamination of unactivated alkenes. Angew Chem Int Ed 122:564–567
76. Turnpenny BW, Hyman KL, Chemler SR (2012) Chiral indole synthesis via enantioselective intramolecular copper-catalyzed alkene hydroamination. Organometallics 31:7819–7892
77. Sherman ES, Fuller PH, Kasi D, Chemler SR (2007) Pyrrolidine and piperidine formation via copper(II)carboxylate-promoted intramolecular carboamination of unactivated olefins: diastereoselectivity and mechanism. J Org Chem 72:3896–3905
78. Coulson DR (1973) Transition metal catalyzed reactions of allenes. J Org Chem 38:1483–1490
79. Padwa A, Filipkowski MA, Meske M, Murphree SS, Watterson SH, Zhijie N (1994) Cyclization reactions of 2,3-bis(phenylsulfonyl)-1,3-butadiene with various carbanions. A [4+1] anionic annulation approach to phenylsulfonyl-substituted cyclopentenes. J Org Chem 59:588–596
80. Zimmer R, Dinesh CU, Nandanan E, Khan FA (2000) Palladium-catalyzed reactions of allenes. Chem Rev 100:3067–3125
81. Morita N, Krause N (2006) Gold-catalyzed cycloisomerization of α-aminoallenes to 3-pyrrolines – optimization and mechanistic studies. Eur J Org Chem 4635–4641
82. Morita N, Krause N (2004) Gold catalysis in organic synthesis: efficient cycloisomerization of α-aminoallenes to 3-pyrrolines. Org Lett 66:4121–4123
83. Meguro M, Yamamoto Y (1998) A new method for the synthesis of nitrogen heterocycles via palladium catalyzed intramolecular hydroamination of allenes. Tetrahedron Lett 39:5421–5424

84. Ha JD, Cha JK (1999) Total synthesis of clavepictines A and B. Diastereoselective cyclization of δ-aminoallenes. J Am Chem Soc 121:10012–10020
85. Kawatsura M, Hartwig JF (2000) Palladium-catalyzed intermolecular hydroamination of vinylarenes using arylamines. J Am Chem Soc 122:9546–9547
86. Qiu S, Wei X, Liu G (2009) Palladium-catalyzed intramolecular hydroamination of allenes coupled to aerobic oxidation. Chem Eur J 15:2751–2754
87. Kadota I, Shibuya A, Lutete LM, Yamamoto Y (1999) Palladium/benzoic acid catalyzed hydroamination of alkynes. J Org Chem 64:4570–4571
88. Patil NT, Pahadi NK, Yamamoto Y (2005) A new route for the synthesis of indolizidine (−)-209D: excellent diastereoselectivity in the intramolecular hydroamination of alkynes. Tetrahedron Lett 46:2101–2103
89. Patil NT, Huo Z, Bajracharya GB, Yamamoto Y (2006) Lactam synthesis via the intramolecular hydroamidation of alkynes catalyzed by palladium complexes. J Org Chem 71:3612–3614
90. Patil NT, Wu H, Yamamoto Y (2007) A route to 2-substituted tetrahydroisoquinolines via palladium-catalyzed intramolecular hydroamination of anilino-alkynes. J Org Chem 72:6577–6579
91. Witham CA, Huang W, Tsung C-K, Kuhn JN, Somorjai GA, Toste FD (2010) Converting homogeneous to heterogeneous in electrophilic catalysis using monodisperse metal nanoparticles. Nat Chem 2:36–39
92. Hashmi ASK, Hutchings GJ (2006) Gold catalysis. Angew Chem Int Ed 45:7896–7935
93. Zhang Z, Liu C, Kinder RE, Han X, Qian H, Widenhoefer RA (2006) Highly active Au(I) catalyst for the intramolecular *exo*-functionalization of allenes with carbon, nitrogen, and oxygen nucleophiles. J Am Chem Soc 128:9066–9073
94. Patil NT, Lutete LM, Nishina N, Yamamoto Y (2006) Gold-catalyzed intramolecular hydroamination of allenes: a case of chirality transfer. Tetrahedron Lett 47:4749–4751
95. Nishina N, Yamamoto Y (2006) Gold-catalyzed intermolecular hydroamination of allenes with arylamines and resulting high chirality transfer. Angew Chem Int Ed 45:3314–3317
96. Broggini G, Borsini E, Fasana A, Poli G, Liron F (2012) Transition-metal-catalyzed hydroamination and carboamination reactions of anthranilic allenamides as a route to 2-vinyl- and 2-(α-styryl)quinazolin-4-one derivatives. Eur J Inorg Chem 3617–3624
97. Tsuhako A, Oikawa D, Sakai K, Okamoto S (2008) Copper-catalyzed intramolecular hydroamination of allenylamines to 3-pyrrolines or 2-alkenylpyrrolidines. Tetrahedron Lett 49:6529–6532
98. Aikawa K, Kojima M, Mikami K (2009) Axial chirality control of gold(biphep) complexes by chiral anions: application in asymmetric catalysis. Angew Chem Int Ed 48:6073–6077
99. Liu L, Wang F, Wang W, Zhao M, Shi M (2011) Synthesis of chiral mono(*N*-heterocyclic carbene) palladium and gold complexes with a 1,1'-biphenyl scaffold and their applications in catalysis. Beilstein J Org Chem 7:555–564
100. Widenhoefer RA (2008) Recent developments in enantioselective gold(I) catalysis. Chem Eur J 14:5382–5391
101. LaLonde RL, Sherry BD, Kang EJ, Toste FD (2007) Gold(I)-catalyzed enantioselective intramolecular hydroamination of allenes. J Am Chem Soc 129:2452–2453
102. LaLonde RL, Wang ZJ, Mba M, Lackner AD, Toste FD (2010) Gold(I)-catalyzed enantioselective synthesis of pyrazolidines, isoxazolidines, and tetrahydrooxazines. Angew Chem Int Ed 49:598–601
103. Hamilton GL, Kang EJ, Mba M, Toste FD (2007) A powerful chiral counterion strategy for asymmetric transition metal catalysis. Science 317:496–499
104. Zhang Z, Bender CF, Widenhoefer RA (2007) Gold(I)-catalyzed enantioselective hydroamination of *N*-allenyl carbamates. Org Lett 9:2887–2889
105. Li H, Lee SD, Widenhoefer RA (2011) Gold(I)-catalyzed enantioselective hydroamination of allenes with ureas. J Organomet Chem 696:316–320

106. Zhang Z, Bender CF, Widenhoefer RA (2007) Gold(I)-catalyzed dynamic kinetic enantioselective intramolecular hydroamination of allenes. J Am Chem Soc 129:14148–14149
107. Sherry BD, Toste FD (2004) Gold(I)-catalyzed propargyl Claisen rearrangement. J Am Chem Soc 126:15978–15979
108. Lutete LM, Kadota I, Yamamoto Y (2004) Palladium-catalyzed intramolecular asymmetric hydroamination of alkynes. J Am Chem Soc 126:1622–1623
109. Patil NT, Lutete LM, Wu H, Pahadi NK, Gridnev ID, Yamamoto Y (2006) Palladium-catalyzed intramolecular asymmetric hydroamination, hydroalkoxylation, and hydrocarbonation of alkynes. J Org Chem 71:4270–4279
110. Zhao P, Incarvito CD, Hartwig JF (2006) Carbon-oxygen bond formation between a terminal alkoxo ligand and a coordinated olefin. Evidence for olefin insertion into a rhodium alkoxide. J Am Chem Soc 128:9642–9643
111. Keith JA, Henry PM (2008) The mechanism of the Wacker reaction: a tale of two hydroxypalladations. Angew Chem Int Ed 48:9038–9049
112. Qian H, Han X, Widenhoefer RA (2004) Platinum-catalyzed intramolecular hydroalkoxylation of λ- and δ-hydroxy olefins to form cyclic ethers. J Am Chem Soc 126:9536–9537
113. Hori K, Kitagawa H, Miyoshi A, Ohta T, Furukawa I (1998) Transition metal-catalyzed cyclization of 2-allylphenol to 2,3-dihydro-2-methylbenzofuran without β-hydride elimination. Chem Lett 27:1083–1084
114. Ohta T, Kataoka Y, Miyishi A, Oe Y, Furukawa I, Ito Y (2007) Ruthenium-catalyzed intramolecular cyclization of hetero-functionalized allylbenzenes. J Organomet Chem 692:671–677
115. Oe Y, Ohta T, Ito Y (2005) Rethenium-catalyzed addition reaction of alcohol across olefins. Synlett 1:179–181
116. Ito Y, Kato R, Hamashima K, Kataoka Y, Oe Y, Ohta T, Furukawa I (2007) Intramolecular cyclization of phenol derivatives with C═C double bond side chains. J Organomet Chem 692:691–697
117. Adrio LA, Quek LS, Taylor JG, Hii KK (2009) Copper-catalysed intramolecular O-H addition to unactivated alkenes. Tetrahedron 65:10334–10338
118. Yang C-G, Reich NW, Shi Z, He C (2005) Intramolecular additions of alcohols and carboxylic acids to inert olefins catalyzed by silver(I) triflate. Org Lett 7:4553–4556
119. Ke F, Li Z, Xiang H, Zhou X (2011) Catalytic hydroalkoxylation of alkenes by iron(III) catalyst. Tetrahedron Lett 52:318–320
120. Komeyama K, Morimoto T, Nakayama Y, Takaki K (2007) Cationic iron-catalyzed intramolecular hydroalkoxylation of unactivated olefins. Tetrahedron Lett 48:3259–3261
121. Yang C-G, He C (2005) Gold(I)-catalyzed intermolecular addition of phenols and carboxylic acids to olefins. J Am Chem Soc 127:6966–6967
122. Chandrasekhar B, Ryu J-S (2012) Gold-catalyzed intramolecular hydroalkoxylation/cyclization of conjugated dienyl alcohols. Tetrahedron 68:4805–4812
123. Kitahara H, Sakurai H (2012) Anti-addition mechanism in the intramolecular hydroalkoxylation of alkenes catalyzed by pvp-stabilized nanogold. Molecules 17:2579–2586
124. Olsson L-I, Claesson A (1979) Synthesis of 2,5-dihydrofurans and 5,6-dihydro-2*H*-pyrans by silver(I)-catalyzed cyclization of allenic alcohols. Synthesis 9:743–745
125. Yamamoto Y, Radhakrishnan U (1999) Palladium catalyzed pronucleophilic addition to unactivated carbon-carbon bonds. Chem Soc Rev 28:199–207
126. Ma S (2005) Some typical advances in the synthetic applications of allenes. Chem Rev 105:2829–2871
127. Hoffmann-Roder A, Krause N (2001) Gold(III) chloride catalyzed cyclization of α-hydroxyallenes to 2,5-dihydrofurans. Org Lett 3:2537–2538
128. Bartolome C, Garcia-Cuadrado D, Ramiro Z, Espinet P (2010) Exploring the scope of nitrogen acyclic carbenes (NACs) in gold-catalyzed reactions. Organometallics 29:3589–3592

129. Wang ZJ, Brown CJ, Bergman RG, Raymond KN, Toste FD (2011) Hydroalkoxylation catalyzed by a gold(I) complex encapsulated in a supramolecular host. J Am Chem Soc 133:7358–7360
130. Kadota I, Lutete LM, Shibuya A, Yamamoto Y (2001) Palladium/benzoic acid-catalyzed hydroalkoxylation of alkynes. Tetrahedron Lett 42:6207–6210
131. Patil NT, Pahadi NK, Yamamoto Y (2005) Pd(0)-PhCOOH catalyzed addition of oxygen pronucleophiles to allenes and internal alkynes. Can J Chem 83:569–574
132. Zhang Z, Widenhoefer RA (2007) Gold(I)-catalyzed intramolecular enantioselective hydroalkoxylation of allenes. Angew Chem Int Ed 46:283–285
133. Aikawa K, Kojima M, Mikami K (2010) Synergistic effect: hydroalkoxylation of allenes through combination of enantiopure BiPHEP-gold complexes and chiral anions. Adv Synth Catal 252:3131–3135

Top Heterocycl Chem (2013) 32: 157–186
DOI: 10.1007/7081_2013_103

Published online: 19 April 2013

Synthesis of Saturated Heterocycles via Metal-Catalyzed Allylic Alkylation Reactions

John M. Ketcham and Aaron Aponick

Abstract Over the past 10 years, significant progress has been made in the field of metal-catalyzed allylic alkylation reactions. Intramolecular variants forming saturated heterocycles comprise a well-known class of reactions that are often utilized in the syntheses of biologically active natural products. Selected recent advances in this area between the years 2002 and 2012 and their applications toward total syntheses are reviewed herein.

Keywords Alkylation · Allylic · Heterocycles · Intramolecular · Metal-catalyzed · Natural products · S_N2' · π-Allyl

Contents

J.M. Ketcham and A. Aponick (✉)
University of Florida, Department of Chemistry, P.O. Box 117200, Gainesville 32611, FL
e-mail: aponick@chem.ufl.edu

Abbreviations

(*R*)-BINAPHANE	(*R*,*R*)-1,2-Bis[(*R*)-4,5-dihydro-3*H*-binaptho-(1,2*c*:2′,1′*e*) phosphino]benzene
Ac	Acetyl
Bn	Benzyl
Boc	*t*-Butoxycarbonyl
Bz	Benzoyl
Cbz	Benzyloxycarbonyl
cod	1,5-Cyclooctadiene
DABCO	1,4-Diazabicyclo[2.2.2]octane
DACH-Ph	1,2-Diaminocyclohexane-*N*,*N*′-bis(2-diphenylphosphinobenzoyl)
dba	Dibenzylideneacetone
DBU	1,8-Diazabicyclo[5.4.0]undec-7-ene
DMPS	Dimethylphenylsilyl
DPPBA	Diphenylphosphinobenzoic acid
dppe	1,2-Bis(diphenylphosphino)ethane
Fmoc	9-Fluorenylmethoxycarbonyl
MeO-BIPHEP	2,2′-Bis(diphenylphosphino)-6,6′-dimethoxy-1,1′-biphenyl
Ns	2-Nitrobenzenesulfonyl
Ph	Phenyl
PMB	*p*-Methoxybenzyl
PMP	*p*-Methoxyphenyl
TBD	1,5,7-Triazabicyclo-[4.4.0]undec-5-ene
Tf	Trifluoromethanesulfonyl
TMS	Trimethylsilyl
Troc	2,2,2-Trichloroethoxycarbonyl
Ts	*p*-Toluenesulfonyl

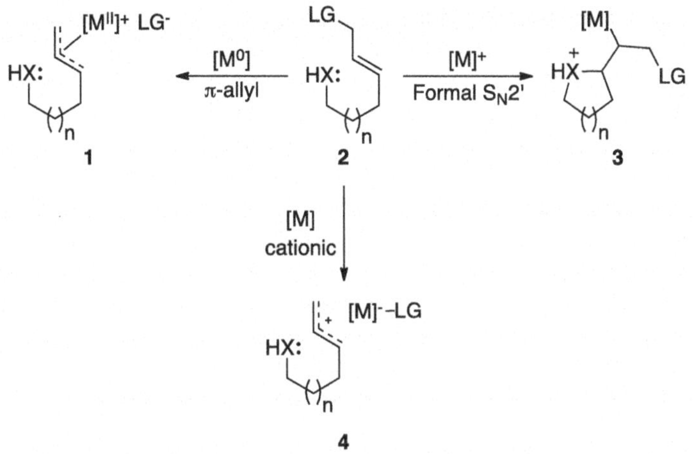

Scheme 1 Activation modes for catalytic π-allyl, formal S_N2', and cationic cyclization reactions

1 Introduction

The ubiquity of heterocycles in biologically active natural products has led to an ever-growing arsenal of methodologies aimed at the production of these cyclic structures. Among cyclization strategies, metal-catalyzed intramolecular allylic alkylation reactions have been particularly fruitful. These facile processes accommodate a broad range of substrates and generally effect ring formation under relatively mild conditions with low catalyst loadings. Mechanistically, metal-catalyzed allylic alkylation reactions can be placed into three distinct categories: formation of π-allyl metal intermediates, direct cyclization by formal S_N2' reactions, and cyclization of cationic systems (Scheme 1). Although these different pathways can in principle give identical products, their mechanistic course varies greatly and is dependent on many factors including: solvent, metal catalyst, leaving group, additives, and ligands. This chapter attempts to categorize the general area by mechanistic class as is dictated by the factors outlined in the ensuing paragraphs.

Reaction conditions for the formation of π-allylmetals (**1**) generally contain nucleophilic/electron-rich metal-catalysts, which act upon allylic systems containing highly reactive leaving groups such as carbonates and halides. During the catalytic cycle, the metal undergoes a redox sequence wherein two electrons are lost and regained during the reaction course. This method has proven to be highly successful and extremely versatile. Recent developments involve an increased number of methods utilizing metals other than palladium and alternative allylic leaving groups.

Formal S_N2' reactions are typically effected by an electrophilic, π-acidic, metal-complex that prefers the formation of a π-complex without undergoing redox during the reaction course. Metal-catalyzed intramolecular formal S_N2' reactions constitute a relatively new class of reaction pathway when compared to the well-known π-allylmetal systems. New methodologies and mechanistic insights are continually being reported and may be beginning a paradigm shift in this area.

Ionization of allylic systems to form cations (**4**) generally employs highly electrophilic metal catalysts that readily ionize the allylic system by abstracting the leaving group. The metal-complex is usually comprised of a hard metal capable of coordinating directly to the leaving group. The cationic nature of these systems adds a significant challenge when enantioenriched products are desired.

Over the past decade, a variety of groups have reported on heterocycle synthesis using cyclization reactions with a diverse set of substrates and catalyst systems. The following chapter is organized first by mechanism and chronologically within each section. Instead of a comprehensive review, this chapter presents selected examples that focus specifically on the formation of a carbon–heteroatom bond via a metal-catalyzed allylic alkylation reaction to form saturated heterocycles.

2 Formation of Saturated Heterocycles via π-Allyl Metal Complexes

The following section covers select examples of π-allyl metal intermediates in the formation of heterocycles over the past 10 years. These processes generally proceed through the classical mechanistic steps: coordination of the catalyst to the allylic olefin, ionization to form a π-allylmetal species, and nucleophilic attack on the complex regenerating the catalyst.

2.1 Heterocycle Synthesis via π-Allyl Palladium Intermediates

Since the initial discoveries of Tsuji [1], Trost [2], and coworkers, the Tsuji–Trost reaction has stood as one of the most versatile synthetic transformations [3–5] for forming both carbon–carbon and carbon–heteroatom bonds. During their syntheses of (±)-desethylibogamine and (+)-ibogamine in the late 1970s, Trost et al. reported some of the earliest examples utilizing this methodology to form a carbon–heteroatom bond in an intramolecular fashion [6, 7]. Over the past 40 years, intramolecular Tsuji–Trost type cyclizations have become commonplace in the synthesis of heterocycles, and this strategy has been used in a myriad of natural product syntheses [8–11].

Driven by their initial studies toward the construction of the core ring structure of vitamin E [12], Trost and coworkers extensively studied a palladium-catalyzed intramolecular asymmetric allylic alkylation (AAA) of phenol-tethered allyl carbonates **5** to form chromans **6** (Scheme 2) [13–15]. These highly useful synthons could be formed in high yield and good enantioselectivities with the use of Pd_2dba_3 and ligand (*R*,*R*)-**L1** under mild conditions. Generally, the *E*-allylic carbonates give

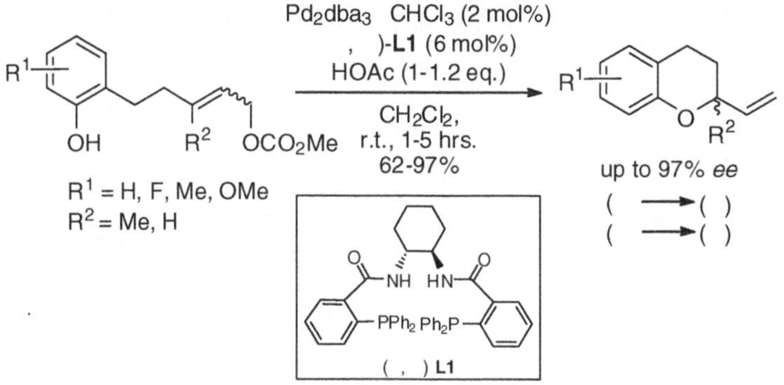

Scheme 2 Pd-catalyzed synthesis of chromans from phenol-tethered allyl carbonates

the (*R*)-chromans while *Z*-allylic carbonates give the (*S*)-chroman products with (*R*,*R*)-**L1**. Interestingly, in most cases the *Z*-allylic systems give higher enantioselectivities. The work culminates in the utilization of these Pd-catalyzed cyclizations in the total syntheses of (+)-clusifoliol [14] and (−)-siccanin [15].

In 2006, the same group found that similar conditions could be used to achieve a one-pot cascade reaction [16, 17], forming piperazinone **9** by a palladium-catalyzed asymmetric allylic alkylation (AAA) reaction between dicarbonate **7** and pyrrole **8** (Scheme 7). This reaction was further used in the formal total synthesis of (−)-agelastatin A (Scheme 3). Additional studies revealed that the regioisomer **10** could be prepared by a sequential palladium-catalyzed process, which was further used in the synthesis of the opposite enantiomer, (+)-agelastatin A (**11**).

In addition to examples using chiral catalysts with achiral substrates, achiral catalyst in combination with chiral substrates can be utilized. Spilling has reported an interesting system using chiral allylic phosphonates [18]. The method was used to synthesize vinyl tetrahydropyran and tetrahydrofuranyl phosphonates and is an extension of their previously reported process for generation of vinyl *N*-heterocyclic phosphonates (Scheme 4) [19]. Although 7- and 8-membered rings could not be formed under the reaction conditions, the 5- and 6-membered ring products **13** were obtained with complete transfer of chirality from carbonate substrates **12** [20]. These vinyl phosphonates are easily transformed into the corresponding β-ketophosphonates via a regioselective Wacker oxidation and can subsequently be used in Horner–Wadsworth–Emmons (HWE) reactions to easily prepare more complex structures. The authors have nicely demonstrated the utility of this method in the formal synthesis of (+)-centrolobine [20], the synthesis of an Amphidinolide F fragment [21], and more recently the synthesis of both diastereomeric nematocidal oxylipids isolated from the Australian sea sponge *Notheia anomala* [22].

As a final example of the versatility of the palladium-catalyzed intramolecular Tsuji–Trost reaction, Comins and coworkers demonstrated the use of a vinylogous amide nucleophile in the synthesis of alkaloid (−)-205B [23]. Isolated in 1987 by

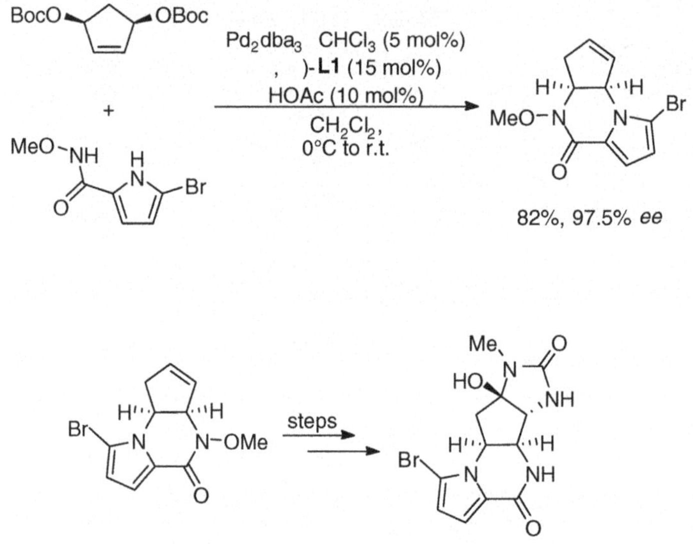

10

11
(+)-agelastatin A

Scheme 3 Synthesis of piperazinone **8** via Pd-catalyzed intramolecular Tsuji–Trost allylation

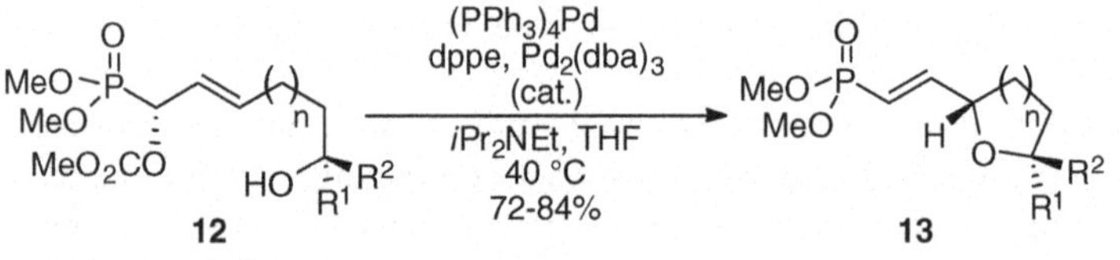

n = 1, 2
R^1, R^2 = H, Me

Scheme 4 Pd-catalyzed synthesis of vinyl tetrahydropyran and tetrahydrofuran phosphonates

Daly et al. [24, 25], alkaloid (−)-205B is structurally unique when compared to other indolizidine alkaloids and its enantiomer has shown selective inhibition for a receptor that is linked with various neurological diseases [26]. Comins' synthesis provides a concise and efficient pathway to **16** in eleven steps [23]. An intramolecular Tsuji–Trost reaction using a vinylogous amide nucleophile **14** gives the product **15** in high diastereoselectivity with the bulky P(^{t}Bu)$_3$ ligand (Scheme 5). The use of Cs_2CO_3 was crucial as other bases led to significant decomposition of the substrate. After this key-step, the total synthesis of the natural product was easily completed from **15** in seven steps.

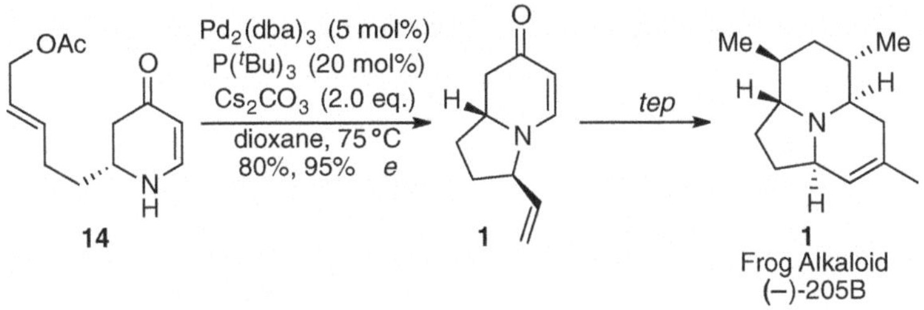

Scheme 5 Comins et al. Total synthesis of alkaloid (−)-205B

2.2 *Heterocycle Synthesis via π-Allyl Iridium Intermediates*

Approximately 40 years after the discovery of the Tsuji–Trost reaction Takeuchi et al. [27, 28] and Helmchen and coworkers [29, 30] reported that iridium complexes were effective catalysts for allylic alkylation reactions. Their pioneering work demonstrated that nucleophilic addition to π-allyl iridium complexes preferentially forms the branched alkylation products, which is in contrast to the linear alkylation products formed by palladium catalysis. Since these initial reports numerous advances have demonstrated the advantages of iridium complexes in allylic alkylation reactions [31, 32]. Given that these iridium-catalyzed processes were developed much more recently than the palladium-catalyzed systems, it is not surprising that intramolecular variants to form heterocycles were not reported until the early 2000s.

In 2003, Takemoto et al. reported the iridium-catalyzed diallylic amination of bis(allylic carbonates) **17** to form various azacycles **19** [33] (Scheme 6). Although the diastereoselectivities were low, the yields and regioselectivities were high. More significantly, this report details the first synthesis of heterocycles via an iridium-catalyzed intramolecular allylic amination strategy.

Soon after this report, Helmchen and coworkers demonstrated the first enantioselective iridium-catalyzed intramolecular allylic amination [34, 35]. After testing various solvents, ligands, additives, etc., it was found that iridium complexes containing phosphoramidite ligands of the general structure **L2** [36] have a dramatic impact on both the reactivity and selectivity of the process (Scheme 7).

Using these conditions, allylic carbonates **20** undergo smooth cyclization to their corresponding azacycles **21** in up to 99% yield and 97% *ee*. Employing similar reaction conditions, they were able to extend this methodology and design systems for the enantioselective formation of chromans and an enantioselective sequential inter-/intramolecular allylic amination reaction [35].

Further exploration of these allylic aminations revealed that the reactions are catalyst controlled and both 2,6-*cis* or 2,6-*trans* piperidines could be formed from the same substrate [37, 38]. For example, the isomeric mixture **22** can be subjected to similar reaction conditions to generate either the *cis*-**23** or *trans*-**24** diastereomeric products depending on which enantiomer of the ligand **L3** is used (Scheme 8).

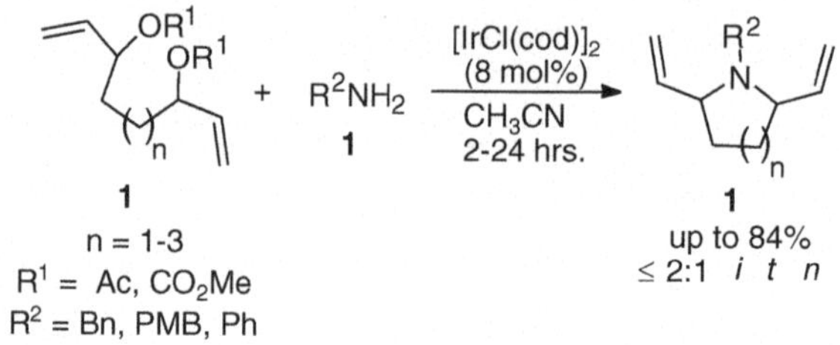

Scheme 6 Iridium-catalyzed sequential allylic amination to form azacycles

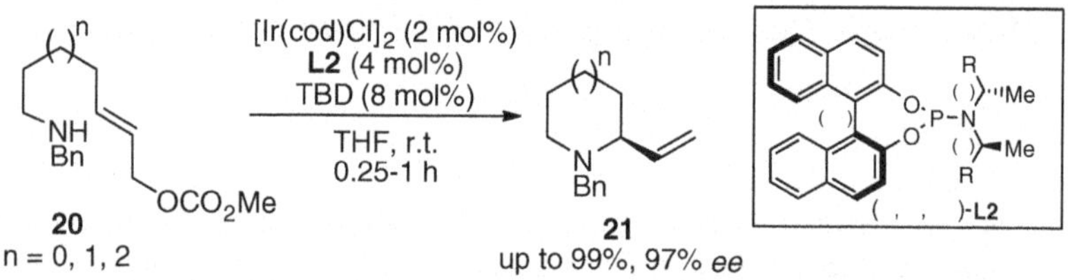

Scheme 7 First enantioselective iridium-catalyzed intramolecular allylic amination

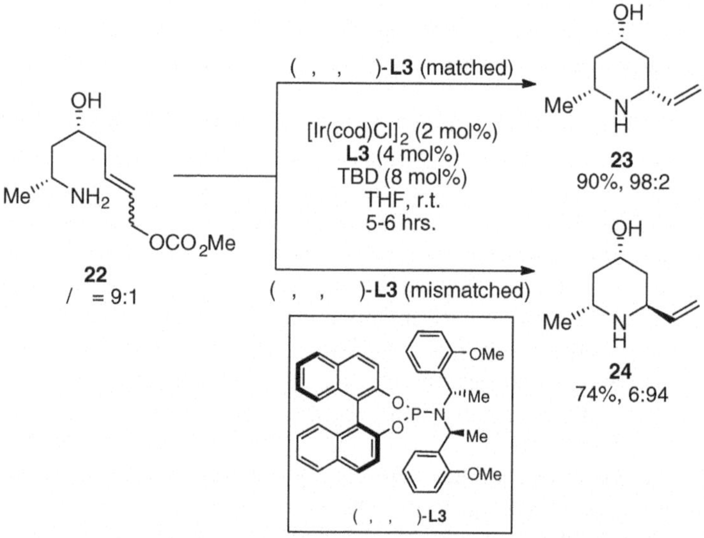

Scheme 8 Iridium-catalyzed allylic alkylations used as a configurational switch

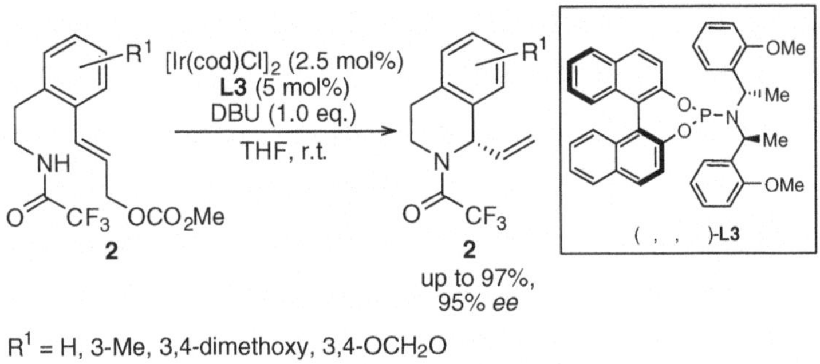

Scheme 9 Iridium-catalyzed formation of tetrahydroisoquinolines

Although only primary amines were used in these cyclization reactions, the yields and selectivities were excellent, and this methodology has been utilized in the total syntheses of prosopis, dendrobate, and spruce alkaloids [38].

More recently, Feringa and coworkers have demonstrated a similar method for constructing tetrahydroisoquinolines [39]. Trifluoroacetylamides **25** readily underwent cyclization to form the corresponding tetrahydroisoquinolines **26** in high yield and enantioselectivity (Scheme 9). This strategy was used to synthesize saturated pyrrolidines and piperidines; however, competing β-hydride elimination was encountered and rendered the formation of azepane derivatives quite challenging. The resulting products could be easily deprotected using K_2CO_3 in $MeOH/H_2O$ without a reduction in *ee*.

2.3 Heterocycle Synthesis via π-Allyl Nickel Intermediates

Heterocyclic formation via π-allyl nickel intermediates are sparse; however, Berkowitz and coworkers undertook an exhaustive study in 2004 [40, 41]. In this report, combinatorial catalysis using an in situ enzymatic screening (ISES) process indicated that nickel complexes could be used to form oxazolidinones via an asymmetric allylic amination reaction.

The report details a screen of more than 25 different bis(phosphine) [40] and P,N-ligands [41] that identify the complex produced from $Ni(cod)_2$ and (*R*)-MeO-BIPHEP as the best catalyst system providing the desired oxazolidinone **28** in 88% yield and 75% *ee* (97% *ee* after one recrystallization) (Scheme 10). The oxazolidinone product **28** was converted to TFA salt of L-glycine **29** in 21% overall yield in five subsequent steps.

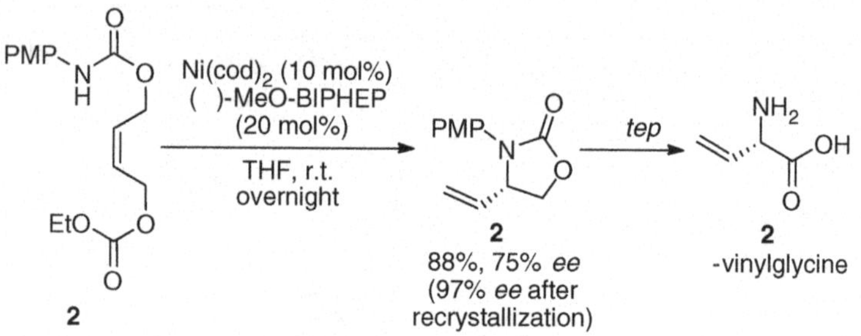

Scheme 10 Nickel-catalyzed formation of oxazolidinones

2.4 Heterocycle Synthesis via π-Allyl Ruthenium Intermediates

Pioneering studies by the Tsuji [42], Watanabe [43], and Trost [44] research groups demonstrated the practicality of ruthenium-complexes for allylic alkylation reactions; however, their application to heterocycle synthesis by intramolecular allylic alkylation has only recently gained popularity.

After their reports detailing the intermolecular dehydrative allylation of alcohols [45, 46] in 2009 Kitamura et al. reported an efficient ruthenium-catalyzed dehydrative cyclization to form cyclic ethers [47]. Reactions were performed in various solvents with very low catalyst loadings (as low as 0.0001 mol%) to provide the products in high yields and enantioselectivities (Scheme 11). Tetrahydropyrans and tetrahydrofurans **31** (as well as chromans) could be efficiently formed from the appropriate diols **30**. However, the preparation of seven-membered cyclic ethers posed a significant challenge, presumably due to the formation of oligomeric side products.

Studies indicate that the chlorine atom in ligand **L4** has a pronounced influence on the reactivity of the complex. The authors suggest that the chlorine atom in **L4** could be playing two distinct roles in obtaining a more energetically favorable transition state. The electronics of the system may be modulated by the chlorine, thereby lowering the LUMO to enable a more facile redox cycle. Additionally, it is proposed that a Cp-H···Cl-R hydrogen bond could further stabilize the transition state. Experimental evidence indicates that the reaction proceeds through an intermediate π-allylruthenium species formed by direct ionization of the allylic alcohol system.

Kitamura has also demonstrated that the same ruthenium-complex could be used in an intramolecular dehydrative cyclization to form azacycles [48]. Various nitrogen heterocycles **33** were synthesized from the corresponding allylic alcohols **32** with catalyst loadings as low as 0.05 mol% (Scheme 12). A variety of protecting groups on the nitrogen were tolerated, and the yields and enantiomeric ratios were excellent. Interestingly, arene-fused azapane **35** could be easily produced under the reaction conditions from allylic alcohol **34**. Conversely, when sulfonamide **36** was subjected to the optimized conditions a competing β-hydride elimination process dominated, producing diene **37** instead of the expected product. In the case of the

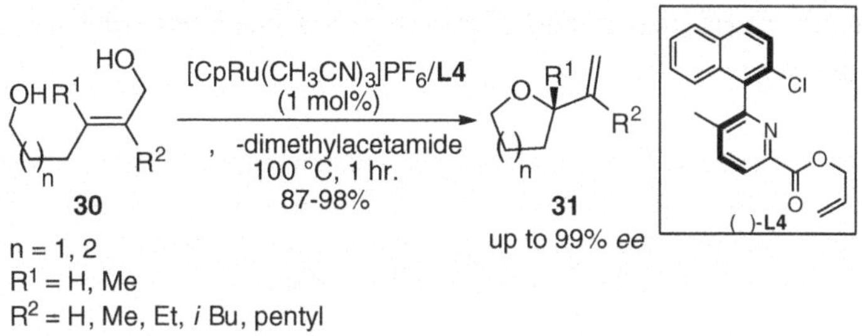

Scheme 11 Ruthenium-catalyzed formation of cyclic ethers

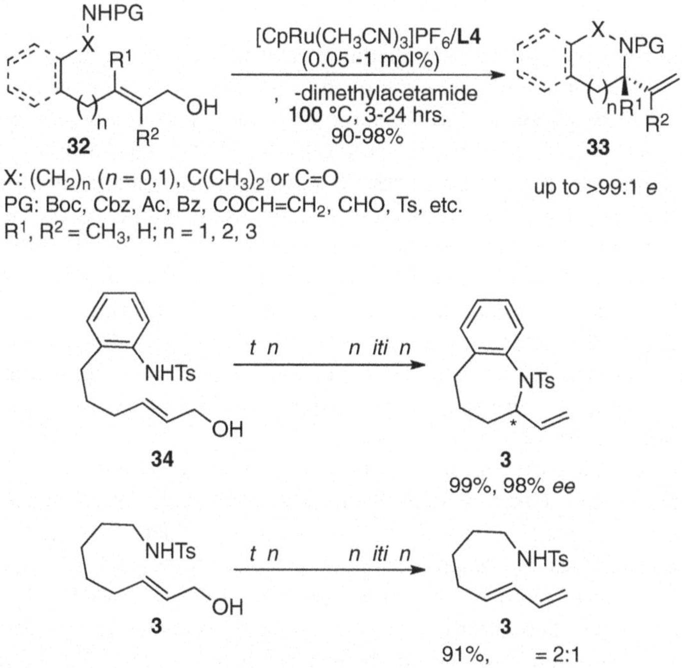

Scheme 12 Ruthenium-catalyzed formation of azacycles

arene-fused azepanes, the authors suggest that the sp^2-carbons of the aniline may permit a better HOMO/LUMO interaction allowing for a higher propensity toward cyclization. However, the conformational effects of these arene-fused sulfonamides may also facilitate a faster cyclization. The aforementioned reactions demonstrate an exceptional methodology for the production of saturated heterocycles and can be carried out on a gram-scale.

3 Formation of Saturated Heterocycles via Formal S_N2' Reactions

Metal-catalyzed formal S_N2' sequences encompass a relatively new strategy for the formation of heterocycles. The reactions are mechanistically distinguished from π-allylmetal chemistry and metal-catalyzed carbocation formation by the fact that a cation (or metal-bound cation) is not generally formed.

3.1 Formal S_N2' Reactions Catalyzed by Palladium Complexes

Given the extensive history of palladium catalysts in the activation of allylic systems it is not surprising that some of the earliest examples of metal-catalyzed formal S_N2' cyclizations to form heterocycles were performed with palladium complexes [49, 50]. To the best of our knowledge, Hirai and coworkers were the first to demonstrate the effectiveness of a palladium(II)-catalyzed formal S_N2' heterocyclization in an enantioselective fashion using allylic ethers and allylic alcohols [49, 50]. The latter report illustrates the successful chirality transfer from the allylic alcohol; substrate **38** was converted to piperidine **39** with complete transfer of chirality. This compound was subsequently transformed to the alkaloid natural product **40**-(+)-coniine (Scheme 13) [50]. Throughout the late nineties, the Hirai group applied these methods to the total synthesis of numerous natural products including: (+)-prosopinine, (+)-palustrine, SS20846A, and 1-deoxymannojirimycin [51–53].

More recently, Uenishi and coworkers have advanced this catalytic methodology to the formation of tetrahydro- and dihydropyrans [54]. These methods were applied directly to the total synthesis of the natural product (−)-laulimalide [55]. During these studies a comparison between the Pd(0)- and Pd(II)-catalyzed cyclizations to form tetrahydropyran **42** and 3,6-dihydropyran **44** indicated that Pd(II) was superior (Scheme 14). For both cyclizations complete chirality transfer was observed; however, in the case of **41** the process was much higher yielding with the Pd(II) source because the use of a Pd(0) catalyst resulted in competing triene formation via β-hydride elimination. Additionally, cyclization of **43** to form pyran **44** did not occur under standard Pd(0) conditions. Mechanistically, the cyclization is assumed to go through a *syn*-addition/*syn*-elimination sequence with respect to the palladium complex [55, 56]. Finally, fragments **42** and **44** were advanced to complete the asymmetric total synthesis of (−)-laulimalide. Uenishi and coworkers have since applied these oxypalladation cyclizations to the construction of several intricate compounds including tetrasubstituted chiral carbon centers [56], and more recently in the total synthesis of (−)-apicularen A and its analogs [57].

In addition to oxygen heterocycles, this synthetic strategy was also applied to the formation of nitrogen heterocycles [58]. Various nitrogen protecting groups (Cbz, Boc, Ts, Fmoc, etc.) were tolerated, but S_N2'-cyclizations of –Cbz protected amines

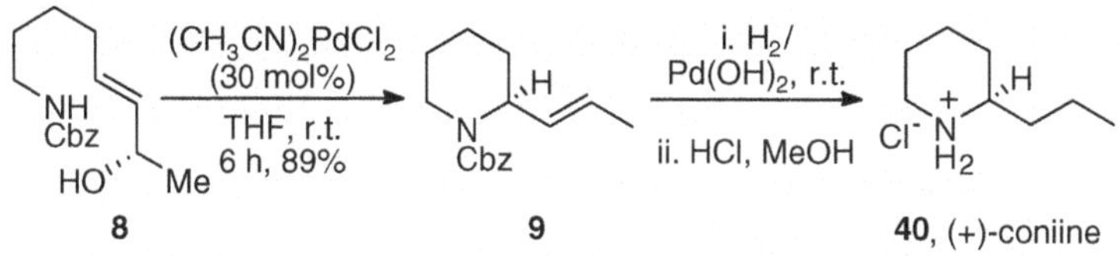

Scheme 13 Pd(II)-catalyzed transfer of chirality

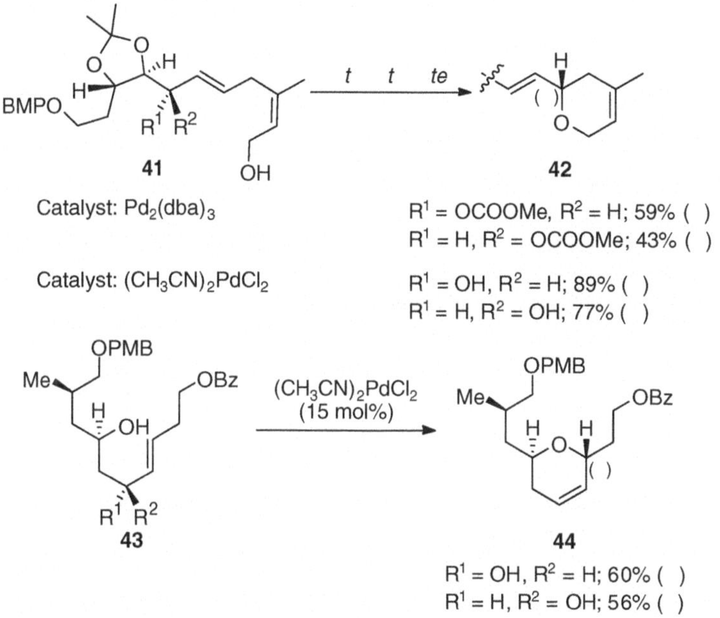

Scheme 14 Studies in the total synthesis (−)-laulimalide

gave the best results. Efficient transfer of chirality was also observed for these transformations. When the enantioenriched allylic alcohols (*R*)- and (*S*)-**45** were treated with 10 mol% of $(CH_3CN)PdCl_2$ the 2-vinylpiperidines were produced in 93% and 92% enantiomeric excess, respectively (Scheme 15). The products (*R*)-**46** and (*S*)-**46** were further used to synthesize the hydrochloride salts of (*S*)-(+)- and (*R*)-(−)-coniine, respectively. These conditions were also found to be highly diastereoselective as demonstrated in the cyclization of the two epimers of **47**. Each substrate epimer was transformed to a different stereoisomer of product **48** in a high diastereomeric ratio for both cases.

Recently, the same group demonstrated a cascade epoxide ring opening to form bis- and tris-tetrahydrofuran rings [59]. Both epimers of epoxide **49** undergo cyclization in under an hour to form the corresponding bis-tetrahydrofuran **50** with good diastereoselectivity (Scheme 16). This method can also accommodate both epimers of diepoxide **51** to give the desired tris-tetrahydrofuran compound **52**

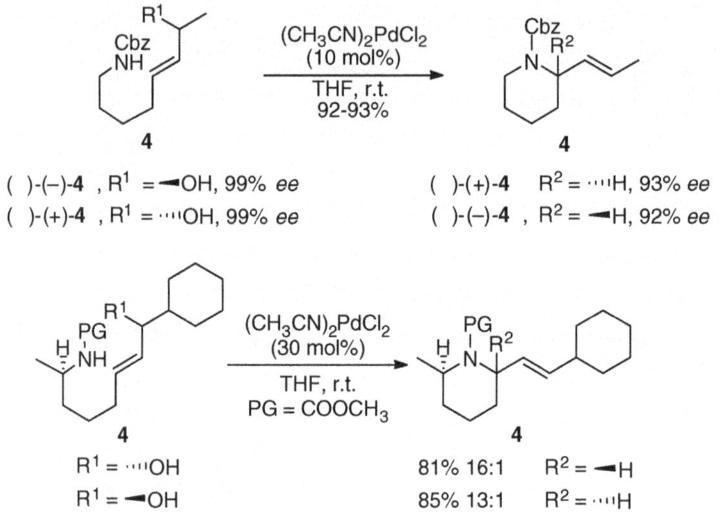

Scheme 15 Pd(II)-catalyzed formation of piperidines

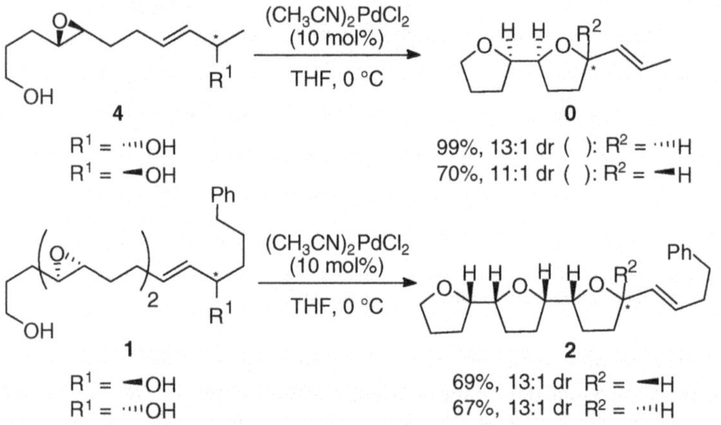

Scheme 16 Pd(II)-catalyzed formation of attached bis- and tris-THF rings

with good selectivities. Preliminary mechanistic studies suggest that the sequence most likely occurs with concerted formation of all C–O bonds rather than a series of stepwise addition reactions.

Recent studies by Aponick and coworkers have revealed a facile spiroketalization methodology utilizing a palladium(II)-catalyzed S_N2' cyclization [60] (Scheme 17). Most notably this method was utilized as the key-step in the total synthesis of acortatarin A [61]. Treatment of allylic ether **55** with 10 mol% of $(CH_3CN)_2PdCl_2$ produced the desired spiroketal **56** in an 87% yield as a 1:1 mixture of epimers which were further elaborated to obtain the desired natural

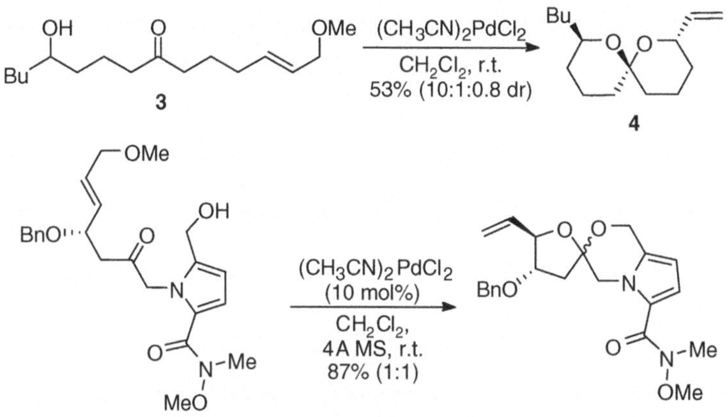

Scheme 17 Pd(II)-catalyzed spiroketalization

product. It should be noted that for the synthesis of spiro *C*-arylglycoribisides a similar spiroketalization strategy was employed by Hirai and coworkers using a hemiacetal nucleophile [62].

3.2 Formal S_N2' Reactions Catalyzed by Gold Complexes

Gold catalysis is a relatively new and ever-expanding field that has provided a wide variety of interesting new methods to the synthetic community [63–66]. With low catalyst loadings and high functional group tolerance, the use of gold complexes as catalysts has become a competitive alternative to some of the traditionally used transition metals. In 2008, Aponick and coworkers were the first to demonstrate a gold-catalyzed dehydrative formal S_N2' cyclization to form tetrahydropyrans and tetrahydrofurans [67, 68]. The process is selective for the formation of *cis*-disubstituted cyclic ethers **58** from monoallylic diols **57** with high diastereoselectivities and yields using very low catalyst loadings (Scheme 18). With the ease of substrate syntheses and catalyst loadings as low as 0.1 mol%, the production of gram-scale quantities of these tetrahydropyrans and furans was readily achieved [68]. Further experimentation demonstrated that the cyclizations did not proceed through a cationic mechanism but rather through a formal S_N2' process.

With respect to heterocycle formation, Aponick [69–71] and others [72, 73] have made significant extensions to these methods including the synthesis of substituted chromenes [71], a stereoselective preparation of 2-vinyl-morpholines [72] and applications to the total synthesis of (+)-isoaltholactone [73], to name a few.

In 2011, the Aponick group reported an efficient transfer of chirality in the cyclization of monoallylic diols **59** to form tetrahydropyrans and morpholines **60** and **61** (Scheme 19) [74]. Selective access to either enantiomer can be achieved from substrates that differ only by the geometry of the olefin, allowing for selective

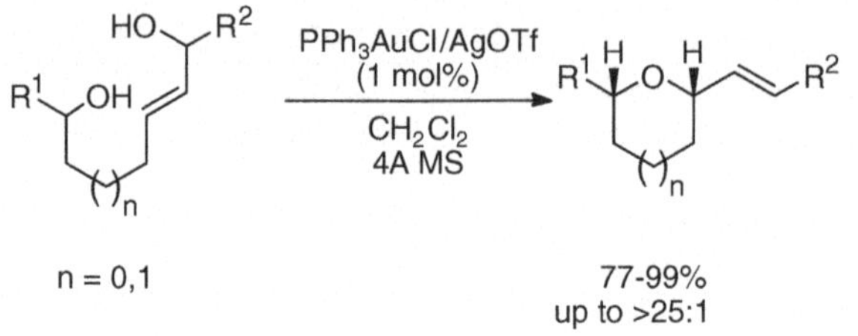

Scheme 18 Gold-catalyzed dehydrative cyclization to form cyclic ethers

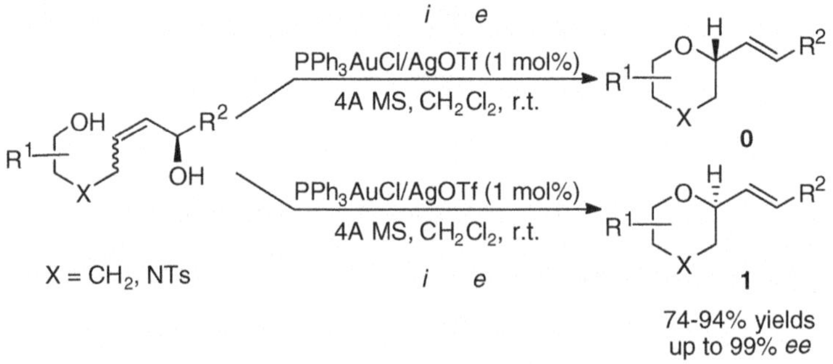

Scheme 19 Gold-catalyzed chirality transfer process for the synthesis of 6-membered heterocycles

access to either stereoisomer. This synthetically practical process provides the desired products in high yields with excellent diastereo- and enantioselectivities. Later that year, a comparative study showed that allylic ethers could be used in place of allylic alcohols to furnish 2-vinyltetrahydropyran products [75].

In 2011, Widenhoefer and coworkers reported a gold-catalyzed intramolecular amination to form azacycles [76]. Secondary amine substrates **62** were demonstrated to undergo cyclization in a high yielding process and, in some examples, with high diastereoselectivity. These basic amines require much higher temperatures (60–100°C) than the corresponding alcohols (Scheme 20). Additionally, subjection of amine **64** under optimized conditions produced the desired piperidine **65** with a complete transfer of chirality. The enantiopurity and absolute configuration of **65** was confirmed by conversion to the hydrochloric acid salt of (*S*)-(+)-coniine.

A highly effective enantioselective intramolecular amination reaction catalyzed by a bis(gold)phosphine complex has also been reported [77]. The new method accommodates a wide range of carbamates **66** to give the desired piperidines and piperazines **67** in high yields and enantioselectivities with the use of a bis(gold) phosphine complex prepared from the bisphosphine ligand **L5** (Scheme 21). Further experiments demonstrated a net *syn*-displacement of the allylic alcohol by the incoming nucleophile.

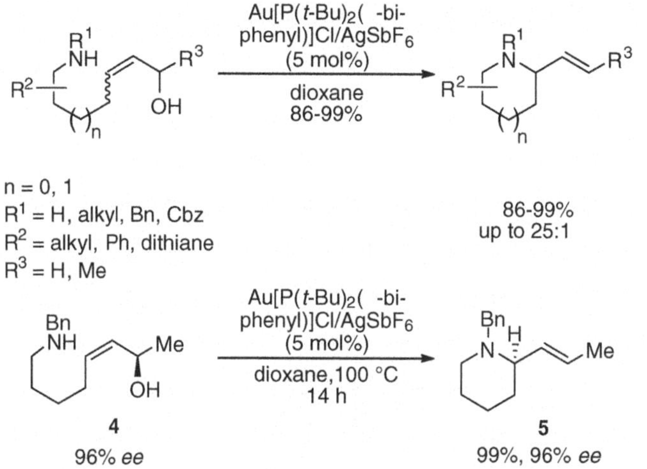

Scheme 20 Gold-catalyzed formation of azacycles

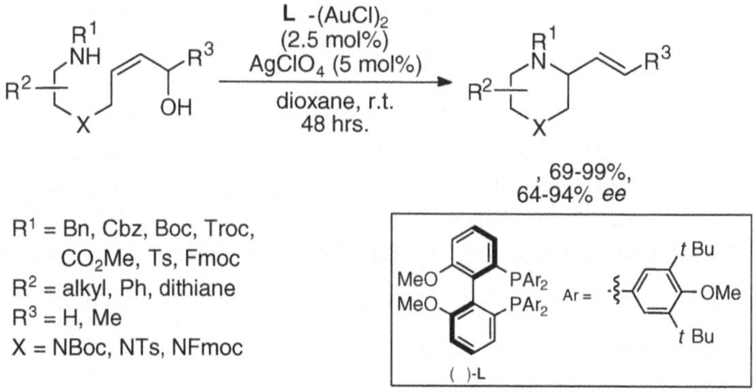

Scheme 21 Formation of azacycles via bis(phosphine)gold complex

Collaborative mechanistic studies by the Aponick and Ess groups have given unequivocal insight into the mechanism of these gold-catalyzed dehydrative cyclization reactions [78]. The experimental and computational studies illustrate the importance of hydrogen bonding with respect to both reactivity and stereoselectivity. Between the three transition states **TS-*anti***, **TS-*syn***, and **TS-*concerted*** the lowest calculated energy state is **TS-*anti*** (Fig. 1). The results suggest that these cyclization reactions must go through a stepwise *anti*-alkoxyauration/*anti*-elimination mechanism in which intramolecular hydrogen bonding between the allylic alcohol and the incoming hydroxyl nucleophile is responsible for both rate acceleration and stereochemical control.

Experimentally, this concept was demonstrated in the gold-catalyzed cyclizations of bicyclic diols **70–72**. The requisite distance for these intramolecular hydrogen bonding interactions cannot be achieved with substrate **70** and

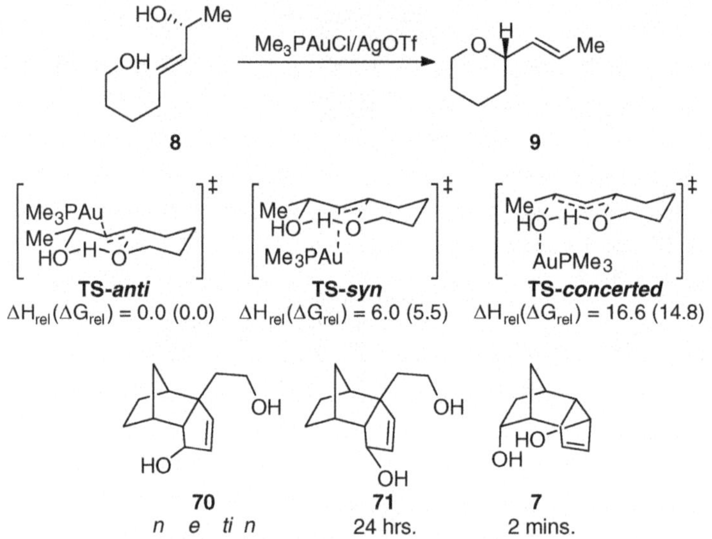

Fig. 1 Pivotal role of hydrogen bonding in gold-catalyzed dehydrative cyclizations of monoallylic diols

consequently no desired cyclization was observed under the optimized conditions. In contrast, as this interaction and the ability of the catalyst to effect an *anti*-addition become more accessible, the desired cyclization reactions become more facile. This was demonstrated in reactions of *endo*- and *exo*-allylic substrates **71** and **72**; **71** undergoes slow reaction whereas **72** is completely converted to cyclized product in only 2 min.

3.3 Formal S_N2' Reactions Catalyzed by Mercury Complexes

In 2008, Nishizawa and coworkers reported an efficient mercury-catalyzed dehydrative cyclization to form various saturated azacycles and indolines [79]. Sulfonamides **73** underwent facile ring closure to form the desired 2-vinylazacycles **74** in high yields with very low catalyst loadings (Scheme 22). Allylic alcohols and ethers also underwent the desired cyclization; however, allylic esters did not cyclize under the optimized conditions.

The same group later established an enantioselective version of these cyclizations using the chiral (R)-BINAPHANE ligand with $Hg(OTf)_2$ [80] (Scheme 23). After screening various ligands and nitrogen protecting groups it was determined that the highest enantioselectivities were achieved with *tert*-butyl substituted sulfonamides. Treating sulfonamides **75** with the chiral mercury-complex at low temperatures gave the desired indolines **76** in high yields with

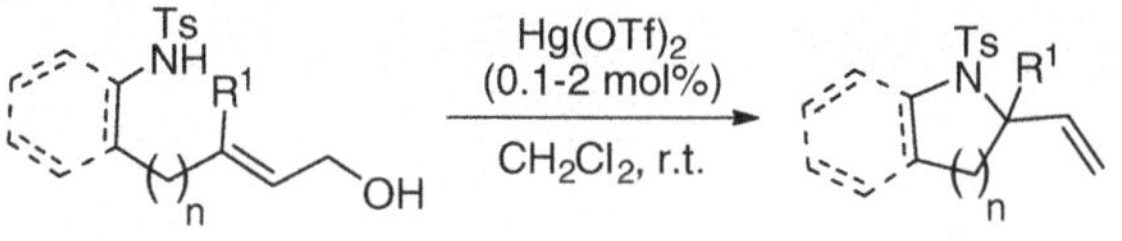

Scheme 22 Mercury-catalyzed dehydrative cyclization to form azacycles

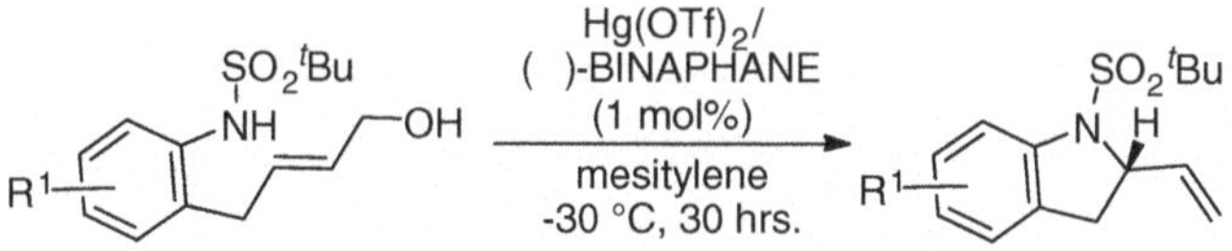

Scheme 23 Mercury-catalyzed enantioselective formation of indolines

moderate to good enantioselectivities. Sulfonamide products **76** could be easily deprotected with anisole in a solution of TFA/CH_2Cl_2, without epimerization of the chiral center.

3.4 Formal S_N2' Reactions Catalyzed by Bismuth Complexes

During their investigations of chirality transfer in metal-catalyzed intramolecular allylic aminations Kawai, Uenishi, and coworkers screened more than ten different metals to find that bismuth(III) triflate gave the best results [81]. Under the optimized, relatively mild bismuth-catalyzed conditions, enantiopure allylic alcohols **77** were transformed to the desired tetrahydroisoquinolines **78** in good yield (Scheme 24). Boc-protected amines provided the highest selectivities, while substituted olefins (R^1 = Me) were converted to cyclic products with significantly lower the enantiomeric ratios. Interestingly, using the conditions optimized for palladium catalysis (Sect. 3.1), high enantioselectivities were observed at −20°C, but the product was produced only in a 20% chemical yield.

To explain these findings a chelated intermediate, **79**, was proposed to rationalize why higher selectivities were obtained using carbamate starting materials (Fig. 2).

In 2011, an extension of the bismuth methodology was reported that expanded the scope to include substituted tetrahydroisoquinolines and gave further insight into the reaction mechanism [82]. The same year a variety of tetrahydroisoquinoline alkaloid natural products (Fig. 3) were prepared to showcase the methodology [83].

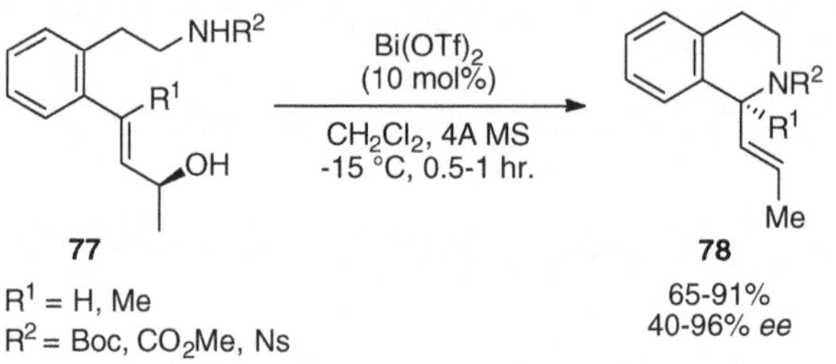

Scheme 24 Bismuth-catalyzed chirality transfer to form tetrahydroisoquinolines

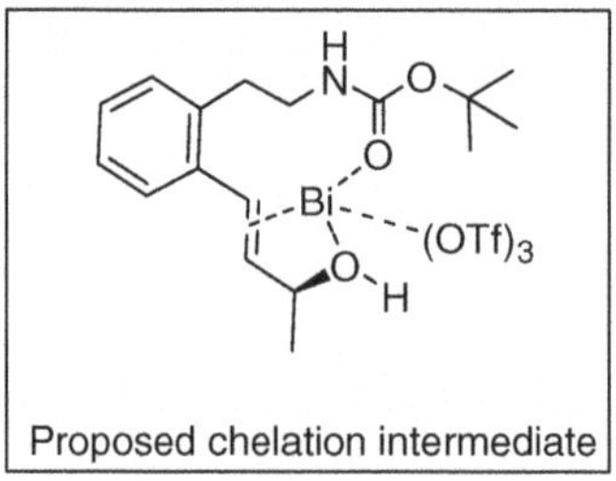

Fig. 2 Proposed bismuth chelation intermediate

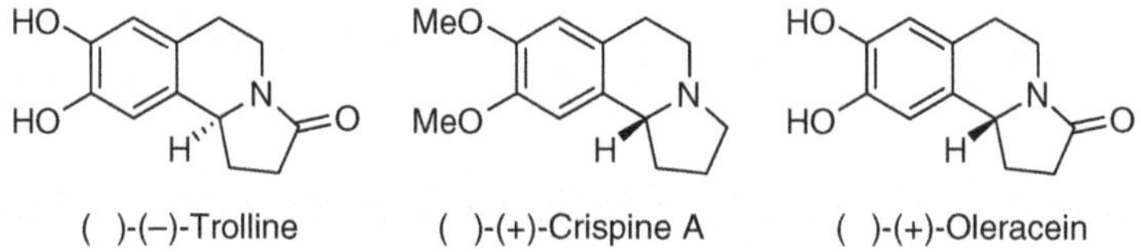

Fig. 3 Tetrahydroisoquinoline natural products prepared via bismuth-catalyzed S_N2' cyclizations

4 Formation of Saturated Heterocycles via Cationic Intermediates

Cationic intermediates in the syntheses of heterocycles have become more prevalent over the past 10 years. Much like π-allylmetal intermediates, an allyl electrophile is produced in these systems. However, the catalyst is not covalently associated with the allyl electrophile, which instead is generated as a free carbocation. This change in mechanism necessitates different types of catalysts to associate with the substrate and remove the allylic leaving group in a different manner.

4.1 Ionization Using Magnesium Complexes

During a total synthesis of (−)-*cis*-clavicipitic acid Jia and coworkers made a serendipitous discovery [84]. Deprotection of the bis(carbamate) **80** using $Mg(ClO_4)_2$ provided the desired azapane **81** instead of the expected deprotection product.

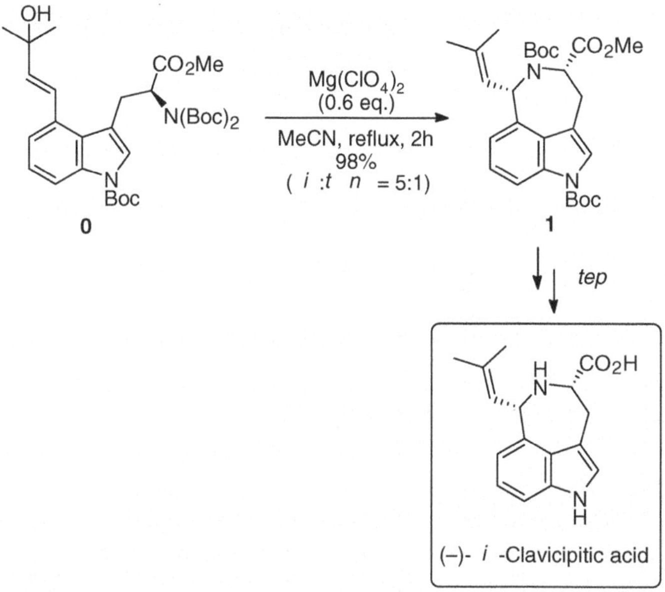

Scheme 25 Magnesium-promoted formation of azapane **81**

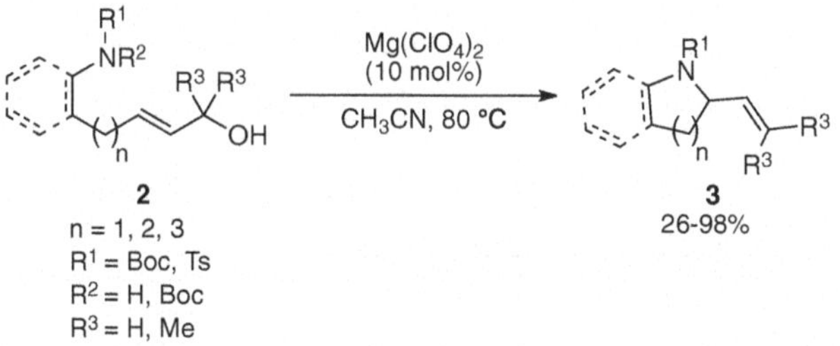

Scheme 26 Formation of azacycles by magnesium-catalyzed dehydrative cyclization

The azapane **81** was obtained with good diastereoselectivity resulting from deprotection followed by magnesium-promoted ionization of the allylic/benzylic system and subsequent cyclization (Scheme 25). This process provided straightforward access to the desired natural product after several steps. The magnesium-promoted process was later used in a one-pot tandem palladium-catalyzed Heck reaction/magnesium promoted dehydrative cyclization in the total syntheses of aurantioclavine and clavicipitic acid [85].

In 2012, the conditions were optimized to allow for a process that is catalytic in magnesium [86]. Treating sulfonamides or carbamates **82** with 10 mol% of $Mg(ClO_4)_2$ at 80°C in acetonitrile gave the desired tetrahydroisoquinolines **83** in reasonable yields for secondary and tertiary allylic alcohols as well as primary allylic acetates (Scheme 26). Piperidines and pyrrolidines could also be prepared; however, substrates containing primary alcohols were shown to be sluggish even with a full equivalent of magnesium. After further optimization, this methodology was then applied to the total synthesis of a known fungal inhibitor demethoxyfumitremorgin C.

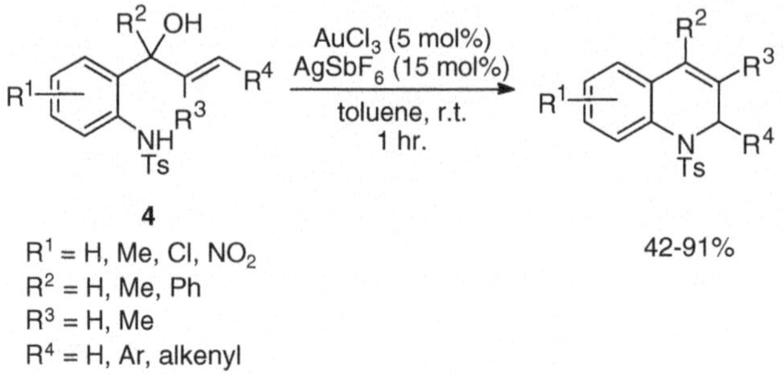

Scheme 27 Gold-catalyzed formation of 1,2-dihydroquinolines

4.2 Ionization Using Gold Complexes

In 2009, Chan et al. described an efficient gold-catalyzed dehydrative cyclization to form 1,2-dihydroquinolines [87]. Under very mild conditions arylsulfonamides **84** were transformed into the desired dihydroquinolines **85** in good yields (Scheme 27). The authors speculate that the process proceeds via a cationic mechanism. Interestingly, this is in contrast to Aponick's chromene synthesis [71], which utilized gold-catalysis but likely does not form a cationic intermediate in many cases. Furthermore, they were able to use this methodology for the total synthesis of the tetrahydroquinoline alkaloid (±)-angustureine.

4.3 Ionization Using Iron Complexes

An attractive method for the diastereoselective formation of *cis*-piperidines and tetrahydropyrans was recently reported by Cossy and coworkers [88, 89]. This iron-catalyzed process provides the desired products **87** from allylic alcohols **86** in high yields and with high diastereoselectivity for the *cis*-products under mild conditions (Scheme 28). Given the cationic nature of the reaction, transposed allylic alcohols were also readily cyclized under the reaction conditions.

Interestingly, this catalyst system can also be applied to the cyclization of ketoalcohol **88** to form the desired spiroketal **89** in a diastereomeric ratio of >99:1. The high stereoselectivity is believed to derive from the epimerization/equilibration of product stereoisomer *trans*-**90** to the more stable *cis*-**90** through an allyl cation intermediate **91** (Scheme 29). Although this iron-catalyzed methodology demonstrates broad functional group tolerance and high diastereoselectivities, the stereochemistry of the allylic system cannot be transferred from the starting material to the product as it can be using Au- and Pd-catalysts as described above in Sect. 3.

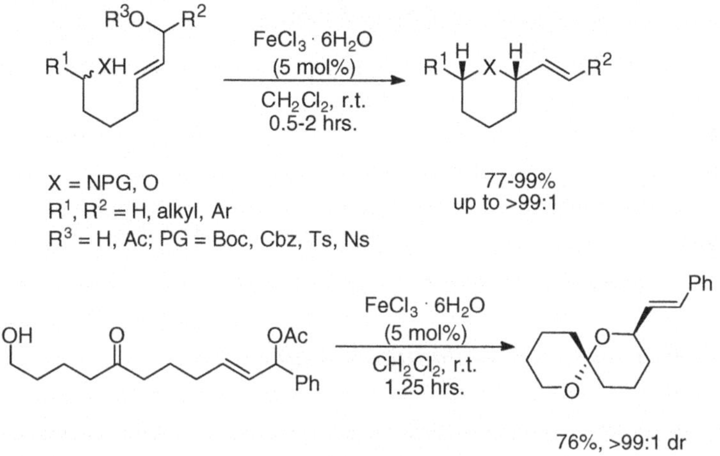

Scheme 28 Iron-catalyzed formation of saturated heterocycles

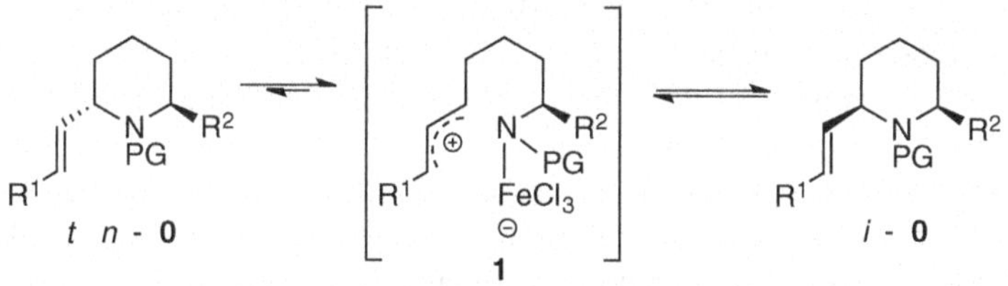

Scheme 29 Rationale for high diastereoselectivity

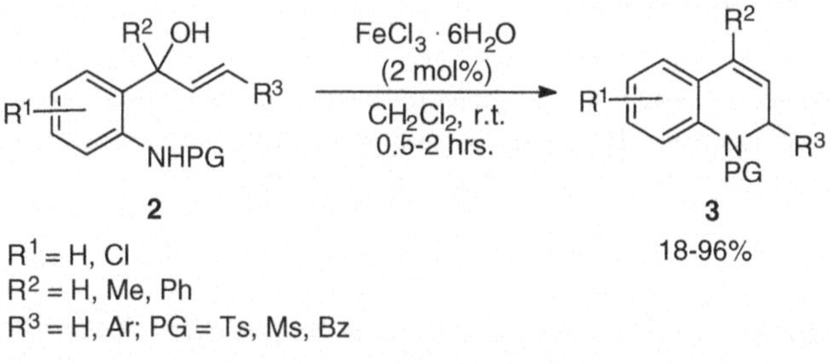

Scheme 30 Iron-catalyzed formation of dihydroquinolines

Sun et al. later used this catalyst system for the formation of substituted dihydroquinolines and quinolines [90]. The process effects the cyclizations of anilines **92** to form dihydroquinolines **93** with low catalysts loadings and good yields in most cases (Scheme 30). When enantiopure allylic alcohols were used they exhibited no transfer of chirality, instead producing a racemic mixture of dihydroquinoline products. Treatment of the products **93** with sodium hydroxide in ethanol at reflux furnished the corresponding quinoline products.

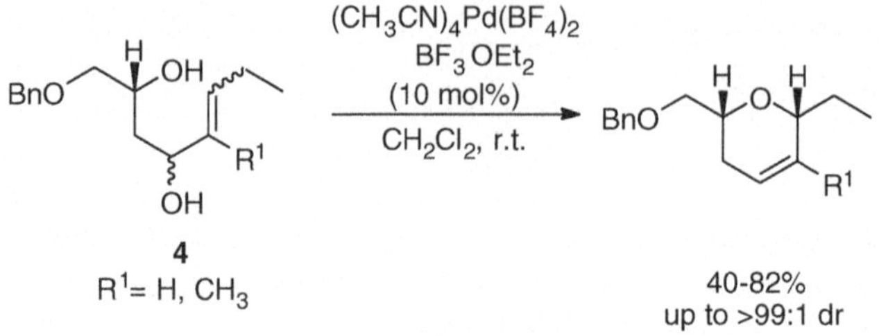

Scheme 31 Stereoselective formation of 2,6-*cis*-dihydropyrans

4.4 Ionization Using Palladium Complexes

During their studies of the total synthesis of jerangolid A, Hanessian and coworkers discovered a highly diastereoselective cyclization of monoallylic diols **94** to form 2,6-*cis*-dihydropyrans **95** with very high diastereoselectivity [91, 92] (Scheme 31). The cyclizations are facile using either the cationic palladium complex $(CH_3CN)_4Pd(BF_4)_2$ or $BF_3{}^{\bullet}OEt_2$ with 10 mol% catalyst loadings. The reaction is similar to those of Uenishi (see Sect. 3.1) but likely proceeds via a cationic intermediate. Regardless of the stereochemical configuration of the alkene and/or the allylic alcohol, the cyclization reactions selectively provided the *cis*-products. This stereochemical outcome likely results from a cationic mechanism. This cycloetherification protocol was used to complete the first total synthesis of jerangolid A in sixteen linear steps from an enantiopure glycidol.

5 Miscellaneous Cases

The following section encompasses selected examples that would not necessarily fit in the previous sections but demonstrate interesting cases for metal-catalyzed formation of saturated heterocycles from allylic systems.

5.1 Formation of Heterocycles via a Sequential Ruthenium Enyne/Palladium Allylation Process

In 2006, Trost and coworkers demonstrated a sequential one-pot ruthenium enyne coupling followed by a palladium-catalyzed allylation to form nitrogen and oxygen heterocycles [93]. Allylic *p*-nitrophenyl ethers **99** generated after the ruthenium enyne coupling of **96** and **97** give the desired heterocycles **98** in moderate to good enantioselectivities (Scheme 32). The process can also be used to form oxygen heterocycles **101** from sequential coupling and cyclization with substrates **96** and **100**.

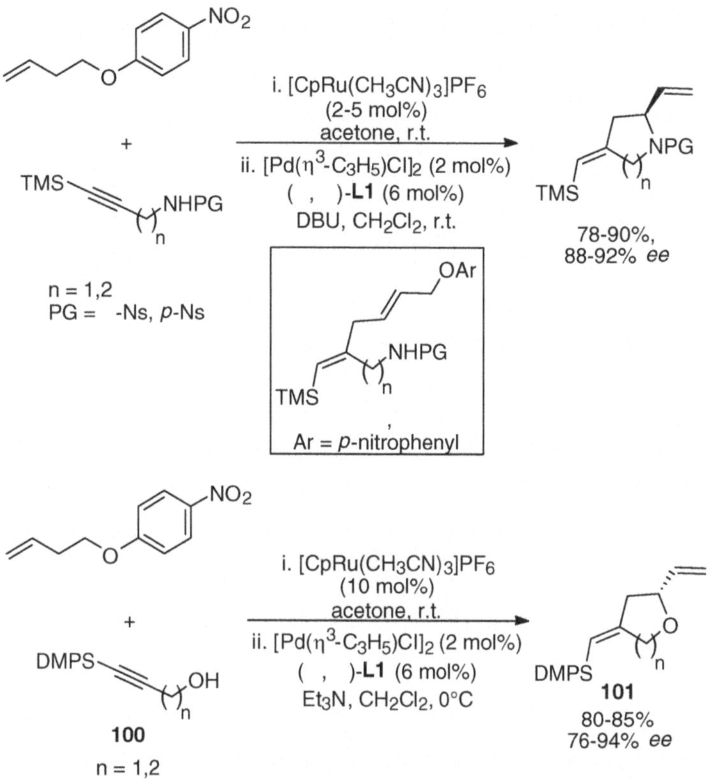

Scheme 32 Sequential Ru/Pd catalysis to form *N*- and *O*-heterocycles

Diastereoselective syntheses of piperidines, tetrahydrofurans, and tetrahydropyrans were also possible through this methodology, generally providing excellent stereoselectivity with the use of chiral ligands. Interestingly, diastereomers that are predominantly thermodynamically disfavored can be obtained through this protocol. Additionally, the stereochemistry seems to be determined by the hard/soft nature of the incoming nucleophiles. For sulfonamides (soft) the initial π-allyl system formed is kinetically trapped, whereas alcohols (hard) go through a slow trapping mechanism allowing for the equilibration/interconversion of the π-allyl diastereomers. Lastly, these methods were applied to the synthesis of the B-ring of the chemotherapeutic natural product bryostatin.

5.2 Formation of Heterocycles via a Tandem Iridium-Catalyzed Vinylation/Allylic Amination Reaction

A short time later, You and coworkers reported an efficient enantioselective iridium-catalyzed tandem allylic vinylation/amination method to form 2,3-dihydro-1*H*-benzo[*b*]azepines [94]. The reaction sequence starts with an allylic

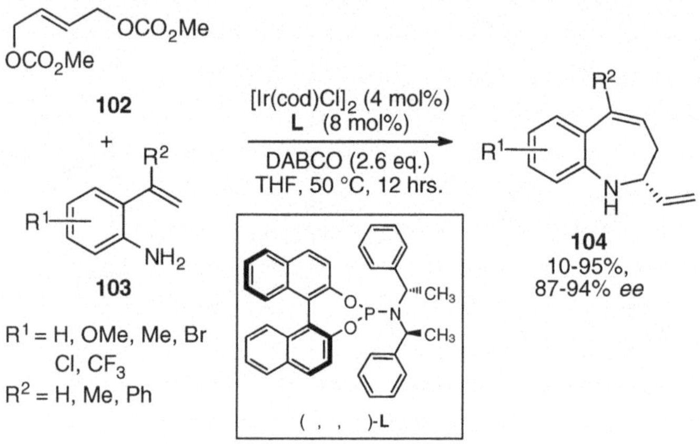

Scheme 33 Enantioselective tandem iridium-catalyzed allylic vinylation/amination reaction to form azepines

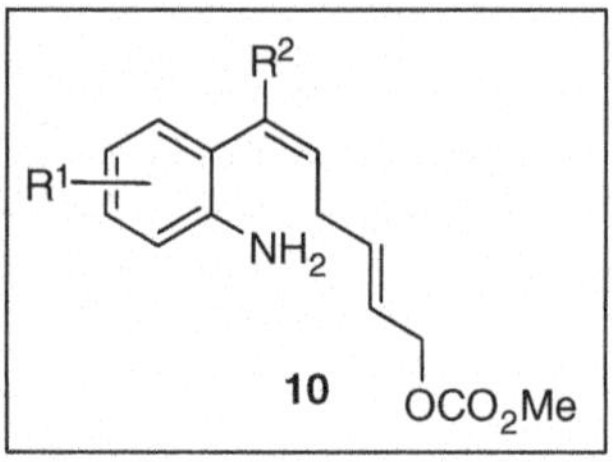

Fig. 4 Allyl carbonate intermediate

vinylation of **102** with **103** thereby creating a monoallylic carbonate intermediate that further undergoes intramolecular allylic amination to give the desired azepines **104** (Scheme 33). In most cases yields were high, and enantioselectivities were consistently good.

Evidence for the proposed pathway was found through experiments that verified the monoallylic carbonate intermediates **105** could both be isolated and cyclized, under optimized conditions (Fig. 4). Moreover, it is interesting to note that this method is one of the few processes in this chapter that gives dependable enantioselective access to 7-membered nitrogen heterocycles.

5.3 Formation of Heterocycles via C–H Activation of Allylic Systems

Metal-catalyzed C–H functionalization has recently become a prominent strategy for the formation of complex structures [95, 96]. The White group has developed various methodologies for the formation of saturated heterocycles via C–H

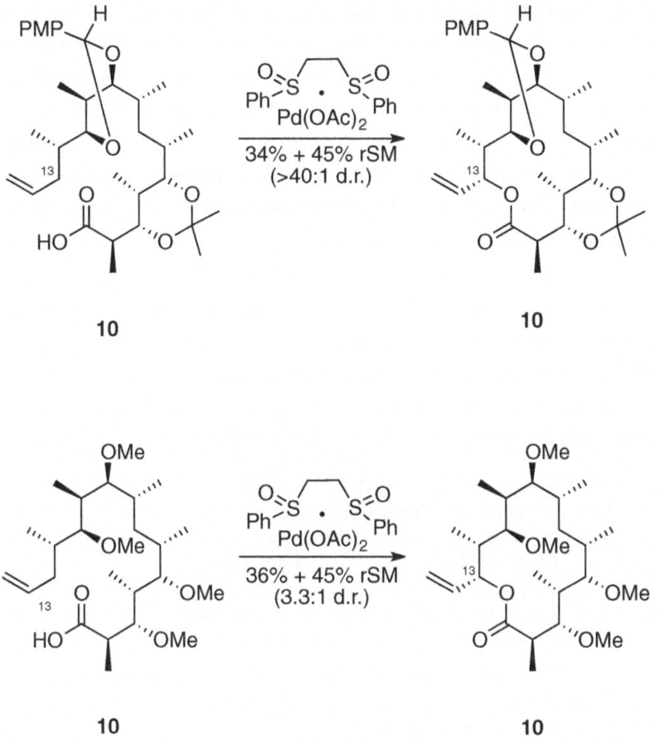

Scheme 34 Studies of the C–H oxidative macrolactonization of erythromycin cores

activation of allylic systems and has applied this approach to the syntheses of biologically relevant compounds [97–103]. While most of the heterocycles formed are intermediates toward 1,2- and 1,3-aminoalcohols or diols these methods can also be used to synthesize heterocycles. For instance, during the total synthesis of 6-deoxyerythronolide B, an efficient macrolactonization was achieved using a palladium-catalyzed C–H oxidation [99, 102] (Scheme 34). Treating compounds **106** and **108** gave the 14-membered macrolides **107** and **109**, respectively. While Yamaguchi macrolactonizations were also performed, the C–H oxidation macrolactonization provides a complementary route without the need for oxidation at the C13 center.

6 Conclusions and Outlook

Over the past 10 years, numerous research groups have demonstrated that a variety of catalysts and reaction pathways can be used to produce structurally unique heterocycles of all varieties. The formation of these compounds through carbon–heteroatom bond forming metal-catalyzed allylic alkylation reactions is an important synthetic strategy that is continually evolving. From the reports discussed in this chapter, one must appreciate the role that these reactions play in the synthesis of various natural products and biologically active compounds. Given that a vast array of structurally diverse heterocycles are found in natural products and biologically important structures, it is likely that these compounds will dictate the need for new methods and that catalytic allylic alkylation will continue to develop and expand in new directions.

References

1. Tsuji J, Takahashi H, Marikawa M (1965) Tetrahedron Lett 6:4387–4388
2. Trost BM, Fullerton TJ (1973) J Am Chem Soc 95:292–294
3. Lu Z, Ma S (2008) Angew Chem Int Ed 47:258–297
4. Trost BM, Zhang T, Sieber JD (2010) Chem Sci 1:427–440
5. Trost BM (2012) Org Process Res Dev 16:185–194
6. Trost BM, Genet JP (1976) J Am Chem Soc 98:8516–8517
7. Trost BM, Godleski SA (1978) J Am Chem Soc 100:3930–3931
8. Trost BM, Oslob JD (1999) J Am Chem Soc 121:3057–3064
9. Williams DR, Meyer KG (1999) Org Lett 1:1303–1305
10. Seki M, Mori Y, Hatsuda M, Yamada S (2002) J Org Chem 67:5527–5536
11. Graenin T, Schmalz HG (2003) Angew Chem Int Ed 42:2580–2584
12. Trost BM, Naoyuki A (1999) Synthesis 1491–1494
13. Trost BM, Shen HC, Dong L, Surivet JP (2003) J Am Chem Soc 125:9276–9277
14. Trost BM, Shen HC, Dong L, Surivet JP, Sylvain C (2004) J Am Chem Soc 126:11966–11983
15. Trost BM, Shen HC, Surivet JP (2004) J Am Chem Soc 126:12565–12579
16. Trost BM, Dong G (2006) J Am Chem Soc 128:6054–6055
17. Trost BM, Dong G (2009) Chem Eur J 15:6910–6919
18. Bisceglia JA, Orelli LR (2012) Curr Org Chem 16:2206–2230
19. Cruz AD, He A, Thanavaro A, Yan B, Spilling CD, Rath NP (2005) J Organomet Chem 690:2577–2592
20. He A, Sutivisedsak N, Spilling CD (2009) Org Lett 11:3124–3127
21. Roy S, Spilling CD (2010) Org Lett 12:5326–5329
22. Roy S, Spilling CD (2012) Org Lett 14:2230–2233
23. Tsukanov SV, Comins DL (2011) Angew Chem Int Ed 50:8626–8628
24. Tokuyama T, Nishimori N, Shimada A, Edwards MW, Daly JW (1987) Tetrahedron 43:643–657
25. Tokuyama T, Garraffo HM, Spande TF, Daly JW (1989) An Asoc Quim Argent 86:291–298
26. Tsuneki H, You Y, Toyooka N, Kagawa S, Kobayashi S, Sasaoka T, Nemoto H, Kimura I, Dani JA (2004) Mol Pharmacol 66:1061–1069
27. Takeuchi R, Kasio M (1997) Angew Chem Int Ed 36:263–265
28. Takeuchi R, Kasio M (1998) J Am Chem Soc 120:8647–8655
29. Janssen JP, Helmchen G (1997) Tetrahedron Lett 38:8025–8026
30. Bartels B, Helmchen G (1999) Chem Commun 741–742
31. Hartwig JF, Stanley LM (2010) Acc Chem Res 42:1461–1465
32. Hartwig JF, Pouy MJ (2011) Top Organomet Chem 34:169–208
33. Miyabe H, Yoshida K, Kobayashi Y, Matsumura A, Takemoto Y (2003) Synlett 1031–1033
34. Welter C, Koch O, Lipowsky F, Helmchen G (2004) Chem Commun 896–897

35. Welter C, Dahnz A, Brunner B, Streiff S, Dubon P, Helmchen G (2005) Org Lett 7:1239–1242
36. Teichert JF, Feringa BL (2010) Angew Chem Int Ed 49:2486–2528
37. Gnamm C, Krauter CM, Brodner K, Helmchen G (2009) Chem Eur J 15:2050–2054
38. Gnamm C, Brodner K, Krauter CM, Helmchen G (2009) Chem Eur J 15:10514–10532
39. Teichert JF, Fananas-Mastral M, Feringa BL (2011) Angew Chem Int Ed 50:688–691
40. Berkowitz DB, Maiti G (2004) Org Lett 6:2661–2664
41. Berkowitz DB, Shen W, Maiti G (2004) Tetrahedron: Asymmetry 15:2845–2851
42. Minami I, Shimizu I, Tsuji J (1985) J Organomet Chem 296:269–280
43. Zhang SW, Mitsudo T, Kondo T, Watanabe Y (1993) J Organomet Chem 450:197–207
44. Trost BM, Fraisse PL, Ball ZT (2002) Angew Chem Int Ed 41:1059–1061
45. Tanaka S, Saburi H, Ishibashi Y, Kitamura M (2004) Org Lett 6:1873–1875
46. Saburi H, Tanaka S, Kitamura M (2005) Angew Chem Int Ed 44:1730–1732
47. Tanaka S, Seki T, Kitamura M (2009) Angew Chem Int Ed 48:8948–8951
48. Seki T, Tanaka S, Kitamura M (2012) Org Lett 14:608–611
49. Hirai Y, Terada T, Amemiya Y, Momose T (1992) Tetrahedron Lett 33:7893–7894
50. Hirai Y, Nagatsu M (1994) Chem Lett 21–22
51. Hirai Y, Watanabe J, Nozaki T, Yokoyama H, Yamauchi S (1997) J Org Chem 62:776–777
52. Yokoyama H, Otaya K, Yamaguchi S, Hirai Y (1998) Tetrahedron Lett 39:5971–5974
53. Yokoyama H, Otaya K, Kobayashi H, Miyazawa M, Yamaguchi S, Hirai Y (2000) Org Lett 2:2427–2429
54. Uenishi J, Ohmi M, Ueda A (2005) Tetrahedron: Asymmetry 16:1299–1303
55. Uenishi J, Ohmi M (2005) Angew Chem Int Ed 44:2756–2760
56. Kawai N, Lagrange JM, Uenishi J (2007) Eur J Org Chem 2808–2814
57. Palimkar SS, Uenishi J, Ii H (2012) J Org Chem 77:388–399
58. Hande SM, Kawai N, Uenishi J (2009) J Org Chem 74:244–253
59. Uenishi J, Fujikura Y, Kawai N (2011) Org Lett 13:2350–2353
60. Palmes JA (2012) Ph.D. Thesis, University of Florida
61. Borrero NV, Aponick A (2012) J Org Chem 77:8410–8416
62. Awasaguchi K, Miyazawa M, Uoya I, Inoue K, Nakamura K, Yokoyama H, Kakuda H, Hirai Y (2010) Synlett 2010:2392–2396
63. Li Z, Brouwer C, He C (2008) Chem Rev 108:3239–3265
64. Corma A, Peyva-Perez A, Sabater MJ (2011) Chem Rev 111:1657–1712
65. Rudolph M, Hashmi ASK (2011) Chem Commun 47:6536–6544
66. Biannic B, Aponick A (2011) Eur J Org Chem 6605–6617
67. Aponick A, Li CY, Biannic B (2008) Org Lett 10:669–671
68. Aponick A, Biannic B (2008) Synthesis 3356–3359
69. Aponick A, Li CY, Palmes JA (2009) Org Lett 11:121–124
70. Aponick A, Li CY, Malinge J, Marques EF (2009) Org Lett 11:4624–4627
71. Aponick A, Biannic B, Jong MR (2010) Chem Commun 46:6849–6851
72. Bandini M, Monari M, Romaniello A, Tragni M (2010) Chem Eur J 16:14272–14277
73. Unsworth WP, Stevens K, Lamont SG, Robertson J (2011) Chem Commun 47:7659–7661
74. Aponick A, Biannic B (2011) Org Lett 13:1330–1333
75. Biannic B, Ghebreghiorgis T, Aponick A (2011) Beilstein J Org Chem 7:802–807
76. Mukherjee P, Widenhoefer RA (2011) Org Lett 13:1334–1337
77. Mukherjee P, Widenhoefer RA (2012) Angew Chem Int Ed 51:1405–1407
78. Ghebreghiorgis T, Biannic B, Kirk BH, Ess DH, Aponick A (2012) J Am Chem Soc 134:16307–16318
79. Namba K, Nakagawa Y, Yamamoto H, Imagawa H, Nishizawa M (2008) Synlett 1719–1723
80. Yamamoto H, Ho E, Namba K, Imagawa H, Nishizawa M (2010) Chem Eur J 16:11271–11274
81. Kawai N, Abe R, Uenishi J (2009) Tetrahedron Lett 50:6580–6583
82. Kawai N, Abe R, Matsuda M, Uenishi J (2011) J Org Chem 76:2102–2114

83. Kawai N, Matsuda M, Uenishi J (2011) Tetrahedron 67:8648–8653
84. Xu Z, Li Q, Zhang L, Jia Y (2009) J Org Chem 74:6859–6862
85. Xu Z, Hu W, Liu Q, Zhang L, Jia Y (2010) J Org Chem 75:7626–7635
86. Jiang D, Xu Z, Jia Y (2012) Tetrahedron 68:4225–4232
87. Kothandaraman P, Foo SJ, Chan PWH (2009) J Org Chem 74:5947–5952
88. Guerinot A, Serra-Muns A, Gnamm C, Bensoussan C, Reymond S, Cossy J (2010) Org Lett 12:1808–1811
89. Guerinot A, Serra-Muns A, Gnamm C, Bensoussan C, Reymond S, Cossy J (2011) Tetrahedron 67:5024–5033
90. Wang Z, Li S, Yu B, Wu H, Wang Y, Sun X (2012) J Org Chem 77:8615–8620
91. Hanessian S, Focken T, Oza R (2010) Org Lett 12:3172–3175
92. Hanessian S, Focken T, Oza R (2011) Tetrahedron 67:9870–9884
93. Trost BM, Machacek MR, Faulk BD (2006) J Am Chem Soc 128:6745–6754
94. He H, Lie WB, Dai LX, You SL (2010) Angew Chem Int Ed 49:1496–1499
95. Davies HML, Du Bois J, Yu JQ (2011) Chem Soc Rev 40:1855–1856
96. Mei TS, Kou L, Ma S, Engle KM, Yu JQ (2012) Synthesis 44:1778–1791
97. Fraunhoffer KJ, White MC (2007) J Am Chem Soc 129:7274–7276
98. Rice GT, White MC (2009) J Am Chem Soc 131:11707–11711
99. Stang EM, White MC (2009) Nat Chem 1:547–551
100. Qi X, Rice GT, Lall MS, Plummer MS, White MC (2010) Tetrahedron 66:4816–4826
101. Gormisky PE, White MC (2011) J Am Chem Soc 133:12584–12589
102. Stang EM, White MC (2011) Angew Chem Int Ed 50:2094–2097
103. Paradine SM, White MC (2012) J Am Chem Soc 134:2036–2039

Top Heterocycl Chem (2013) 32: 187–224
DOI: 10.1007/7081_2013_106

Published online: 24 April 2013

Synthesis of Heterocycles via Metal-Catalyzed Domino/One-Pot Reactions That Generate a C−N or C−O Bond

Juliane Keilitz, Hasnain A. Malik, and Mark Lautens

Abstract This chapter focuses on transition metal-catalyzed domino (cascade) or one-pot syntheses of heterocycles via the formation of a carbon–nitrogen, –oxygen, or –sulfur bond. A precise classification of domino, one-pot, and tandem reactions is given. However, despite that rather strict definition, the chapter includes a variety of processes that are important from a mechanistic and synthetic point of view. These are methods which showcase both ingenious and efficient reaction design while simultaneously aiming to minimize deleterious byproduct formation as well as uneconomical workup and purification steps. While there are several types of protocols highlighted within this section, there is a larger emphasis on transition metal-catalyzed cycloisomerization methods, the utility of *gem*-dihaloolefins, and C−H functionalization protocols within the framework of domino catalysis.

Keywords C−H functionalization · Cycloisomerization · *gem*-Dihaloolefins · Green chemistry · Heterocycles · Transition metal catalysis

Contents

J. Keilitz and M. Lautens (✉)
Department of Chemistry, Davenport Research Laboratories, University of Toronto, 80 St. George Street, Toronto, ON, Canada M5S 3H6
e-mail: mlautens@chem.utoronto.ca

H.A. Malik
Novartis Institutes for Biomedical Research, 250 Massachusetts Avenue, Cambridge, MA 02142, USA

1 Introduction

Transition metal-catalyzed reactions have gained increasing importance in synthetic organic chemistry over the past few decades. There has been a heightened focus on the ability to perform multiple chemical transformations utilizing one or more catalysts in a single reaction vessel. While the ability to achieve several chemical reactions in a one-pot fashion is obviously attractive from the perspective of synthetic efficiency, the potential for cost-savings and positive environmental impact that result from the elimination of time-consuming workup and purification protocols cannot be understated. Indeed, it is highly advantageous for the continued development of chemical processes to be even more effective and robust while resulting in an overall low environmental footprint.

It should be noted that while there has been an effort by some to rigorously classify and define the many types of cascade/domino/tandem transformations [1–4], there remains a lack of consensus in literature. As a consequence, the terms "cascade," "domino," "tandem," "one-pot," "sequential," among many others, are at times routinely and casually interchanged. For the purposes of this review we will take an inclusive view to highlight and illustrate processes that we believe to be important from a mechanistic and synthetic point of view. As a result, while some processes we may describe will not fall within the strictest definition of cascade, domino, or tandem reactions, we did not exclude any processes that display creative reaction design.

There have been early efforts to put forth a clear and unified means to define the variety of reactions that involve multiple sequential chemical transformations that occur in a single vessel. One effective descriptor has been developed by Fogg and dos Santos (Fig. 1) [3].

The content of this chapter will be organized in the following manner:

Section 2: Metal-Catalyzed Cascade Reactions That Result in the Generation of a C−N or C−O Bond

Section 2.1: Synthesis of Heterocycles via Domino/One-Pot Cycloisomerization Sequences
Section 2.2: Synthesis of Heterocycles via Use of *gem*-Dihaloolefins
Section 2.3: Synthesis of Heterocycles via C−H Functionalization
Section 2.4: Miscellaneous Domino Methods for the Synthesis of Heterocycles

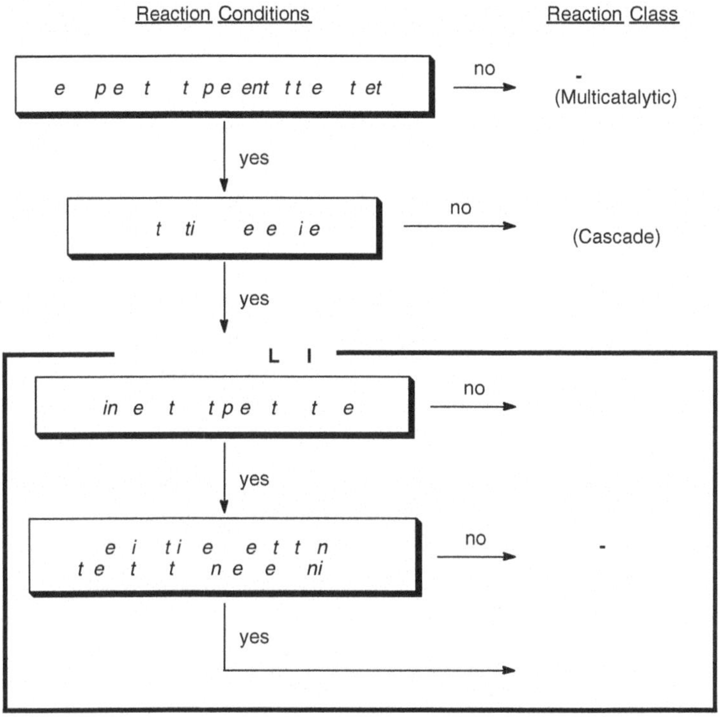

Fig. 1 Flowchart for classification of one-pot processes involving sequential elaboration of an organic substrate via multiple catalytic transformations

2 Metal-Catalyzed Cascade Reactions That Result in the Generation of a C–N or C–O Bond

2.1 Synthesis of Heterocycles via Domino/One-Pot Cycloisomerization Sequences

The importance of indole and benzofuran scaffolds as intermediates, natural products, and pharmaceuticals cannot be understated [5–14]. While benzofurans are a common motif in many natural and pharmaceutically relevant compounds, indoles are even more ubiquitous. The most common synthetic approach to these types of structures is illustrated below – usually involving the cycloisomerization of an *ortho*-alkynyl phenol or aniline starting material to the corresponding benzofuran or indole, respectively (Scheme 1). In this section, we will attempt to highlight a variety of approaches to this cycloisomerization in a domino (cascade) or one-pot fashion.

Fürstner and coworkers reported a platinum-catalyzed cycloisomerization/formal allyl transfer method of the synthesis of benzofurans, indoles, and isochromene-1-ones (Scheme 2) [12]. Reaction times vary from 1 to 12 h for complete conversion of starting

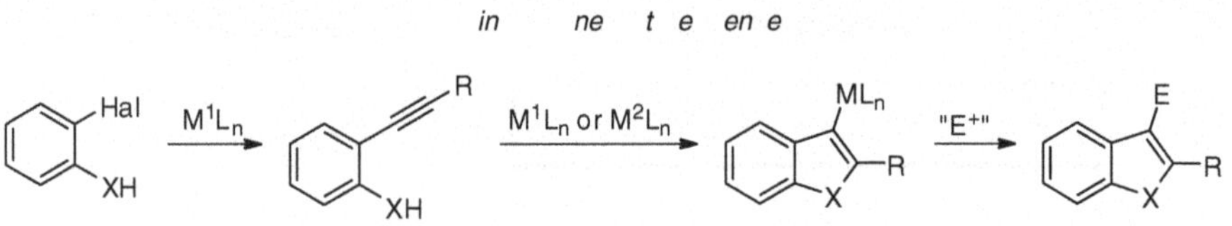

Scheme 1 The general synthetic approach to the domino or one-pot synthesis of benzofuran and indole heteroaromatics

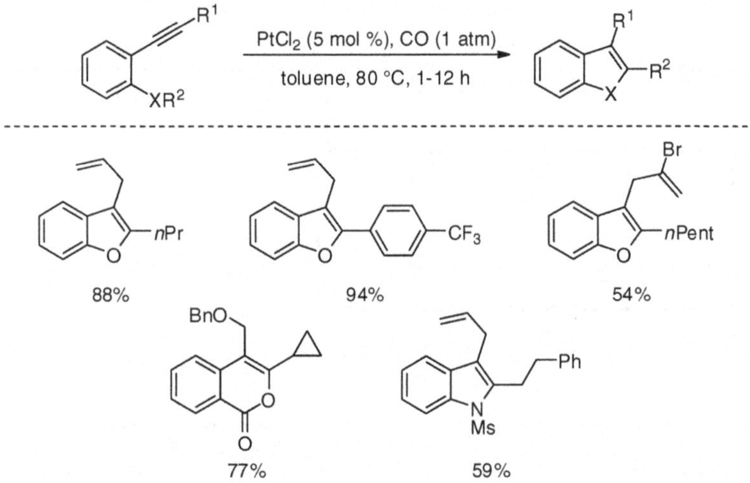

Scheme 2 Platinum-catalyzed domino synthesis of heteroaromatic compounds through a cyclization/allyl transfer pathway

materials. The reaction undergoes a domino sequence and is completely atom-economical, save the requirement of carbon monoxide gas, which is necessary and has been empirically found to accelerate Pt-catalyzed rearrangement reactions. It is noteworthy that vinyl halides are tolerated under the reaction conditions. A closely related transformation was independently discovered and reported by Yamamoto and coworkers whereby the use of cyclooctadiene as a ligand offsets the requirement of CO gas [15].

Zhang and coworkers later expanded and diversified the utility of this type of platinum-catalyzed strategy by employing *N*-(2-alkynylphenyl)lactams as substrates for the synthesis of fused indole products (Scheme 3) [16]. The putative mechanism involves cycloisomerization, ring-expansion/rearrangement, and a 1,2-shift. Levels of selectivity range from good to excellent for the rearrangement product.

Venkataraman and coworkers reported the copper-catalyzed domino Sonogashira/cycloisomerization reaction for the synthesis of a variety of benzofurans (Scheme 4) [17]. Yields are generally good to excellent and the method displays high functional group tolerance. It is worth mentioning that aryl bromides and chlorides remain untouched throughout the catalytic cycle. The reaction also has the added benefit of

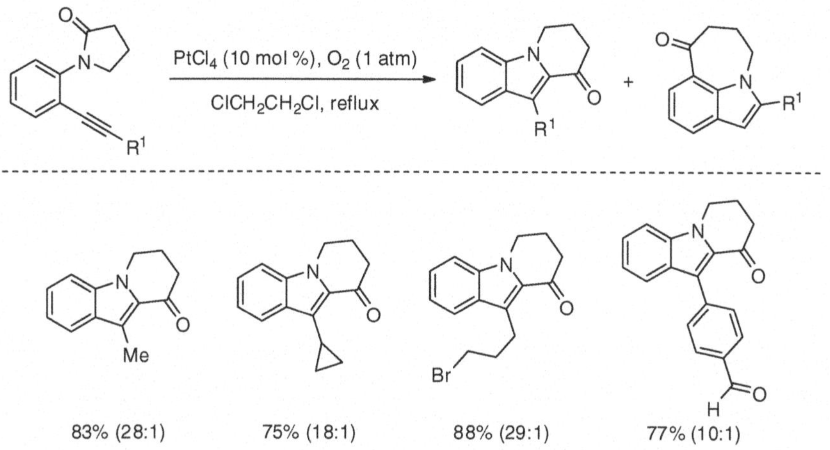

Scheme 3 Platinum-catalyzed domino cycloisomerization/ring-expansion/alkyl transfer reactions to form cyclic-ketone-fused indoles

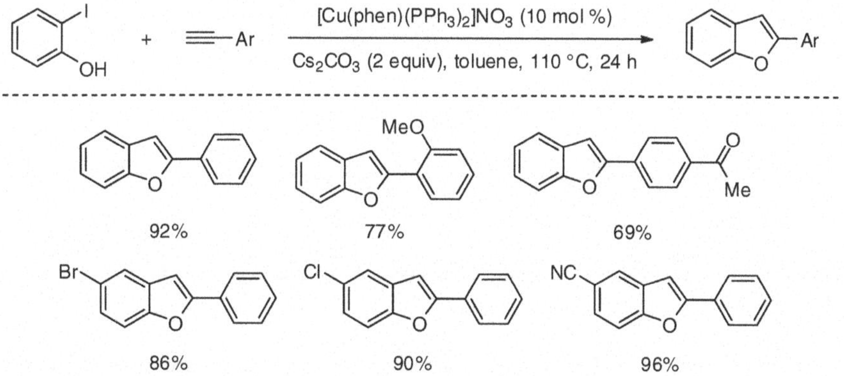

Scheme 4 Copper-catalyzed domino Sonogashira/cycloisomerization sequence for the synthesis of 2-substituted benzofurans

being palladium-free which renders the method more attractive for scale-up with respect to cost.

The iron-catalyzed Sonogashira reaction of 2-iodophenol with terminal aromatic alkynes results in the formation of benzofuran products (Scheme 5) [18–20].[1] Interestingly, the simple Sonogashira product, namely the newly formed internal alkyne, is obtained when *N*-benzyl 2-substituted iodoanilines are employed. The use of an inexpensive and environmentally benign iron catalyst is noteworthy.

[1] There have been recorded instances in literature by Bolm, Buchwald, and others where it has been determined that trace metal impurities are the catalytically active species in transition metal-catalyzed reactions. This is especially relevant in many iron-catalyzed methods.

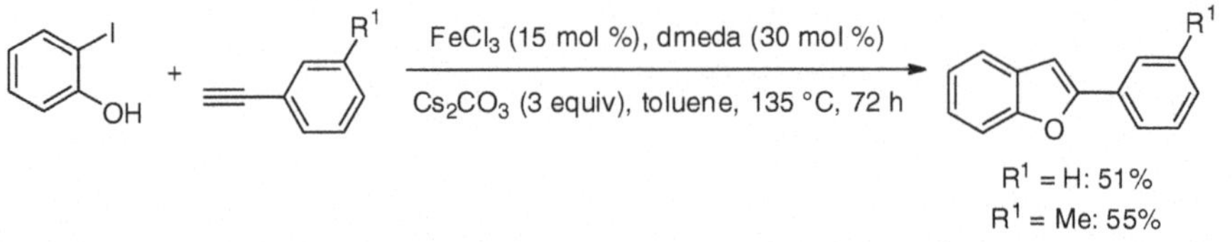

Scheme 5 Iron-catalyzed domino Sonogashira/cycloisomerization for the synthesis of 2-substituted benzofurans

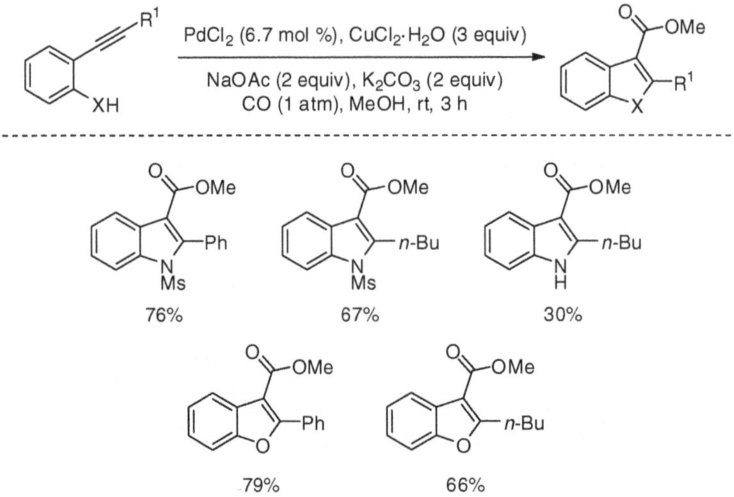

Scheme 6 Palladium-catalyzed domino carbonylative cycloisomerization for the synthesis of heteroaromatics

Sakamoto and coworkers reported a palladium-catalyzed domino carbonylative cyclization for the synthesis of a variety of heteroaromatic compounds (Scheme 6) [21]. Under an atmosphere of carbon monoxide gas in methanol, 2,3-disubstituted indoles and benzofurans could be furnished. Unprotected aniline starting materials were poor substrates and yields were considerably lower than the mesyl-protected variants.

Lu and coworkers disclosed a useful means to synthesize unprotected 2,3-disubstituted indoles in one-pot reaction sequence (Scheme 7) [22]. This palladium-catalyzed reaction incorporates a sequential Sonogashira reaction followed by cycloisomerization and deprotection to afford a variety of 2,3-disubstituted indole scaffolds. While there are currently a limited number of examples that could be carried out in a domino fashion (where all starting materials are present at the beginning of the reaction), high yields can be obtained when the aryl halide is added portionwise at the completion of the Sonogashira coupling step.

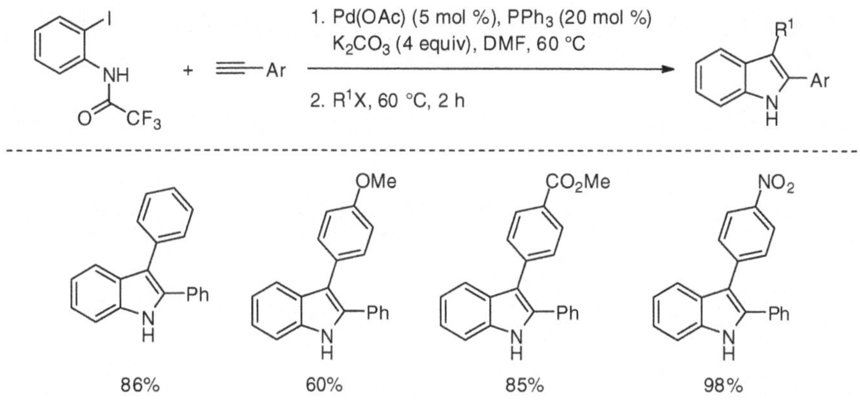

Scheme 7 One-pot palladium-catalyzed Sonogashira/cyclization sequence for the synthesis of 2,3-disubstituted indoles

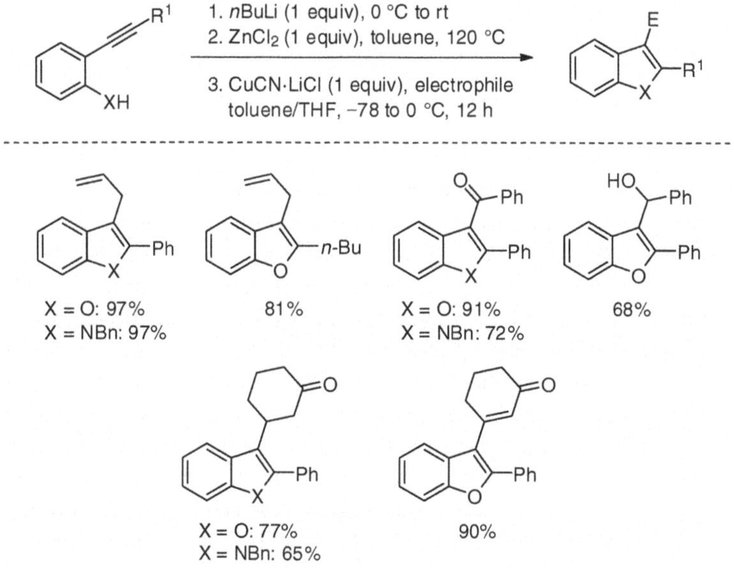

Scheme 8 One-pot synthesis of 2,3-disubstituted indoles and benzofurans

Nakamura and coworkers have reported a one-pot method for the synthesis of a diverse array of 2,3-disubstituted indoles and benzofurans (Scheme 8) [23]. Although the reaction is not catalytic in nature, it has several advantages over similar catalytic methods. Specifically, this method represents one of the rare examples of this type of transformation where allyl halides, acyl chlorides, aldehydes, α,β-unsaturated carbonyls, and vinyl halides are all shown to be competent electrophile partners and yields range from moderate to excellent in all the cases described.

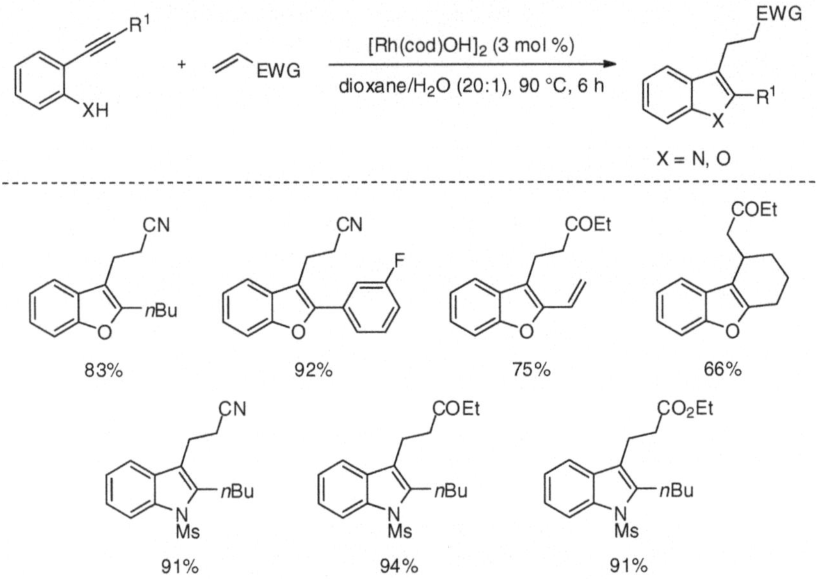

Scheme 9 Rhodium-catalyzed domino cycloisomerization/1-4-addition to afford 2,3-disubstituted benzofurans and indoles

Our own group's interest in tandem and domino processes has led us to develop the rhodium-catalyzed cycloisomerization of *ortho*-alkynyl phenols and anilines followed by electrophile trapping to afford benzofuran and indole products (Scheme 9) [24]. We were able to capitalize on the stability of a rhodium(I)-intermediate that could undergo facile migratory insertion reactions with a variety of π-electrophiles. Deleterious β-hydride elimination by-product formation (usually a minor by-product in this transformation) can be minimized and in many cases completely eliminated through the use of BINAP as a ligand. This method provides an efficient and expedient route to a variety of heteroaromatic compounds under relatively mild conditions. If this domino process is engineered to undergo two intramolecular steps (cycloisomerization followed by intramolecular 1,4-addition), tricyclic compounds can be synthesized in synthetically useful yields.

The scope of this rhodium-catalyzed domino process was later expanded to include reactions with internal alkynes as electrophiles (Scheme 10) [25]. The regioselectivity for alkyne insertion varies from low to high, where the highest levels of regiocontrol are hypothesized to be dependent on a putative heteroatom chelation to the rhodium(I) intermediate (see Scheme 10). Indeed, there seems to be some experimental support for this phenomenon as *ortho*-alkynyl phenols provide a ca. 80:20 mixture of regioisomers and protected *ortho*-alkynyl phenols result in a complete loss of regioselectivity.

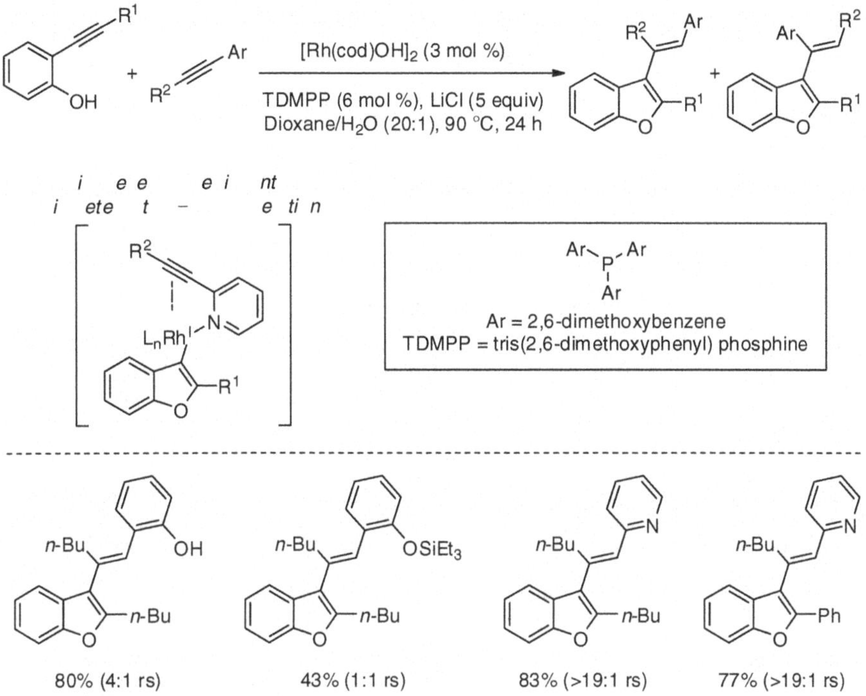

Scheme 10 Rhodium-catalyzed domino cycloisomerization/alkyne migratory insertion to afford 2,3-disubstituted benzofurans

2.2 *Synthesis of Benzofurans and Indoles via Use of* gem-*Dihaloolefins*

In recent years, *gem*-dihaloolefins have attracted attention as versatile substrates for the synthesis of heterocycles via tandem sequences [26–33]. They can be readily obtained through a Ramirez olefination [34–37] of a suitably *ortho*-substituted aniline, phenol, or thiophenol which allows for modular syntheses of indoles, benzofurans, or benzothiophenes, respectively. Our group was the first to employ *ortho-gem*-dihalo vinyl substrates in transition metal-catalyzed tandem reactions [37–44]. While there is uncertainty as to which step occurs first as the substrate is varied, it is generally the case that an initial C–N, C–O, or C–S coupling leads to a 2-bromoindole, -benzofuran, or -benzothiophene [45] moiety which can further react in a separate but tandem transition metal-catalyzed coupling reaction (Scheme 11).

The palladium-catalyzed intramolecular C–N bond formation and intermolecular Suzuki–Miyaura cross-coupling of *ortho-gem*-dibromoolefins with organoboron reagents was first reported by Bisseret with limited substrates [46] and fully developed by our group (Scheme 12) [38]. Specifically, we showed that $Pd(OAc)_2$, with the use

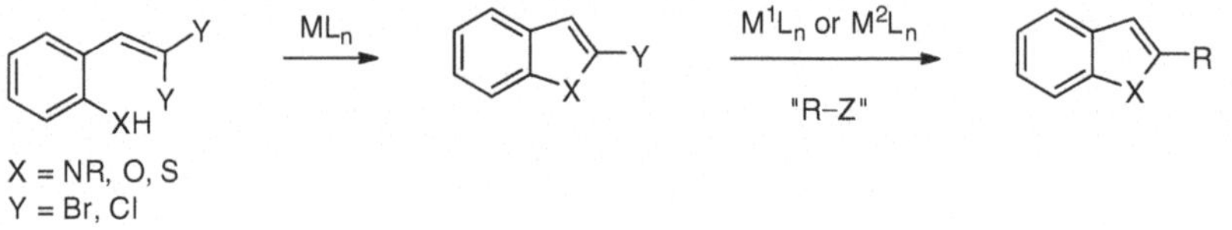

Scheme 11 General approach toward the synthesis of indoles, benzofurans, or benzothiophenes using *ortho-gem*-dihaloolefins

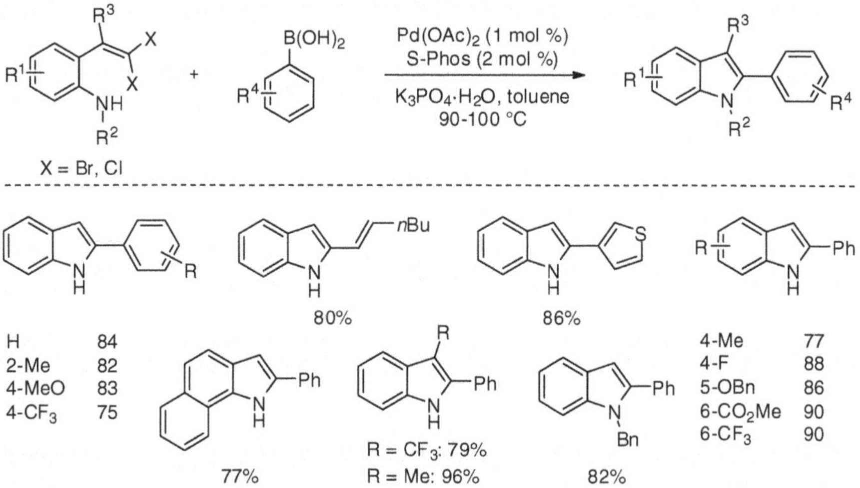

Scheme 12 Tandem Buchwald–Hartwig amination/Suzuki–Miyaura cross-coupling of *ortho-gem*-dihalovinylanilines with organoboron reagents

of Buchwald's SPhos ligand, provides access to a variety of 2- and 2,3-substituted indoles [38, 39]. Substitution at a variety of positions on the indole heterocycle is tolerated and the products are obtained in good to excellent yields (72–96%) within 1–14 h. Interestingly, the use of *ortho-gem*-dichlorovinylanilines provides almost quantitative yields, which is hypothesized to occur due to a higher level of chemoselectivity.

Mechanistic investigations revealed that, in the parent substrate, the Buchwald–Hartwig coupling occurs first. What is not known is if selective insertion into the (*Z*)-C–X bond is responsible or if isomerization of the (*E*)-inserted product to the more reactive (*Z*)-isomer can occur [39]. The generality and practicality of this tandem Buchwald–Hartwig amination/Suzuki–Miyaura cross-coupling sequence were demonstrated through the synthesis of four different KDR kinase inhibitors which are potential therapeutics (Fig. 2) [37] and with the synthesis of various azaindoles and thienopyrroles which were previously not accessible by such a modular and general approach [41].

Alper and coworkers extended the reaction by developing a tandem C–N coupling followed by a carbonylation (Scheme 13) [47]. The reaction is performed

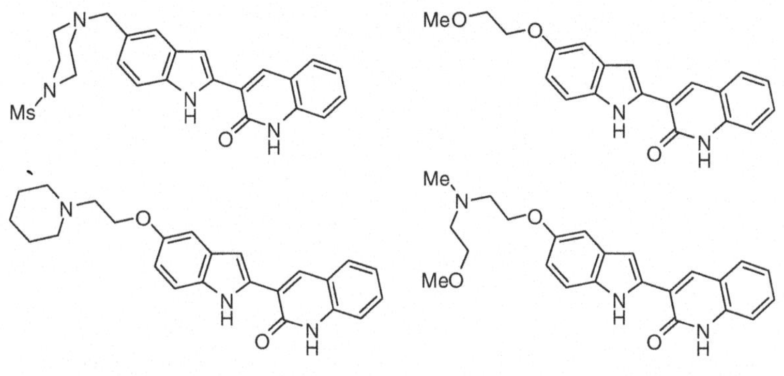

Fig. 2 Targets synthesized via tandem Buchwald–Hartwig amination/Suzuki–Miyaura cross-coupling sequence

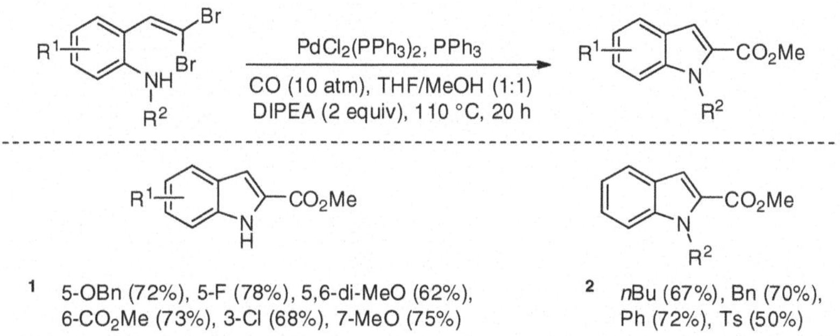

Scheme 13 Palladium-catalyzed tandem intramolecular amination/carbonylation sequence

under a CO atmosphere (10 atm) in a THF/MeOH mixture. Various functional groups are tolerated both at the amine moiety and on the aryl ring, including halogens such as chlorine or fluorine. However, when bromine atoms are present on the aromatic ring, a second carbonylation reaction takes place at this position.

In 2009, the group of Pontikis and Florent reported a tandem C–N coupling/carbonylation/C–C coupling sequence employing *gem*-dibromoolefins that furnish the synthesis of 2-aroyl- or 2-heteroaroyl indoles, respectively (Scheme 14) [48]. A range of substituents is tolerated on the aromatic ring of the *gem*-dibromovinyl substrates, but no substitution in the 3-position of the indole has been reported. Sterically demanding substituents at the boronic acid reduce overall reactivity; the use of 2-methoxyphenylboronic acid delivers the corresponding product in a modest yield of 40%, while 2,6-di-methylphenylboronic acid provides no observable product formation.

2-Vinylic indoles and their tricyclic derivatives can be obtained through a tandem Buchwald–Hartwig coupling followed by a Heck–Mizoroki cross-coupling sequence (Scheme 15) [40]. The only limitation of this reaction is in the formation of 3-substituted derivatives, where poor yields were observed when the corresponding

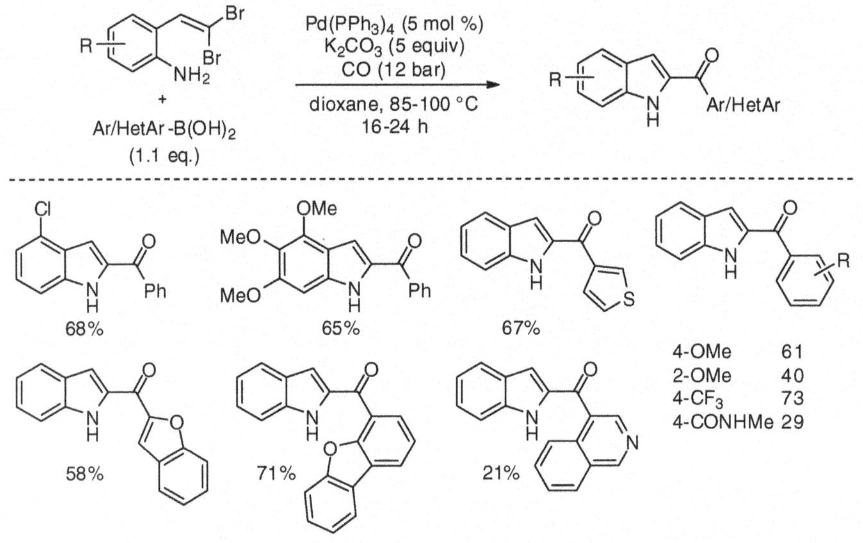

Scheme 14 Palladium-catalyzed C–N coupling/carbonylation/C–C coupling sequence

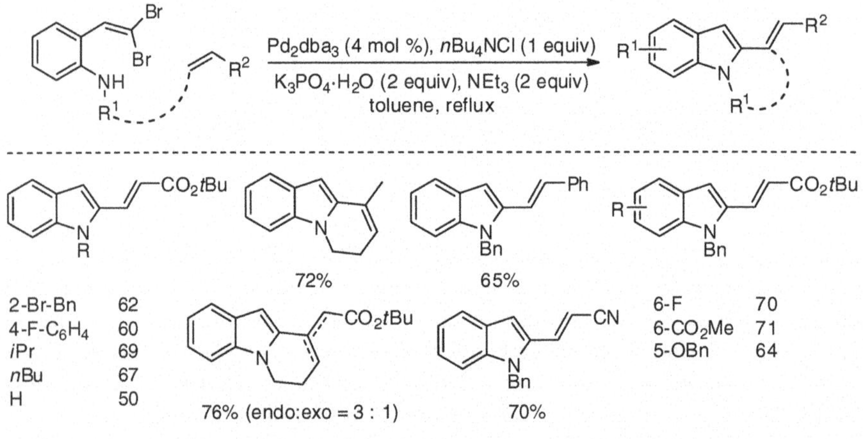

Scheme 15 Palladium-catalyzed tandem Buchwald–Hartwig/Heck–Mizoroki reaction

substituted dibromovinylanilines are utilized. An intramolecular variant of this method was realized by tethering the alkene moiety to the nitrogen atom of *ortho-gem*-dibromovinylaniline. The tandem reaction yields the corresponding tricyclic adducts as mixtures of two easily separable isomers and even a non-activated alkene could be employed.

For the synthesis of 2-alkynyl indoles and benzofurans, a tandem copper- and palladium-catalyzed cross-coupling reaction was developed involving an Ullmann-type reaction and a Sonogashira cross-coupling tandem reaction [43]. Interestingly, heterogeneous Pd/C (2 mol%) in conjunction with 4 mol% CuI was found to be the

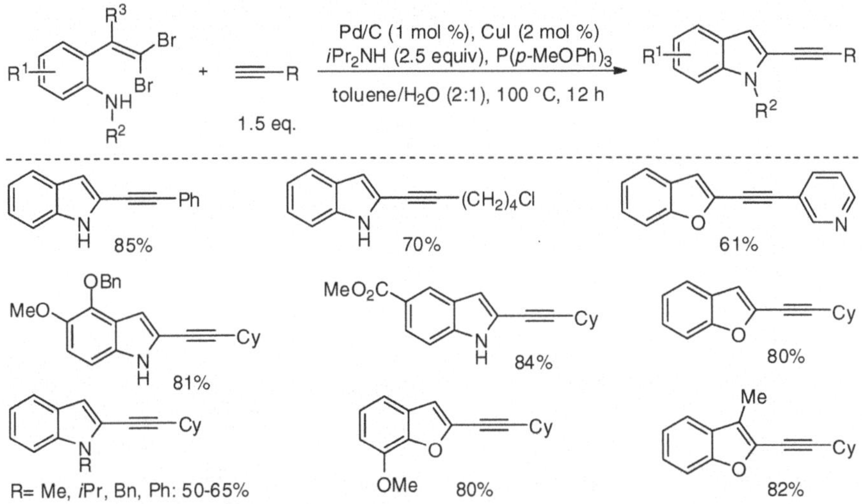

Scheme 16 Sequence of Ullmann-type reaction and Sonogashira cross-coupling for the synthesis of 2-alkynyl indoles and benzofurans

best co-catalyst combination (Scheme 16). However, since the solid support did not negatively affect the efficacy of the catalytic system, it was assumed that the reaction itself occurs in the homogeneous organic phase with trace amounts of leached palladium(0). Different aromatic and aliphatic alkynes as well as a variety of substituted *ortho-gem*-dibromovinyl derivatives could be utilized and the corresponding anilines and benzofurans were obtained in moderate to good yields (40–98%). Substitution on the aniline nitrogen atom generally lowers the yield and the synthesis of 3-substituted indoles has not been reported.

ortho-gem-Dibromoolefins tethered to amino acids were utilized for the synthesis of imidazoindolones via a double amidation reaction (Scheme 17) [42]. The reactions require 12–49 h for completion and a range of substituents is tolerated. However, in case of the 3-substituted *gem*-dibromoolefin, the catalyst loading had to be increased to obtain the corresponding imidazoindolone in a reasonable yield. The preservation of the chiral center originating from the amino acid was highly variable and the extent of epimerization was assumed to depend on a variety of factors. The rate of conversion of the 2-bromoindole intermediate to the product is vital as the proton at the stereocenter is much more acidic than in the starting *gem*-dibromoolefin and therefore much more susceptible to epimerization under the reaction conditions. Thus the extent of epimerization is highly dependent on the rate of the second amidation step.

In 2012, Wang, Lv, and coworkers reported a Cu_2O-catalyzed C–N/C–X (X = N, O, S) coupling for the formation of oxazino[3,2-*a*]indole, thiazino[3,2-*a*]indole, and indolo[2,1-*b*]quinazoline derivatives (Scheme 18) [49]. This transformation operates under ambient conditions and it was shown that the copper catalyst is necessary for both steps to occur. Although substitution is possible at most positions, when strongly

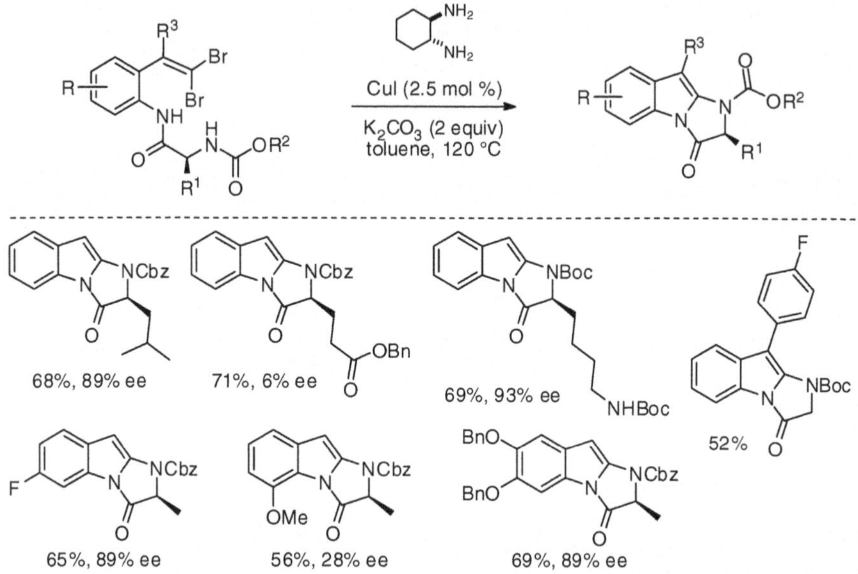

Scheme 17 Copper-catalyzed tandem intramolecular amidation for the synthesis of imidazoindolones

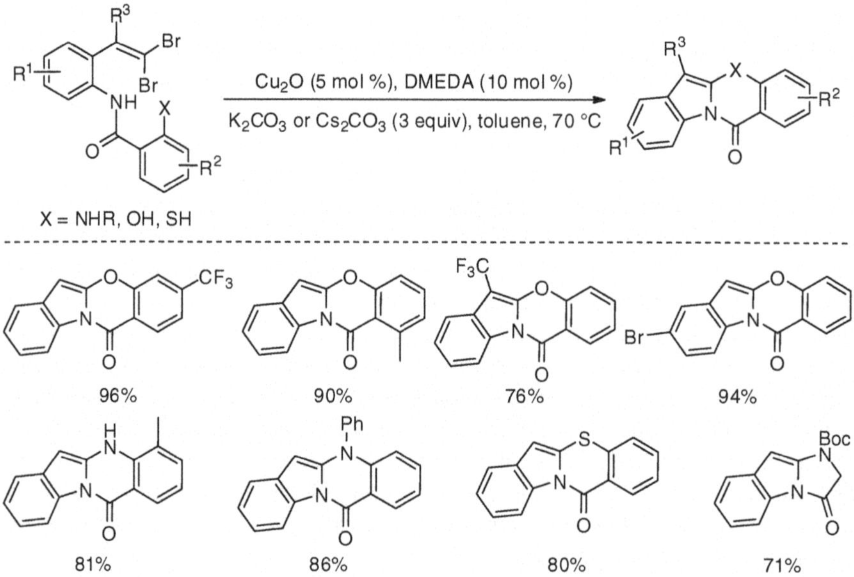

Scheme 18 Copper-catalyzed tandem C–N/C–X (X = N, O, S) coupling for the formation of polycyclic indole derivatives

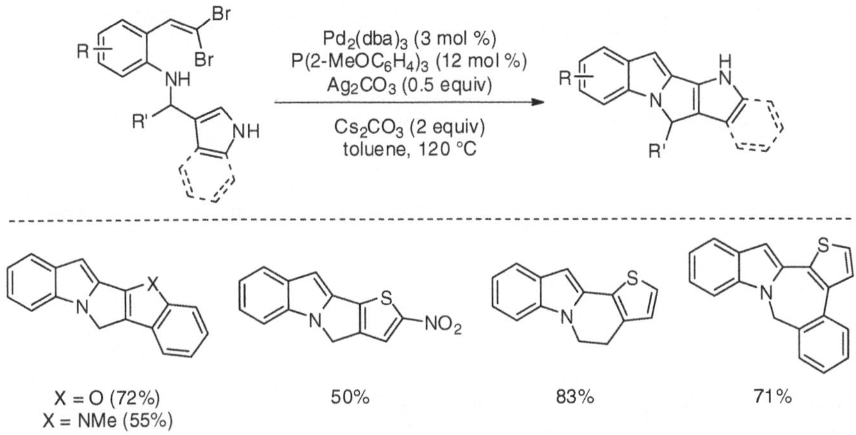

Scheme 19 Palladium-catalyzed Buchwald–Hartwig amination/direct arylation sequence toward tetra- and pentacyclic indole derivatives

electron-withdrawing substituents are present on the aromatic ring of the *gem*-dibromovinyl aniline moiety, only trace amounts of product formation are observed. In order to obtain the corresponding indolo[2,1-*b*]quinazolines, a change of base selection from K_2CO_3 to Cs_2CO_3 was required.

Our group was also able to combine a tandem Buchwald–Hartwig cross-coupling followed by a direct arylation reaction yielding tetracyclic and pentacyclic indole derivatives (Scheme 19) [44]. A variety of indoles with substituents of different electron character are easily accessible and by using a higher catalyst loading even seven-membered rings can be obtained. An apparent limitation is found to be substitution in the 3-position of the indole, which completely inhibits product formation.

Bao and coworkers employed isocyanates as nucleophilic acceptors by introducing them *ortho* on *gem*-dibromovinylbenzene (Scheme 20) [30]. Those *ortho*-*gem*-dibromovinyl isocyanates were reacted with *N*-alkylanilines to provide pyrimido[1,2-*a*]indol-1(2*H*)-one derivatives through a sequence of nucleophilic addition of the aniline group to the isocyanate moiety, copper-catalyzed *N*-arylation, and palladium-catalyzed C–H functionalization. Although this sequence is only a one-pot process (requiring sequential addition of transition metal catalysts) it is a rare example of this type of sequences involving a C–H functionalization step. The scope with regard to substitution is broad and even 3-substituted indoles could be obtained from the corresponding isocyanate substrates. A limitation of the method is that anilines with strongly electron-withdrawing groups on the phenyl ring fail to display reactivity. Additionally, high steric hindrance on the amine prevents addition onto the isocyanate and unprotected amines only furnish urea intermediates where the subsequent Cu-catalyzed *N*-arylation step does not occur.

Bao and coworkers then applied the same method to the synthesis of unsymmetrical 1,1′-carbonyl-2,2′-biindolyls, but their initial experiments proved unsuccessful.

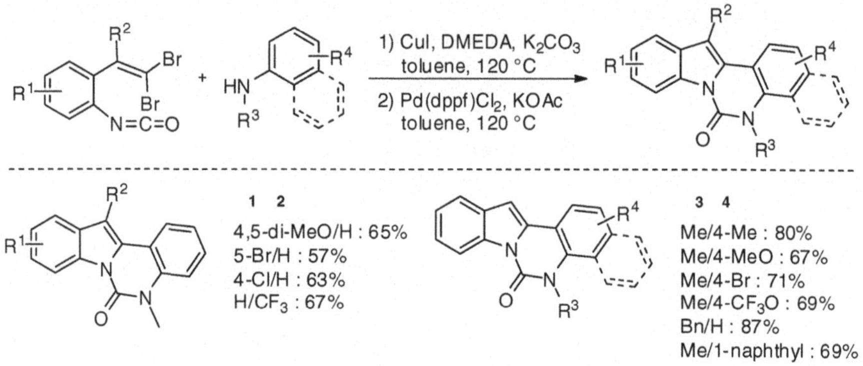

Scheme 20 One-pot synthesis of pyrimido[1,2-*a*]indol-1(2*H*)-ones via nucleophilic addition/copper-catalyzed *N*-arylation/palladium-catalyzed C–H functionalization

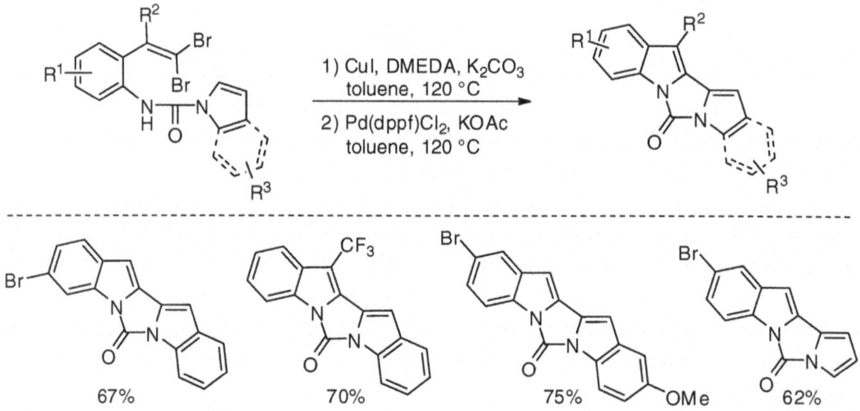

Scheme 21 One-pot sequence for the copper- and palladium-catalyzed formation of unsymmetrical 1,1′-carbonyl-2,2′-biindolyl derivatives

Therefore, indole-1-carboxylic acid and *ortho-gem*-dibromovinyl aniline were coupled and under the established reaction conditions led to the formation of the desired products (Scheme 21) [50]. The method efficiently provides moderate to good yields. A limitation was found to be that the synthesis of unsymmetrical products bearing electron-deficient groups on both indole rings was unsuccessful. Attempts to employ only one metal catalyst and one base failed.

Bao’s most recent contribution in this area is a two component sequence where an aromatic acid chloride and the well-studied *ortho-gem*-dibromovinyl aniline react via amide formation/Cu-catalyzed intramolecular C–N coupling/C–H activation to form 6*H*-isoindolo[2,1-*a*]indol-6-ones [51].

In 2009, our group published a tandem process for the synthesis of benzothiophenes consisting of an intramolecular *S*-vinylation followed by intermolecular carbon–carbon bond formation either through a Suzuki–Miyaura, Heck, or Sonogashira reaction (Scheme 22) [52]. Although sulfur has a long-standing

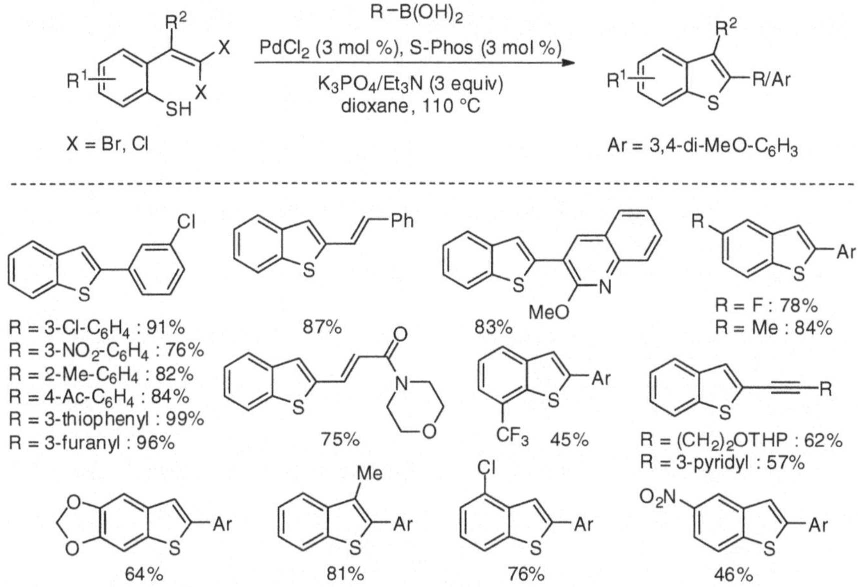

Scheme 22 Synthesis of substituted benzothiophenes via intramolecular *S*-vinylation and intermolecular carbon–carbon cross-coupling

reputation as a catalyst poison, various substituted benzothiophenes can be obtained in good to excellent yields via the *S*-vinylation/Suzuki–Miyaura sequence. A variety of boronic acids and different boron nucleophiles (e.g., boronic esters, trifluoroborate salts, and trialkylboranes) are compatible with this process. In contrast, the nature of the thiophenol fragment has significant influence and the presence of strongly electron-withdrawing substituents provides low yields or the failure of the tandem reaction. The method was also extended to Heck- and Sonogashira-coupling reactions. It is noteworthy that the Sonogashira sequence can be catalyzed by Pd/C.

In 2010, Chen and coworkers utilized aryl- and alkyl-substituted *gem*-dibromovinyl derivatives for the preparation of imidazo[2,1-*b*]-thiazoles and related *N*-fused heterocycles via copper-catalyzed 1,2-aminothiolation (Scheme 23) [31]. For aryl-substituted *gem*-dibromovinyl compounds the 3-substituted imidazo[2,1-*b*]-thiazoles are obtained exclusively while for alkyl-substituted *gem*-dibromovinyl compounds a mixture of the 2- and 3-substituted products are obtained of which the 2-substituted one is the major isomer. The method was also applicable to the aminothiolation of unsubstituted and substituted 2-mercaptoimidazole, perimidine, and pyrimidine derivatives.

Alper and coworkers reported the synthesis of 2-carbonylbenzo[*b*]thiophene derivatives via a selective palladium-catalyzed tandem procedure (Scheme 24) [33]. An intramolecular C–S coupling/intermolecular carbonylation sequence yields various highly functionalized benzo[*b*]thiophenes in moderate yields. The strategy was also applicable for *gem*-dichlorovinyl derivatives, although the desired product was obtained in a lower yield.

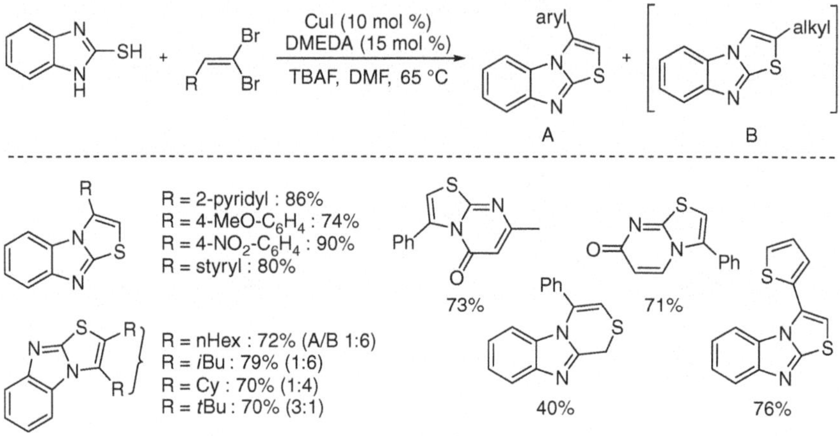

Scheme 23 Synthesis of imidazo[2,1-*b*]-thiazoles via copper-catalyzed 1,2-aminothiolation

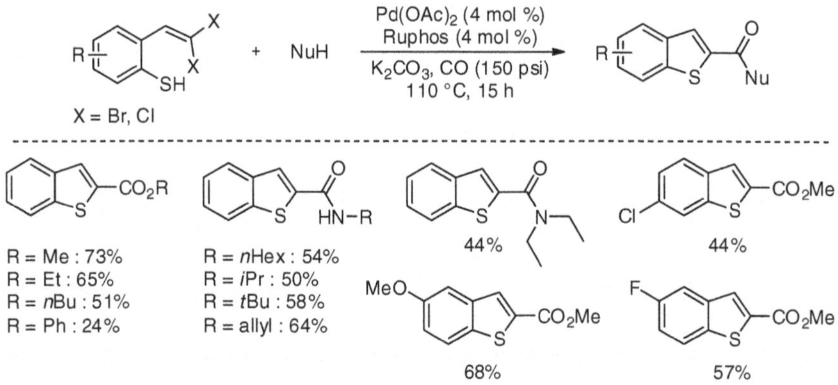

Scheme 24 Tandem palladium-catalyzed intramolecular C–S coupling/intermolecular carbonylation for the synthesis of 2-carbonylbenzo[*b*]thiophenes

2.3 Synthesis of Heterocycles via C–H Functionalization

Direct arylation (or C–H functionalization) offers several advantages such as the use of simplified/unfunctionalized starting materials and a higher degree of atom economy when compared to "traditional" cross-coupling methods [53, 54].

In 2003, Zhu and coworkers reported the synthesis of polyheterocycles by a palladium-catalyzed intramolecular *N*-arylation/C–H functionalization/aryl–aryl bond forming tandem process (Scheme 25) [55, 56]. Interestingly, the authors were able to access medium-sized and even macrocyclic ring systems by their method which was applied to the synthesis of azaphenanthrenes fused with an 8-, 10-, 11-, and 13-membered lactam. The reaction temperature was found to be important, with higher temperatures providing higher yields. It was assumed that

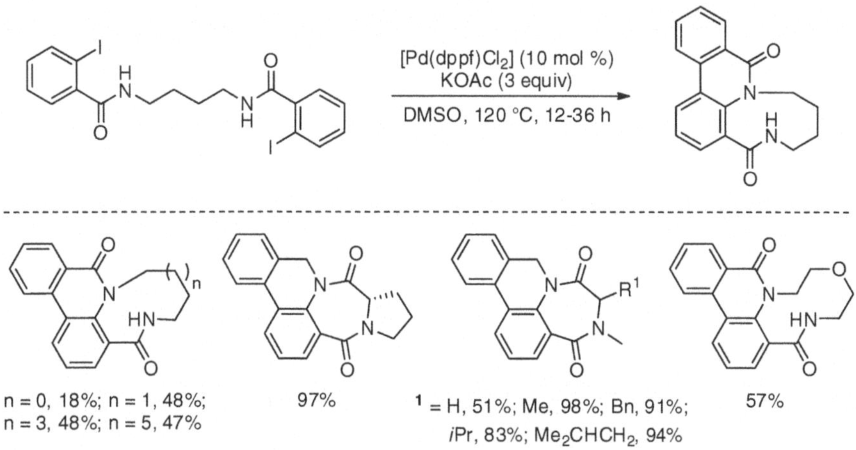

Scheme 25 Palladium-catalyzed intramolecular *N*-arylation/C–H functionalization/aryl–aryl bond forming tandem reaction toward polyheterocycles

a template effect, due to chelation of the transition metal to the two amido groups, leads to conformational pre-orientation which might be the reason for the high efficiency of this method.

The approach of Ackermann and coworkers for the synthesis of annulated heterocycles involves an amination step and a direct arylation sequence by using anilines and 1,2-dihalo-(hetero)aryls (Scheme 26) [57]. It is noteworthy that easily available and inexpensive chloro-substituted starting materials can be employed. Additionally, primary anilines are applicable for the synthesis of carbazoles which avoids complex protection/deprotection procedures. The authors also demonstrated the efficiency of their approach to the synthesis of naturally occurring murrayafoline A [58].

A route to 3-substituted indoles from *ortho*-dihalobenzenes and allylic amines via intermolecular aryl amination and Heck cyclization was reported by Jørgensen and coworkers in 2008 (Scheme 27) [59]. In consideration of previous results, it was postulated that aryl amination is the first step in the sequence. The regiochemistry of the final product is controlled by the chemoselective amination of the aryl iodide position, and therefore the preparation of functionalized products is limited by the availability of the corresponding 1,2-dihaloarene starting materials. Substituents other than a methyl or benzyl at the 3-position have not been yet reported. Conveniently, the addition of an aryl bromide or aryl iodide after completion of the first two steps generates the corresponding *N*-arylated indole product.

An intermolecular *N*-arylation/intermolecular carbopalladation/C–H functionalization/C–C bond formation sequence was realized by Neuville, Zhu, and coworkers for the synthesis of 3-(diarylmethylene)oxindoles (Scheme 28) [60]. This procedure allows for the formation of one C–N and two C–C bonds by way of three different catalytic cycles in a one-pot fashion. The procedure requires addition of the aryl iodide after the *N*-arylation step is completed, and it is important to use an excess quantity of

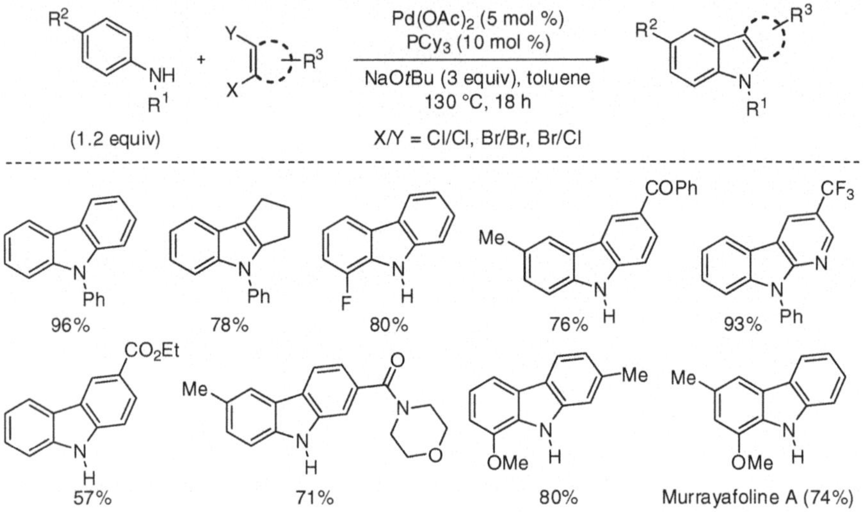

Scheme 26 Palladium-catalyzed synthesis of annulated heterocycles by an amination/direct arylation sequence

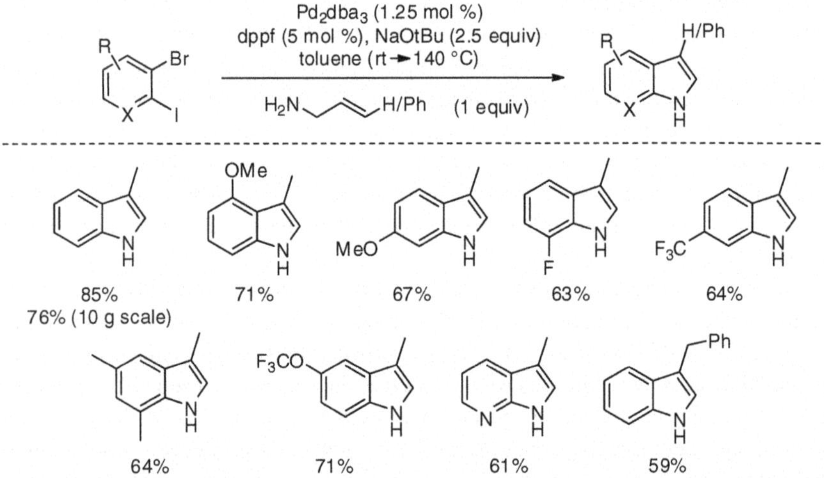

Scheme 27 Palladium-catalyzed synthesis of 3-substituted indoles via intermolecular aryl amination and Heck cyclization

palladium relative to the ligand, as Xantphos is necessary for the initial step while it serves to later inhibit the carbopalladation sequence. The scope of this transformation is somewhat limited since the *N*-arylating agent requires an electron-withdrawing group in the *para*-position (*ortho*- or *meta*-substituted aryl bromides were unsuitable) and yields are reduced when the aryl iodide bears an electron-donating group.

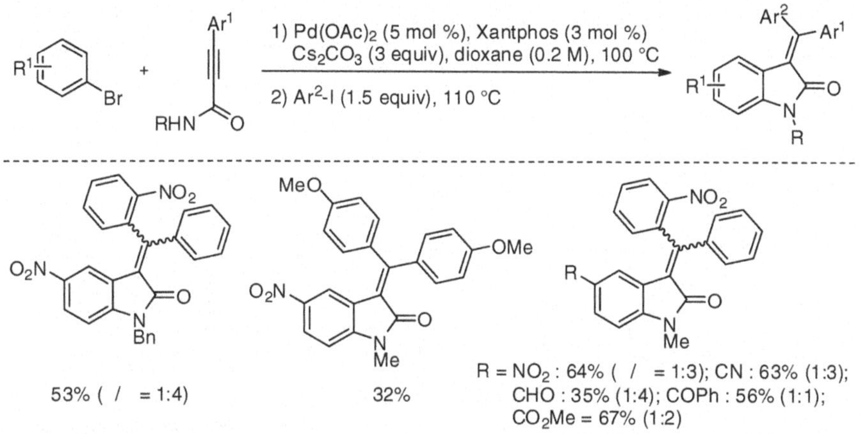

Scheme 28 Synthesis of 3-(diarylmethylene)oxindoles by an intermolecular *N*-arylation/intermolecular carbopalladation/C–H functionalization/C–C bond formation sequence

A powerful example of C–H functionalization in a domino process was reported by Catellani and coworkers who used norbornene as an organic co-catalyst and accomplished a sequence of domino *ortho*-functionalization terminated by cross-coupling [61–64]. Our group successfully implemented this norbornene-mediated C–H functionalization process in domino reactions for the synthesis of various substituted heterocycles [65–69].

The efficiency of this methodology is illustrated by a domino reaction developed by our group in 2007 (Scheme 29) [65]. An intermolecular alkylation at the *ortho*-position of an aryl iodide is followed by an intramolecular amination to afford functionalized indolines and tetrahydroquinolines from simple precursors. The protecting group on nitrogen proved to be important since Boc, Bz, and Ts functional groups only led to decomposition of the starting material. Ethyl carbamate, phenyl, or 4-nitrophenyl protected anilines provided the corresponding functionalized indolines in moderate to good yields. Strongly electron-donating groups at the 2-position are not generally tolerated. However, the use of 2-chloroiodobenzene is possible and the Cl-substituent can be easily converted into electron-rich alcohols, amines, or thiols. Extension of this methodology to the synthesis of tetrahydroquinolines was also shown.

A major drawback of this norbornene-mediated methodology is the requirement of a substituent in the second *ortho*-position of the starting aryl halide, which is necessary to exert regiocontrol over the C–H functionalization step and to avoid double alkylation.

For the synthesis of indoles, our group utilized azirines as the coupling partner in a domino C–H activation/*N*-arylation reaction [66]. Initially, the use of α-haloimines as coupling partners was intended, but during our studies it became apparent that

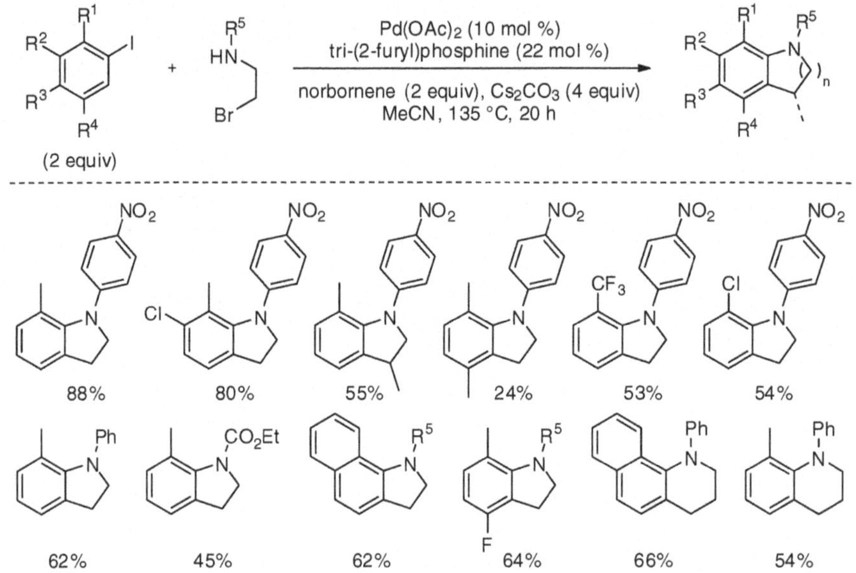

Scheme 29 Palladium-catalyzed domino C–C/C–N coupling of bromoalkylamines for the synthesis of benzannulated *N*-heterocycles

their synthesis is low yielding and often accompanied by decomposition. Therefore, we turned to strained 2*H*-azirines as the 1,3-dipole (Scheme 30). Most substituted indoles are obtained in moderate to excellent yields, except when substituents are placed at the 2-position or when an alkyl or carbonyl group is present at the 3-position of the azirine ring system, which leads to azirine decomposition. During optimization, an unusual tetracyclic by-product was observed that contains two equivalents of the azirine and can be avoided or produced selectively by adjusting the reaction conditions (see Scheme 30).

A similar methodology was used for the synthesis of phenanthridines from aryl iodides and *N*-unsubstituted or *N*-silylimines (Scheme 31) [67]. The key step in this transformation is the cleavage of the N–H or N–Si bond in the catalytic cycle which is necessary for the formation of a palladium–imido intermediate which releases the product upon reductive elimination. A mechanistic constraint is that the imine derivative must carry a group on the nitrogen atom which can be cleaved in the catalytic cycle. Another requirement is the presence of an *ortho*-substituent on the aryl iodide. The reaction tolerates a number of substituents on aryl iodide and the azirine, and our group later also showed that instead of aryl iodides the corresponding aryl triflates can be used which are more easily accessible [68]. We were able to demonstrate the applicability of this methodology in the formal syntheses of nitidine and NK190 starting from the corresponding aryl triflates.

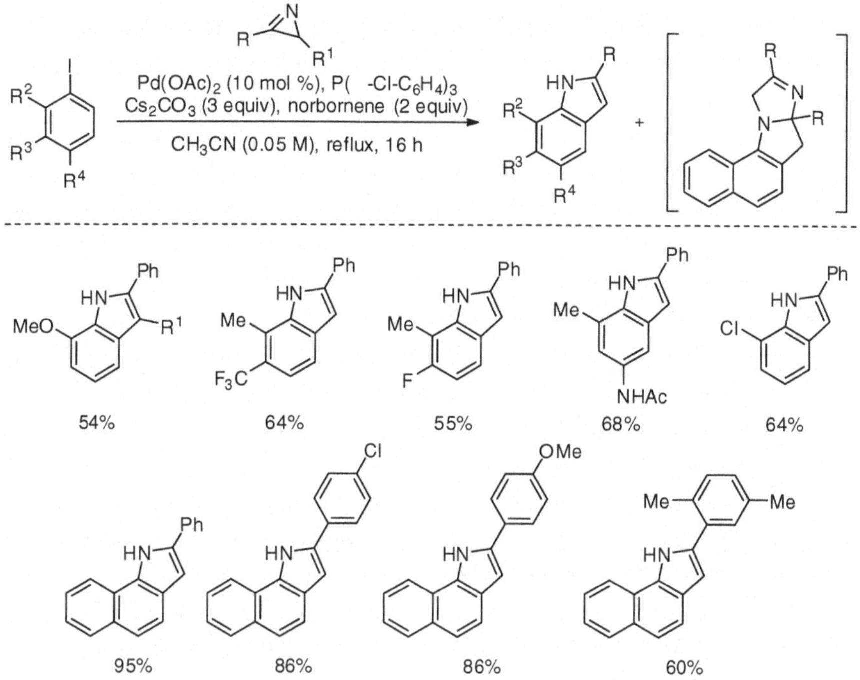

Scheme 30 Palladium-catalyzed domino reaction of azirines with aryl iodides

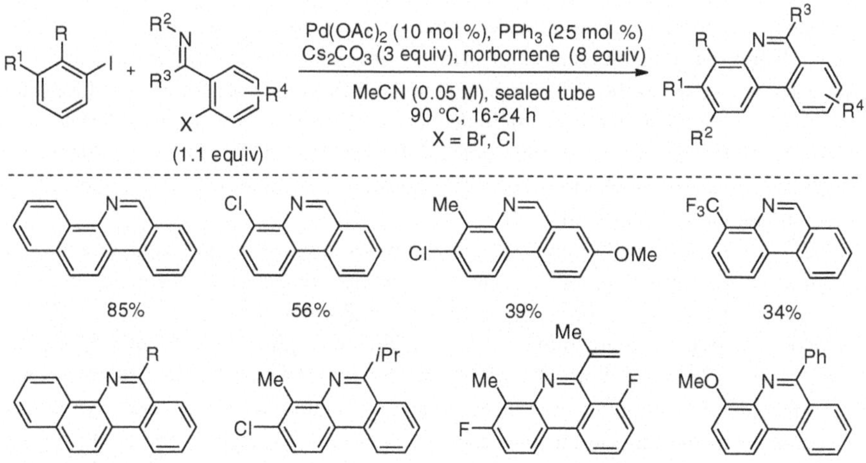

Scheme 31 Palladium-catalyzed domino direct arylation/N-arylation for the synthesis of phenanthridines

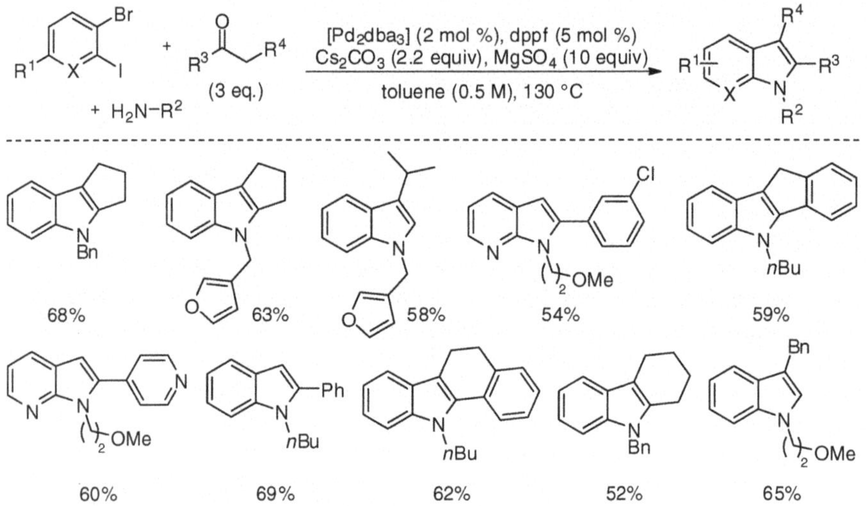

Scheme 32 Palladium-catalyzed tandem synthesis of substituted indoles via Buchwald–Hartwig amination/condensation/arene–alkene coupling

2.4 *Miscellaneous Domino Methods for the Synthesis of Heterocycles*

A recent example of a domino indole synthesis is a three-component palladium-catalyzed process reported by Kurth and coworkers (Scheme 32) [70]. This three-step process involves a Buchwald–Hartwig reaction, a condensation, and an arene–alkene coupling. A variety of primary amines, carbocycles, an anisole, or a pyridine can be used as the aryl compound, and the carbonyl compounds can be cyclic and acyclic ketones as well as aldehydes. Several experiments were undertaken to determine the sequence of events, and it was concluded that Buchwald–Hartwig coupling initiates the catalytic cycle. The postulated mechanism was supported by quantum chemical calculations.

An example of an orthogonal tandem catalysis is the rhodium-catalyzed alkyne arylation/palladium-catalyzed *N*-arylation that was presented by our group in 2011 (Scheme 33) [71]. We reported the successful implementation of a catalyst system consisting of two different metals with two different phosphine ligands in which both catalysts coexist and preferentially promote two out of three possible reactions to produce 1,2-dihydroquinoline derivatives in moderate to good yields. An initial optimization of the individual steps led to conditions that yielded the final product in 69% yield (versus 71% yield over two steps) by using preformed catalysts. An extensive investigation of the reactivity of the possible metal–ligand combinations showed that [Rh(BINAP)] does not reversibly bind XPhos, while palladium can reversibly bind to both ligands. Since [Pd(BINAP)] is catalytically inactive in the

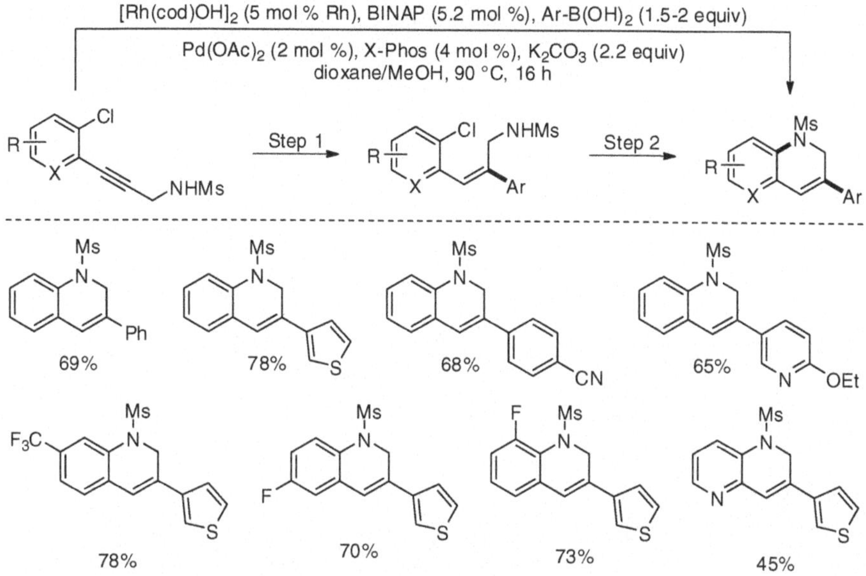

Scheme 33 Orthogonal tandem catalysis for the synthesis of 1,2-dihydroquinoline derivatives

C–N coupling, the amount of BINAP or [Rh(BINAP)], which is a source of trace amounts of free BINAP, had to be carefully adjusted in order to avoid inhibition of the C–N coupling step.

Barluenga and coworkers utilized the bidentate nature of the azaallylic anion as a synthon for palladium-catalyzed construction of various substituted indoles (Scheme 34) [72]. An azaallylic anion can be easily generated in situ through the deprotonation of an imine with α-hydrogen atoms which can then participate in an intermolecular α-arylation reaction. The authors developed a sequence which includes the imine formation, thereby achieving a three-component reaction where the same palladium catalyst promotes three different and independent reactions: (1) the formation of the imine by alkenyl amination, (2) α-arylation of the (deprotonated) imine, and (3) intramolecular *N*-arylation. The reaction conditions of the imine formation are very similar to those of the tandem *C*-arylation/*N*-arylation process, and a couple of successful examples with moderate to good yields were reported. While two different regioisomeric indoles can theoretically be obtained when unsymmetrical 1-bromo-2-chlorobenzene derivatives are employed, only one isomer is ever observed. This regioselectivity may be explained through the different rates of oxidative addition of the palladium catalyst into aryl bromides versus aryl chlorides.

The group of Willis combined the palladium-catalyzed urea arylation with a base-promoted ester amidation to synthesize 3-alkylated 2,4-quinazolinediones (Scheme 35) [73]. This transformation requires relatively high amounts of catalyst loading and long reaction times. An interesting aspect is the fact that for all unsymmetrical urea derivatives studied, the 3-alkyl regioisomer was obtained selectively. This regioselectivity is assumed to arise from the fact that the initial

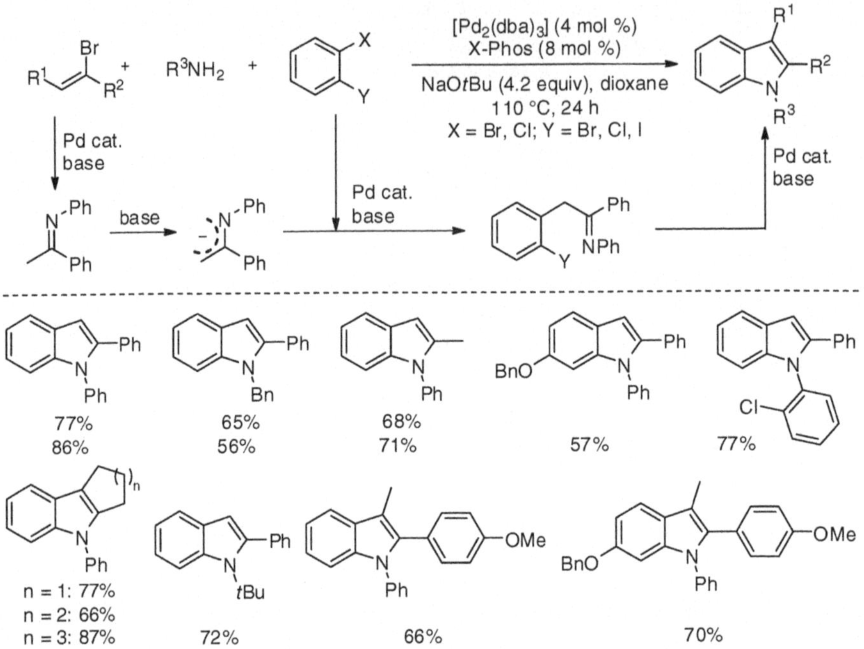

Scheme 34 Use of the azaallylic anion as synthon in palladium-catalyzed tandem reactions ([a]products obtained directly from the preformed imine)

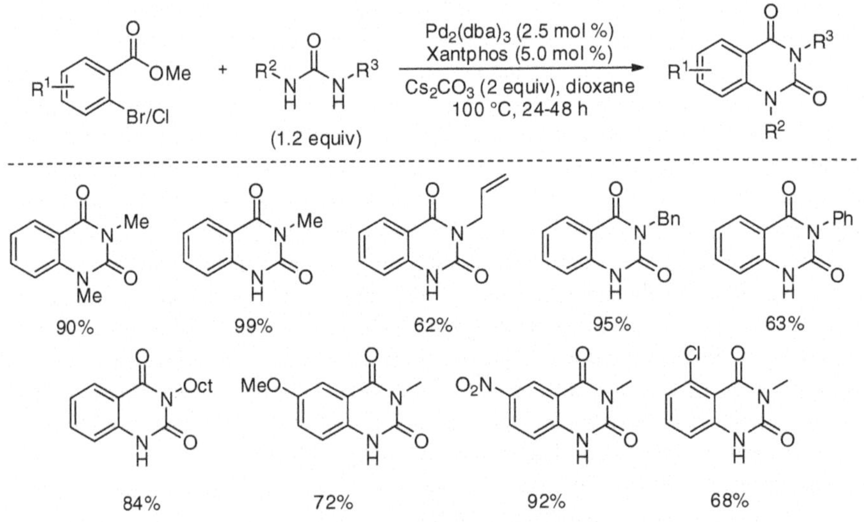

Scheme 35 Tandem palladium-catalyzed urea arylation/intramolecular ester amidation for the regioselective synthesis of 3-alkylated 2,4-quinazolinediones

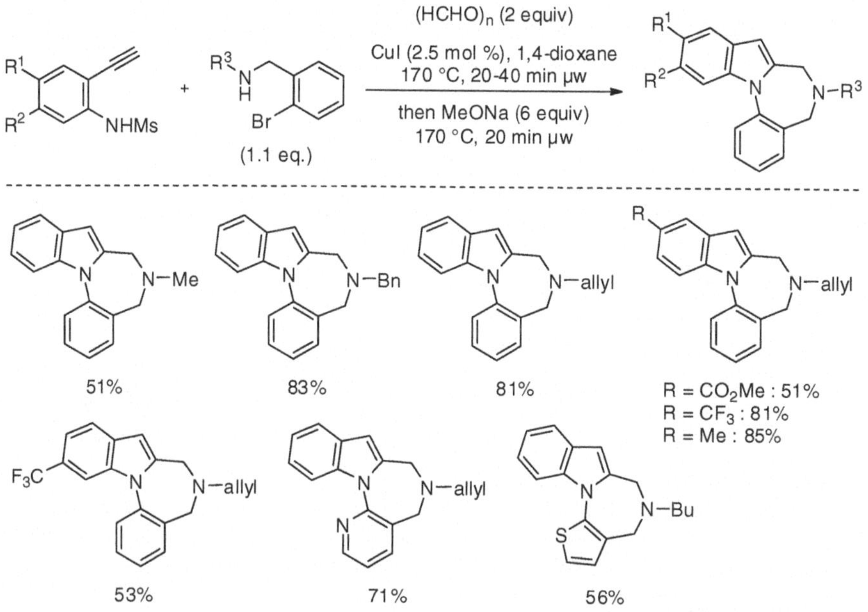

Scheme 36 Synthesis of indole-fused 1,4-diazepines via a ligand-free copper-catalyzed three-component coupling/cyclization/*N*-arylation sequence

arylation reaction occurs on the least hindered, unsubstituted *N*-atom of the urea and is then followed by the ring-closing amidation.

In 2008, Fujii, Ohno, and coworkers reported a ligand-free copper-catalyzed three-component coupling sequence during which four bonds and two rings are formed (Scheme 36) [74]. The sequence is initiated by a Mannich-type reaction followed by intramolecular indole formation. After indole formation is complete, addition of base initiates amine deprotection and the final *N*-arylation can proceed to form indole-fused 1,4-diazepines. The addition of base at a later stage is necessary to avoid decomposition of the starting material. Various *N*-substituted *ortho*-bromobenzylamines and 2-ethynylanilines (with electron-donating or electron-withdrawing groups) as well as heterocyclic secondary amines can be employed to produce the corresponding products in moderate to good yields.

Barluenga, Valdés, and coworkers utilized *N*-tosylhydrazones in a new type of cross-coupling process (Scheme 37) [75]. An intermolecular arylation between tosylhydrazone and a 1-bromo-2-chlorobenzene derivative followed by an intramolecular amination yields substituted tetrahydroquinolines in moderate to good yields. Microwave heating promoted the reaction in one pot, and the tosylhydrazone can be generated in situ from the corresponding carbonyl compound and tosylhydrazine, making the overall process an efficient three-component coupling sequence. A limitation is the failure of the cyclization step when electron-withdrawing substituents are present on the nitrogen atom. The authors were also able to show that chiral substrates can be transformed without loss of enantiomeric excess.

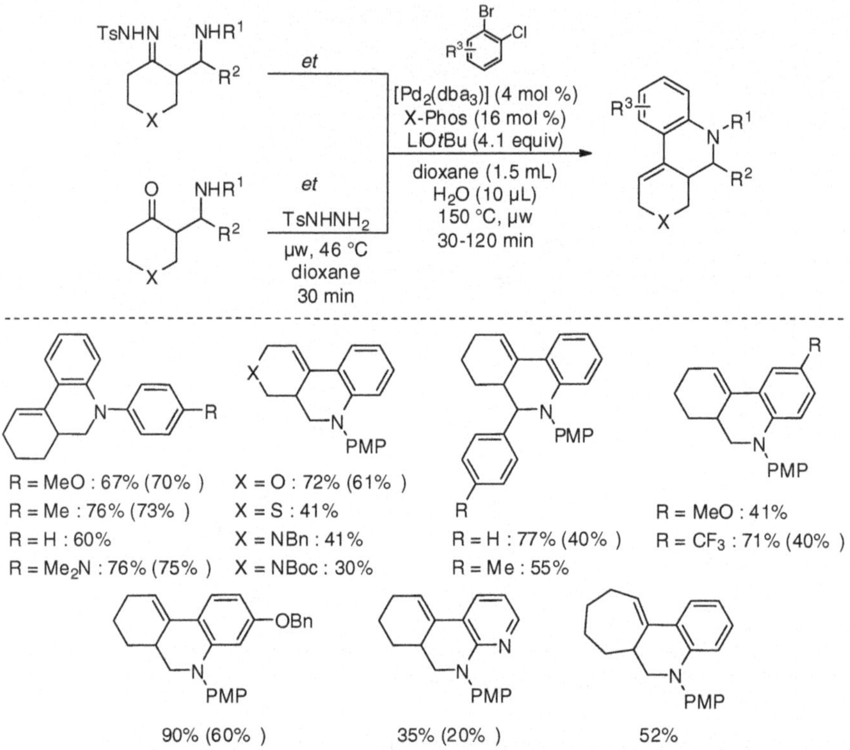

Scheme 37 Tosylhydrazide-promoted palladium-catalyzed synthesis of tetrahydroquinolines via intermolecular arylation/intramolecular amination ([a]results obtained with method B)

Jiang, Ma, and coworkers developed a CuI/L-proline-catalyzed tandem process that generates 3-methyleneisoindolin-1-ones from readily available 2-bromobenzamides and terminal alkynes (Scheme 38) [76]. A variety of functionalized arylacetylenes, aliphatic alkynes, substituted aryl bromides, and a wide range of *N*-substituents were tolerated. In most cases, only the *Z*-isomer was observed. The authors hypothesize that the Sonogashira coupling of aryl bromides with 1-alkynes occurs first. After deprotonation of the amide moiety, the CuI-mediated additive cyclization takes place in a 5-*exo* manner exclusively, which is different for base- or Lewis acid-mediated cyclizations.

Very recently, Nakamura and coworkers utilized in situ generated *N*-allenylimines for the construction of azepine derivatives (Scheme 39) [77]. Starting from *ortho*-propargylic cyclopropylcarbaldoximes, a rhodium catalyst, and TPPMS (sodium diphenylphosphinobenzene-3-sulfonate) the corresponding azepine oxide derivatives are obtained in good yields through a tandem 2,3-rearrangement/heterocyclization reaction. The rhodium catalyst serves a dual role as both π-acidic and redox catalyst. All products are obtained with a *Z*-configuration at the alkylidene moiety, regardless of the configuration of the starting material. For the

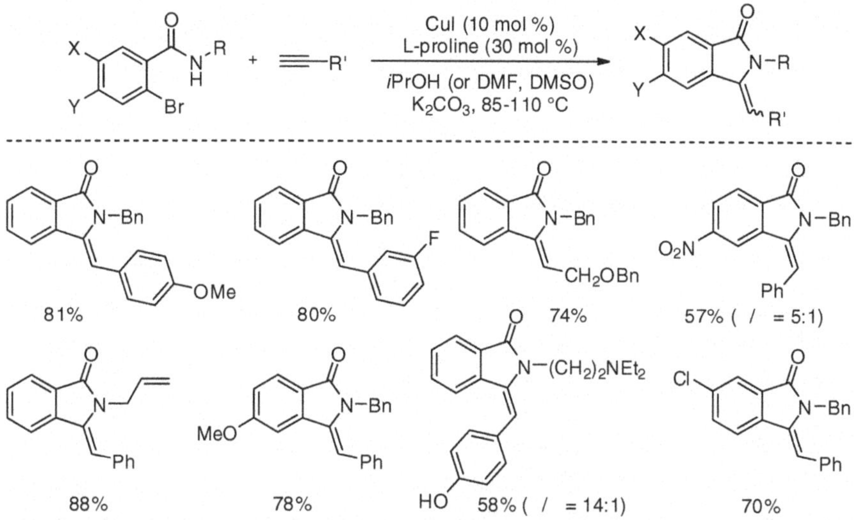

Scheme 38 CuI/L-proline-catalyzed tandem approach toward 3-methyleneisoindol-1-ones

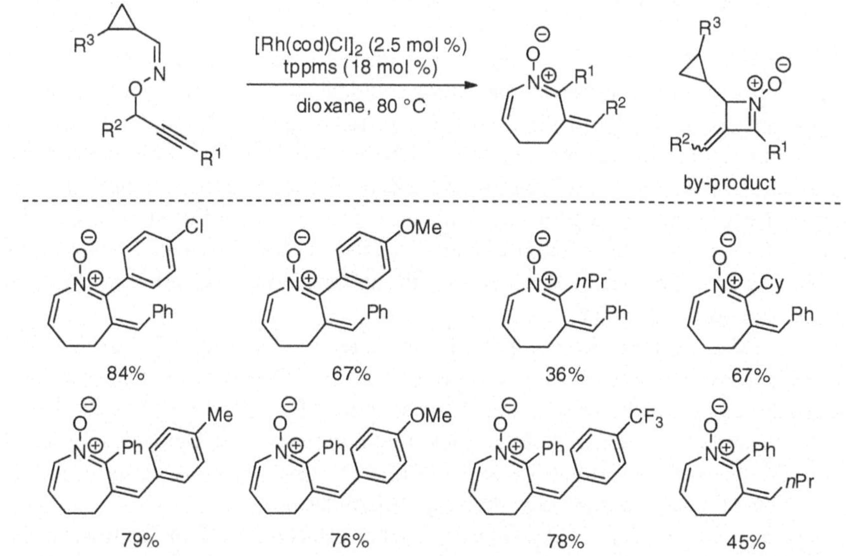

Scheme 39 Rhodium-catalyzed tandem 2,3-rearrangement/heterocyclization for the synthesis of azepine derivatives

(*E*)-isomer of the starting material, reaction conditions had to be re-optimized. In some cases the four-membered cyclic nitrone was obtained as a by product (see Scheme 39). It was shown to be stable under the reaction conditions and is not converted to the product.

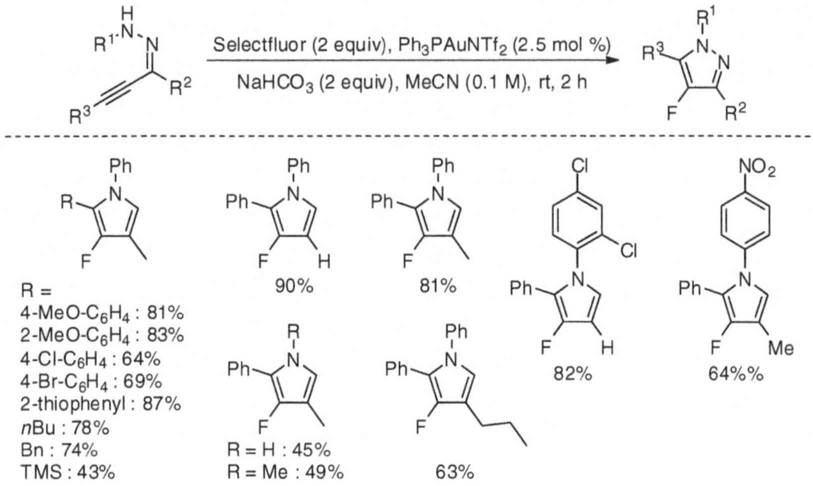

Scheme 40 Gold-catalyzed tandem aminofluorination of alkynes for the synthesis of fluorinated pyrazoles

In 2011, Liu, Xu, and coworkers reported a novel synthesis of fluorinated pyrazoles via a gold-catalyzed tandem aminofluorination of alkynes in the presence of Selectfluor [78]. This methodology was designed to overcome limitations of known approaches to fluoropyrazoles among which are low yields, multiple steps, harsh reaction conditions, or the use of dangerous reagents. The method works at room temperature and has broad scope (Scheme 40). The authors proposed the coordination of an Au^{I} or Au^{III} salt to the alkyne as the key mechanistic step. It is unclear at which step Selectfluor participates in transferring a fluorine atom to the final product. Under the reaction conditions, when a non-fluorinated analogue is obtained as a by-product of the reaction, it can be readily converted to the final fluorine-containing product.

Buchwald and coworkers reported the tandem synthesis of pyrroles and pyrazoles from haloenynes by reaction of either a Boc-protected amine or a bis(Boc)hydrazide (Scheme 41) [79]. The sequence of copper-catalyzed amidation and hydroamidation yields various substituted pyrroles and pyrazoles in good yields. Mechanistic investigations showed that the reaction most likely proceeds via an initial C–N coupling followed by hydroamidation.

Tang, Fan, and coworkers developed a copper-catalyzed tandem reaction for the synthesis of *N*-heteroarylated indoles and benzimidazoles which involves a conjugate addition, two cyclizations, and an aromatization (Scheme 42) [80].

Willis and coworkers reported the synthesis of 2-quinolones via a palladium-catalyzed alkenyl aminocarbonylation followed by intramolecular amidation (Scheme 43) [81]. For the 2-quinolone synthesis it is important at which site the initial reaction takes place (aryl halide versus alkenyl halide, see Scheme 43) and which of the two catalytic reactions occurs first (amination or carbonylation). It is postulated that the alkenyl halide is the first site of reaction and carbonylation is

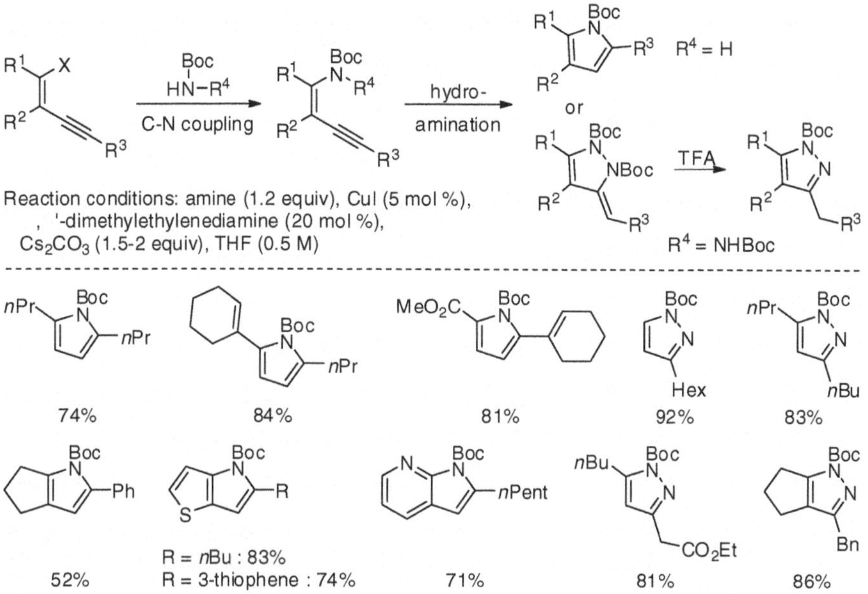

Scheme 41 Tandem synthesis of pyrroles and pyrazoles by copper-catalyzed amidation/hydroamidation of haloenynes

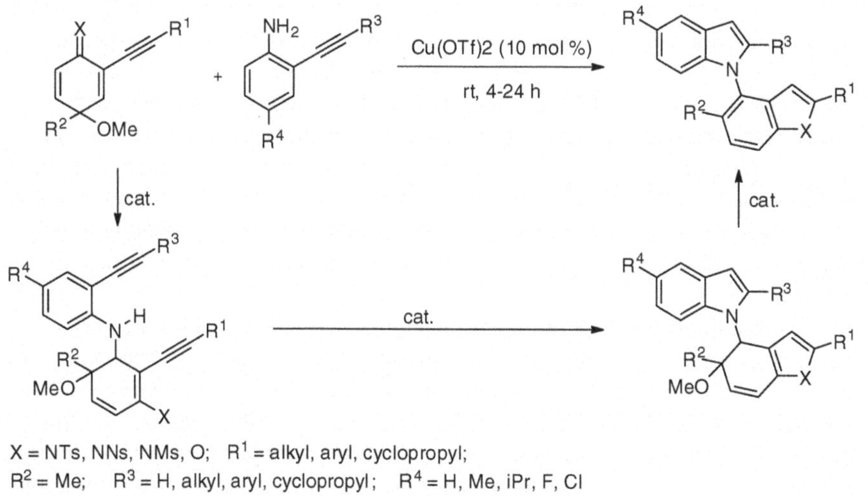

Scheme 42 Copper-catalyzed tandem reaction for the synthesis of *N*-heteroarylated indoles and benzimidazoles

the faster of the two processes. In some cases it was beneficial to remove the CO atmosphere after 3 h. In order to obtain the corresponding isoquinolone, it was necessary to change the order of reagent addition which was done by applying the CO atmosphere at a later stage in the reaction. A limitation of this approach is the

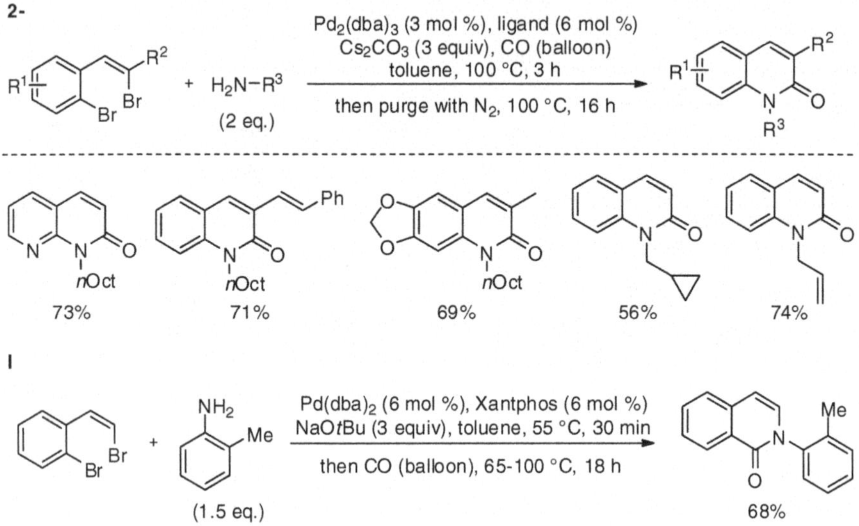

Scheme 43 Palladium-catalyzed alkenyl aminocarbonylation/intramolecular aryl amidation for the synthesis of 2-quinolones

requirement for a sterically demanding *N*-nucleophile since less hindered amines lead to competing indole formation.

In 2008, Shin and coworkers reported a gold-catalyzed generation of an azomethine ylide via an internal redox reaction between a tethered nitrone and an alkyne. The ylide then undergoes an efficient diastereoselective cycloaddition cascade (Scheme 44) [82]. Various platinum(II), silver(I), and gold(III) salts were found to be effective catalysts, while $AuCl_3$ provided the best results. This methodology is particularly attractive because it avoids the use of explosive diazo derivatives and is 100% atom economic. Metal-catalyzed cycloaddition reactions that result in the generation of a C–N or C–O bond will be discussed in length in another chapter of this book and consequently will not be visited further in this section.

The versatility of incorporating the Sonogashira reaction has been exploited by Ohe and coworkers for the synthesis of hetero α,α′-dimers of heteroaromatic compounds (Scheme 45) [83]. By utilizing bimetallic palladium/copper catalysis, a tandem process can be achieved for the synthesis of a variety of dimeric compounds. Due to the fact that this is a three-component coupling process, chemical diversity can be established very quickly under mild reaction conditions.

Wang and coworkers recently reported an elegant illustration of a copper-catalyzed domino coupling of a diverse array of *N*-tosylhydrazones and a series of terminal alkynes for the synthesis of 2-substituted benzofurans and indole heterocycles (Scheme 46) [84].

Glorius and coworkers published a report that describes the copper-catalyzed domino reaction of 1,2-dihalo carbo- and heterocycles with primary amides for the

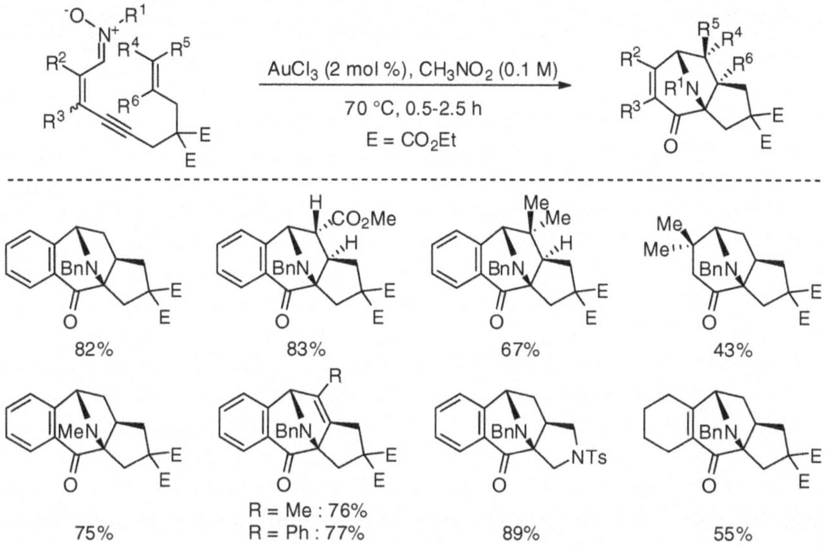

Scheme 44 Gold-catalyzed internal redox/dipolar cycloaddition cascade

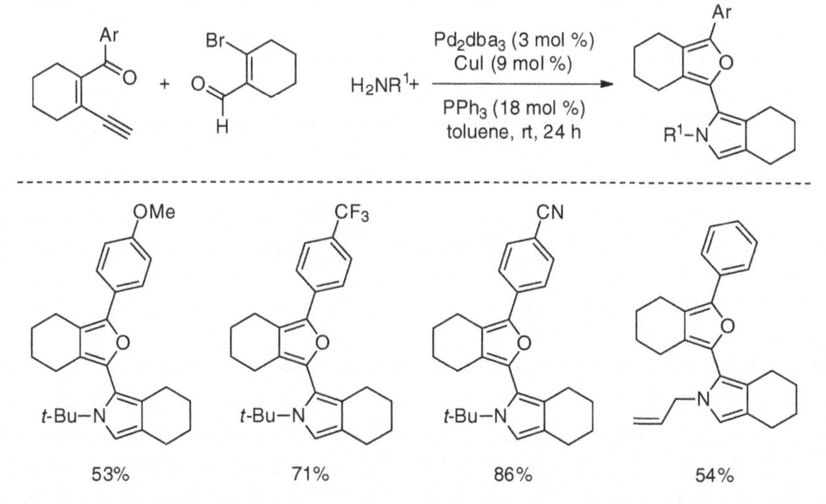

Scheme 45 Palladium/copper-catalyzed tandem multicomponent coupling for the synthesis of α,α′-heteroaromatic dimers

synthesis of benzoxazole products (Scheme 47) [85]. A variety of chloro-, bromo-, and iodo-containing 1,2-dihaloarene starting materials are shown to be competent coupling partners in this methodology.

The Tsuji–Trost reaction has been utilized in a variety of transformations to achieve complex target structures and intermediates [86]. To this end, the

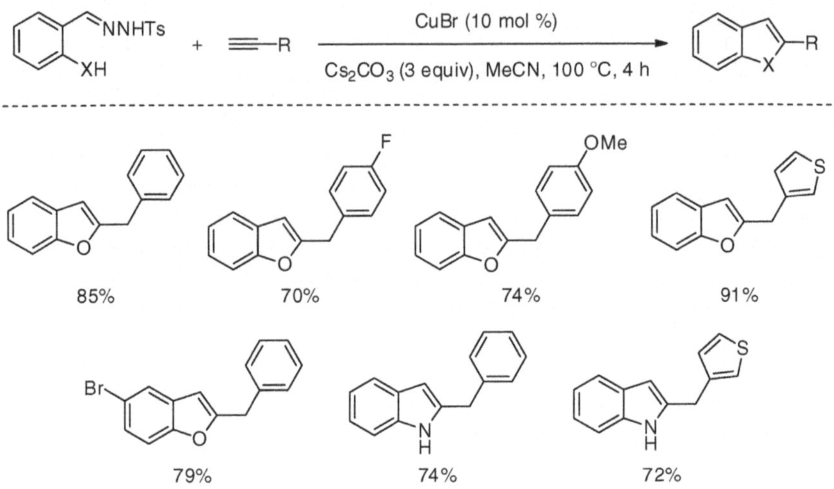

Scheme 46 Copper-catalyzed domino coupling of *N*-tosylhydrazones and terminal alkynes for the synthesis of 2-substituted benzofurans and indoles

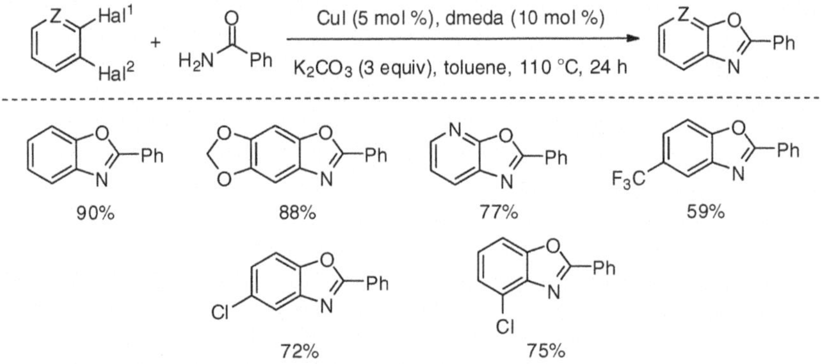

Scheme 47 Copper-catalyzed domino synthesis of benzoxazole heterocycles

Tsuji–Trost reaction has been successfully coupled in tandem or domino processes for the synthesis of heterocycles [87]. Menche and coworkers reported in 2010 the concise Pd-catalyzed diastereoselective domino synthesis of tetrahydropyran heterocycles via an oxa-Michael addition/Tsuji–Trost reaction (Scheme 48) [88]. By employing a chiral alcohol with a pendant allyl carbonate moiety, tetrahydropyrans could be constructed with up to four non-contiguous stereocenters. Yields were generally moderate and level of diastereoselectivity ranged from low to high, dependent on the substrate combination.

Menche and coworkers recently disclosed an expansion of their previous method for the synthesis of masked 1,3-diols via a domino Pd-catalyzed diastereoselective hemiacetal formation/Tsuji–Trost reaction sequence (Scheme 49) [89]. The scope

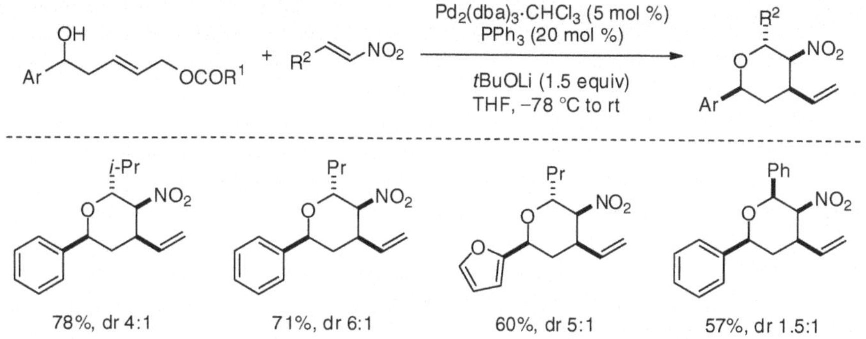

Scheme 48 Domino synthesis of tetrahydropyrans via an oxa-Michael addition/Tsuji–Trost reaction

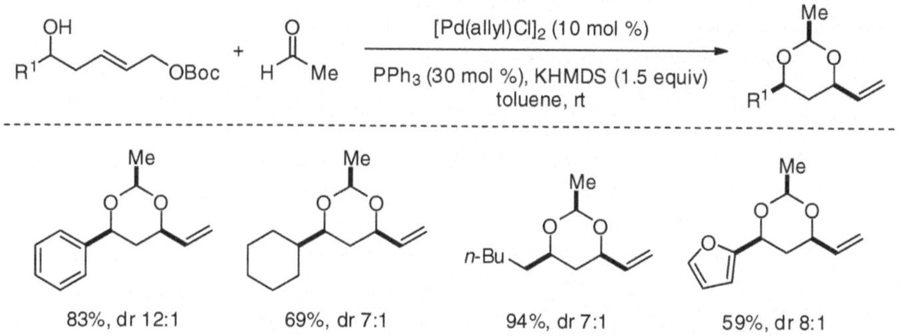

Scheme 49 Domino synthesis of protected 1,3-diols via hemiacetal formation/Tsuji–Trost reaction

of the transformation is quite broad and yields range from moderate to excellent with generally high levels of diastereoselectivity.

3 Conclusion

While this particular chapter was specifically focused on the synthesis of heterocycles via metal-catalyzed domino reactions that result in the generation of a C–N or C–O bond, the general field of these types of "domino" transformations represents one of the most efficient, elegant, and atom-economical means to construct complex target structures. It can be expected that efficiencies in these transformations will only increase in the coming years. The ability to reduce the environmental impact represents a potential advantage of this approach. Indeed, the reduction of numerous workup steps such as extractions, purifications (chromatography, recrystallization, distillation, etc.), and the lessening/elimination of the requirement of toxic reagents is an attractive and important goal. It is to be expected that research groups will continue to pursue advances in the field of domino catalysis.

Acknowledgments The authors would like to thank the Natural Sciences and Engineering Council of Canada (NSERC) and the University of Toronto for financial support. NSERC/Merck-Frosst Canada is thanked for an Industrial Research Chair. JK thanks the Deutsche Forschungsgemeinschaft for a postdoctoral fellowship.

References

1. Tietze LF, Beifuss U (1993) Angew Chem Int Ed 32:131
2. Tietze LF (1996) Chem Rev 96:115
3. Fogg DE, dos Santos EN (2004) Coord Chem Rev 248:2365
4. Chapman CJ, Frost CG (2007) Synthesis 1
5. Singh BN, Vaughan Williams EM (1970) Br J Pharmacol 39:657
6. Baba K, Takeuchi K, Hamasaki F, Kozawa M (1986) Chem Pharm Bull 34:595
7. Lounasmaa M, Tolvanen A (2000) Nat Prod Rep 17:175
8. Katritzky A, Rees C, Scriven E (eds) (1996) Indoles. In: Gribble GW (ed) Comprehensive heterocyclic chemistry II. Pergamon, Oxford, p 207
9. Kadieva MG, Oganesyan ÉT (1997) Chem Heterocycl Compd 33:1245
10. Krüger K, Tillack A, Beller M (2008) Adv Synth Catal 350:2153
11. Zeni G, Larock RC (2004) Chem Rev 104:2285
12. Fürstner A, Davies PW (2005) J Am Chem Soc 127:15024
13. Li G, Huang X, Zhang L (2008) Angew Chem Int Ed 47:346
14. Koradin C, Dohle W, Rodriguez AL, Schmid B, Knochel P (2003) Tetrahedron 59:1571
15. Shimada T, Nakamura I, Yamamoto Y (2004) J Am Chem Soc 126:10546
16. Nakamura I, Mizushima Y, Yamamoto Y (2005) J Am Chem Soc 127:15022
17. Bates CG, Saejueng P, Murphy JM, Venkataraman D (2002) Org Lett 4:4727
18. Carril M, Correa A, Bolm C (2008) Angew Chem Int Ed 47:4862
19. Buchwald SL, Bolm C (2009) Angew Chem Int Ed 48:5586
20. Thomé I, Nijs A, Bolm C (2012) Chem Soc Rev 41:979
21. Kondo Y, Shiga F, Murata N, Sakamoto T, Yamanaka H (1994) Tetrahedron 50:11803
22. Lu BZ, Zhao W, Wei H-X, Dufour M, Farina V, Senanayake CH (2006) Org Lett 8:3271
23. Nakamura M, Ilies L, Otsubo S, Nakamura E (2006) Org Lett 8:2803
24. Isono N, Lautens M (2009) Org Lett 12:1329
25. Boyer A, Isono N, Lackner S, Lautens M (2010) Tetrahedron 66:6468
26. Rao MLN, Jadhav DN, Dasgupta P (2010) Org Lett 12:2048
27. Berciano BP, Lebrequier S, Besselièvre F, Pigue S (2010) Org Lett 12:4038
28. Coste A, Karthikeyan G, Couty F, Evano G (2009) Angew Chem Int Ed 48:4381
29. Coste A, Couty F, Evano G (2009) Org Lett 11:4454
30. Wang Z-J, Yang J-G, Yang F, Bao W (2010) Org Lett 12:3034
31. Xu H, Zhang Y, Huang J, Chen W (2010) Org Lett 12:3704
32. Qin X-R, Cong X-F, Zhao D-B, You J-S, Lan J-B (2011) Chem Commun 47:5611
33. Zeng F, Alper H (2011) Org Lett 13:2868
34. Ramirez F, Desal NB, McKelvie N (1962) J Am Chem Soc 84:1745
35. Corey EJ, Fuchs PL (1972) Tetrahedron Lett 36:3769
36. Eymery F, Iorga B, Savignac P (2000) Synthesis 85
37. Fang Y-Q, Karisch R, Lautens M (2007) J Org Chem 72:1341
38. Fang Y-Q, Lautens M (2005) Org Lett 7:3549
39. Fang Y-Q, Lautens M (2008) J Org Chem 73:538
40. Fayol A, Fang Y-Q, Lautens M (2006) Org Lett 8:4203
41. Fang Y-Q, Yuen J, Lautens M (2007) J Org Chem 72:5152
42. Yuen J, Fang Y-Q, Lautens M (2006) Org Lett 8:653

43. Nagamochi M, Fang Y-Q, Lautens M (2007) Org Lett 9:2955
44. Bryan CS, Lautens M (2008) Org Lett 10:4633
45. Newman SG, Aureggi V, Bryan CS, Lautens M (2009) Chem Commun 5236
46. Thielges S, Meddah E, Bisseret P, Eustache J (2004) Tetrahedron Lett 45:907
47. Vieira TO, Meaney LA, Shi Y-L, Alper H (2008) Org Lett 10:4899
48. Arthuis M, Pontikis R, Florent J-C (2009) Org Lett 11:4608
49. Xia Z, Wang K, Zheng J, Ma Z, Jiang Z, Wang X, Lv X (2012) Org Biomol Chem 10:1602
50. Wang Z-J, Yang F, Lv X, Bao W (2011) J Org Chem 76:967
51. He H-F, Dong S, Chen Y, Yang Y, Le Y, Bao W (2012) Tetrahedron 68:3112
52. Bryan CS, Braunger JA, Lautens M (2009) Angew Chem Int Ed 48:7064
53. Alberico D, Scott ME, Lautens M (2007) Chem Rev 107:174
54. Kuhl N, Hopkinson MN, Wencel-Delord J, Glorius F (2012) Angew Chem Int Ed 51:10236
55. Cuny G, Bois-Choussy M, Zhu J (2003) Angew Chem Int Ed 42:4774
56. Cuny G, Bois-Choussy M, Zhu J (2004) J Am Chem Soc 126:14475
57. Ackermann L, Althammer A (2007) Angew Chem Int Ed 46:1627
58. Knölker H-J, Reddy KR (2002) Chem Rev 102:4303
59. Jensen T, Pedersen H, Bang-Andersen B, Madsen R, Jørgensen M (2008) Angew Chem Int Ed 47:888
60. Pinto A, Neuville L, Zhu J (2009) Tetrahedron Lett 50:3602
61. Catellani M, Fagnola MC (1994) Angew Chem Int Ed Engl 33:2421
62. Catellani M (2003) Synlett 298
63. Motti E, Ippomei G, Deledda S, Catellani M (2003) Synthesis 2671
64. Faccini F, Motti E, Catellani M (2004) J Am Chem Soc 126:78
65. Thansandote P, Raemy M, Rudolph A, Lautens M (2007) Org Lett 9:5255
66. Candito DA, Lautens M (2010) Org Lett 12:3312
67. Candito DA, Lautens M (2009) Angew Chem Int Ed 48:6713
68. Blanchot M, Candito DA, Larnaud F, Lautens M (2011) Org Lett 13:1486
69. Thansandote P, Chong E, Feldmann K-O, Lautens M (2010) J Org Chem 75:3495
70. Knapp JM, Zhu JS, Tantillo DJ, Kurth MJ (2012) Angew Chem Int Ed 51:10588
71. Panteleev J, Zhang L, Lautens M (2011) Angew Chem Int Ed 50:9089
72. Barluenga J, Jiménez-Aquino A, Valdés C, Aznar F (2007) Angew Chem Int Ed 46:1529
73. Willis MC, Snell RH, Fletcher AJ, Woodward RL (2006) Org Lett 8:5089
74. Ohta Y, Chiba H, Oishi S, Fujii N, Ohno H (2008) Org Lett 10:3535
75. Barluenga J, Quiñones, Cabal M-P, Aznar F, Valdés C (2011) Angew Chem Int Ed 50:2350
76. Li L, Wang M, Zhang X, Jiang Y, Ma D (2009) Org Lett 11:1309
77. Nakamura I, Okamoto M, Sato Y, Terada M (2012) Angew Chem Int Ed 51:10816
78. Qian J, Liu Y, Zhu J, Jiang B, Xu Z (2011) Org Lett 13:4220
79. Martín R, Rodríguez Rivero M, Buchwald SL (2006) Angew Chem Int Ed 45:7079
80. Yang M, Tang J, Fan R (2012) Chem Commun 48:11775
81. Tadd AC, Matsuno A, Fielding MR, Willis MC (2009) Org Lett 11:583
82. Yeom H-S, Lee J-E, Shin S (2008) Angew Chem Int Ed 47:7040
83. Murata T, Murai M, Ikeda Y, Miki K, Ohe K (2012) Org Lett 14:2296
84. Zhou L, Shi Y, Xiao Q, Liu Y, Ye F, Zhang Y, Wang J (2011) Org Lett 13:968
85. Altenhoff G, Glorius F (2004) Adv Synth Catal 346:1661
86. Trost BM, Crawley ML (2003) Chem Rev 103:2921
87. Balme G, Bouyssi D, Monteiro N (2006) Pure Appl Chem 78:231
88. Wang L, Li P, Menche D (2010) Angew Chem Int Ed 49:9270
89. Wang L, Menche D (2012) Angew Chem Int Ed 51:9425

Top Heterocycl Chem (2013) 32: 225–270
DOI: 10.1007/7081_2013_108

Published online: 30 April 2013

Synthesis of Saturated Heterocycles via Metal-Catalyzed Formal Cycloaddition Reactions That Generate a C–N or C–O Bond

Jerome Waser

Abstract In this section, the synthesis of saturated N- and O-heterocycles via formal cycloaddition is presented. The main focus is on metal-catalyzed reactions involving C–C or C–X σ bond cleavage in three- or four-membered rings. After a fast presentation of pioneering works, the important breakthroughs of the last two decades are presented. The section starts with reactions involving three-membered rings. Formal [3+2] cycloadditions of donor–acceptor-substituted cyclopropanes and methylenecyclopropanes with carbonyls and imines are important methods to access tetrahydrofuran and pyrrolidine heterocycles. Formal [3+3] cycloadditions have emerged more recently. On the other hand, reactions of epoxides and aziridines with carbon monoxide or cumulenes are now well-established methods to access heterocycles. These processes have been completed more recently with cycloaddition with olefins, carbonyls, and imines. The section ends with the emerging field of four-membered ring activation for cycloaddition with π systems.

Keywords Aziridines · Cycloaddition · Cyclobutanes · Cyclopropanes · Epoxides · Heterocycles · Pyrrolidines · Tetrahydrofurans

Contents

J. Waser (✉)
Ecole Polytechnique Fédérale de Lausanne Institute of Chemical Sciences and Engineering, EPFL SB ISIC LCSO, BCH 4306, 1015 Lausanne, Switzerland
e-mail: Jerome.waser@epfl.ch

1 Introduction: Definitions and Scope of the Section

The discovery of classical cycloaddition reactions, such as the (hetero) Diels–Alder and 1,3-dipolar cycloadditions, has contributed tremendously to a more efficient access toward both carbocycles and heterocycles. The introduction of the term cycloaddition was necessary to distinguish these new types of reactions from previously discovered processes leading to cyclic structures, such as the famous Robinson annulation. In principle, each cycloaddition can be considered as a special case of the more general annulation process, but from which point on an annulation can be called a cycloaddition has been the topic of intensive discussions for decades, and is still not settled today. In 1968, Huisgen proposed a set of rules for the definition of cycloaddition, and the two first are still largely recognized as prerequisite [1]:

- Huisgen Rule 1: "Cycloadditions are ring closures in which the number of σ bonds increases."
- Huisgen Rule 2: "Cycloadditions are not associated with the elimination of small molecules or ions. The cycloadduct corresponds to the sum of the components."

The current official definition of cycloaddition by IUPAC is very close to these first two rules of Huisgen:

> A reaction in which two or more unsaturated molecules (or parts of the same molecule) combine with the formation of a cyclic adduct in which there is a net reduction of the bond multiplicity.

Although the rule that all the atoms of the starting materials have to be included in the product is not explicitly included in the definition, this requirement is usually recognized by most organic chemists. Even if electrocyclic cyclization processes were included in the original definition of Huisgen, the term cycloaddition is mostly used today for those reactions proceeding via the formation of at least two new bonds. Nevertheless, several researchers think that the term cycloaddition should be more strictly limited to reactions involving a continuous overlap of π electrons, and consequently allowing a concerted process. In fact, in his seminal publication, Huisgen already introduced further rules, in particular rule number 3, which explicitly stated that cycloadditions should not involve the cleavage of sigma bonds:

- Huisgen Rule 3: "Cycloadditions do not involve the cleavage of σ bonds."

Unfortunately, in the same publication, Huisgen also described several reactions proceeding via σ-bond cleavage as cycloaddition.

To solve this definition dilemma, several researchers have used the term "formal cycloaddition." Although this term has not yet been strongly defined, we propose to use it here for those reactions following the rules 1 and 2 of Huisgen and the IUPAC definition, but not the more strict criteria of rule 3 and the non-interrupted π system of electrons (Scheme 1). In contrast to annulation reactions, the formation of small molecules or changes in the connectivity of atoms not involved in the formation of the new bonds in the ring are not allowed in this case. For example, even if the

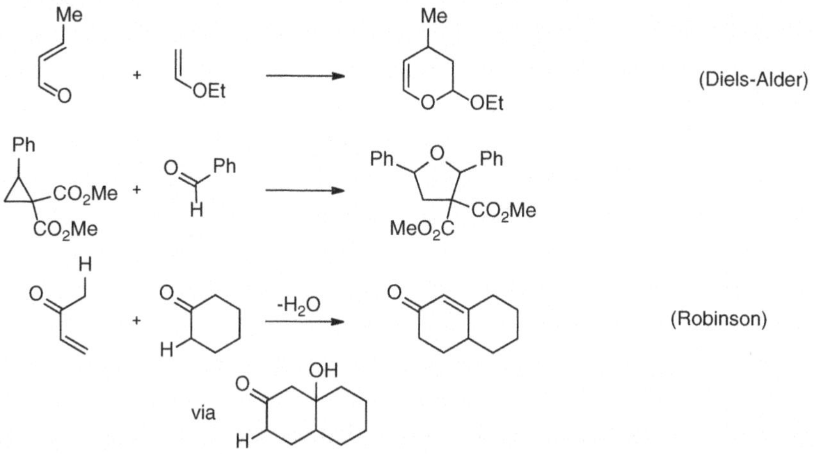

Scheme 1 Examples of cycloaddition, formal cycloaddition, and annulation as defined in this section

alcohol intermediate formed first in the Robinson annulation formally contains all the atoms of the starting materials, the position of the indicated hydrogen has changed. Although also highly useful, such processes will not be included in this section. Furthermore, we will limit ourselves to reactions for which the definition is valid for the used starting materials, and not on transiently generated reactive intermediates:

Definition of "formal cycloaddition" in this section: "A reaction in which two or more molecules (or parts of the same molecule) combine with the formation of a cyclic adduct, involving the formation of at least two new σ bonds and the cleavage of at least one σ bond, but not associated with the elimination of small molecules or changes in the connectivity of atoms except for ring formation."

This type of reaction is highly useful for the synthesis of heterocycles, as it gives a direct access to more saturated derivatives, in contrast to classical cycloadditions involving only π systems, but still conserves the perfect atom-economy of the process. On the other hand, the cleavage of σ bonds is much more difficult than the rearrangement of π electrons. To increase the reactivity of the substrates, the use of ring strain often together with the further polarization of σ bonds with functional groups has been the most successful, and this section will be limited to this approach. In order especially to highlight the synthetic complementarity with cycloadditions of conjugated systems, we will limit the discussion to reactions giving access to heterocycles with no more than one unsaturated center and leading to O- or N-containing heterocycles.

In order to give a better systematic overview of this fast growing field, the section has been organized according to the following criteria (Scheme 2):

(a) Ring size of the formal cycloaddition substrate (three, four or larger).
(b) Structure of the ring: all carbons, with one oxygen or one nitrogen and with more than one heteroatom.

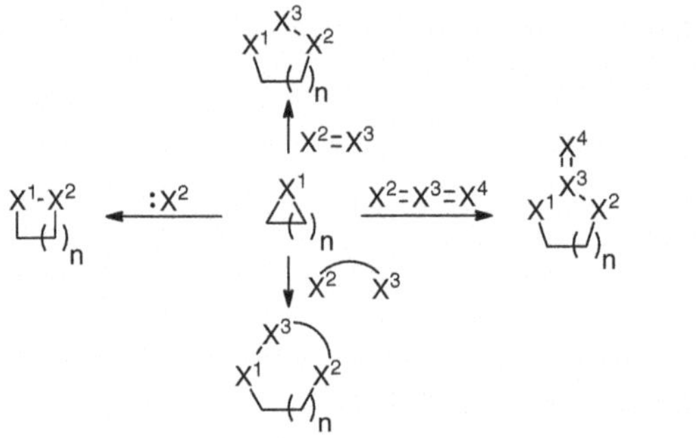

Scheme 2 Classification criteria for formal cycloaddition reactions

(c) Number of added atoms during the cycloaddition process: one, two (divided in isolated π systems and cumulenes), and larger.
(d) Structure of the reacting partner: all carbon, with one oxygen or one nitrogen and with more than one heteroatom.

2 Reactions Involving Three-Membered Rings

Three-membered rings have been by far most often used in formal cycloaddition reactions. This is probably due first to the activation of the σ-bond originating from ring strain, which is essential to allow cycloaddition under mild conditions. Secondly, there are numerous synthetic methods to access three-membered rings, especially cyclopropanes, epoxides, and aziridines. This has led to a widespread use of these substrates in cycloaddition and annulation reactions.

2.1 Reactions with Cyclopropanes

Cyclopropanes are very important in organic chemistry, both as structural elements of synthetic and bioactive compounds and as platforms for further functionalization. They are also interesting from the theoretical point of view and are best described by the use of Walsh orbitals, which explain their partial π character. For these reasons, they can be considered as one-carbon homologues of olefins. Despite their high strain energy (26 Kcal/mol), cyclopropanes are still stable compounds, and most useful formal cycloadditions of cyclopropanes have relied on further activation of the C–C bond via polarization, especially through the introduction of vicinal donor and acceptor groups (donor–acceptor-substituted cyclopropanes) [2–8]. A second possibility for further activation is the introduction of unsaturation, which further increases ring strain and stabilizes potential reactive intermediate, as exemplified by the rich chemistry of alkylidenecyclopropanes [9].

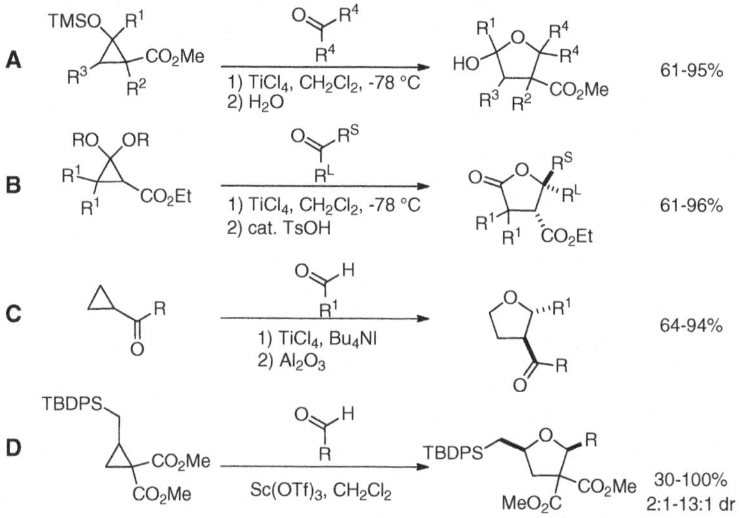

Scheme 3 Pioneering examples of formal [3+2] cycloaddition of cyclopropanes with carbonyl compounds

2.1.1 Formal [3+2] Cycloadditions with Isolated π Systems

C–O Bond Formation

The formal cycloaddition of cyclopropanes with carbonyl compounds gives a fast and atom-economical access to important tetrahydrofuran derivatives. Pioneering works of Reissig and co-workers in the 1980s have already shown the potential of oxygen-substituted cyclopropanes to access either tetrahydrofurans or lactones (Scheme 3**A**, **B**) [10–15]. Oshima and co-workers later showed that unsubstituted cyclopropanes could also be used for the cycloaddition (Scheme 3**C**) [16], and Yadav introduced in 2006 silyl activated cyclopropanes as another alternative (Scheme 3**D**) [17]. In 2011, Dobbs and co-workers demonstrated that cycloaddition of silylmethyl-substituted cyclopropanes was also possible in the absence of the diester activating group [18].

Nevertheless, despite these promising studies, the interest in [3+2] cycloadditions remained limited for several decades, probably because the factors controlling the stereoselectivity of the reaction were poorly understood. The situation changed dramatically when Johnson and co-workers demonstrated in 2005 that the Lewis acid-catalyzed [3+2] cycloaddition of aryl-diester-substituted cyclopropanes with carbonyl compounds was not only diastereoselective, but also highly enantiospecific (Scheme 4) [19, 20]. The reaction was successfully extended to alkenyl- and alkyl-substituted cyclopropanes. To rationalize the observed enantiospeficity, Johnson and co-workers proposed that the reaction proceeded via a tight ion pair **II** [21, 22]. The existence of such "intimate ion pairs" has also been proposed by other authors [23–25]. A stereoselective anti-attack of the aldehyde followed by a fast bond rotation

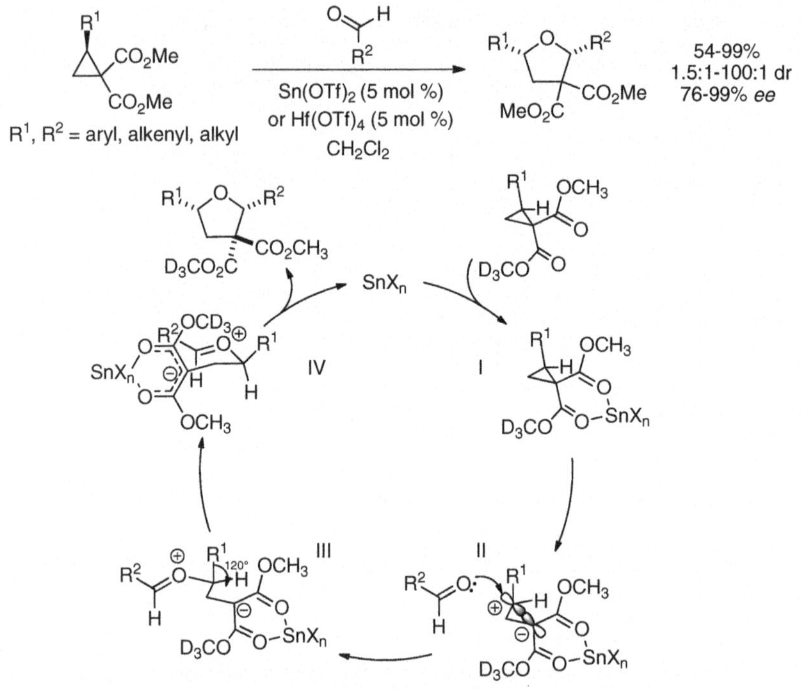

Scheme 4 Enantiospecific formal [3+2] cycloaddition reported by Johnson and co-workers

of 120°C would lead to the favored envelope conformation **IV**, in which all groups are in favorable pseudo-equatorial positions. Finally, C–C bond formation would give the observed tetrahydrofuran. The proposed mechanism was further confirmed by the stereospecificity observed when a deuterium label was introduced on one of the two ester groups of the cyclopropane.

With electron-rich aryl substituents, racemization of the starting material was observed. This result opened the way for the development of the first dynamic kinetic asymmetric formal [3+2] cycloaddition of aldehydes and cyclopropanes, using a magnesium PYBOX catalyst (Eq. 1) [26].

Equation 1. Dynamic kinetic asymmetric [3+2] formal cycloaddition

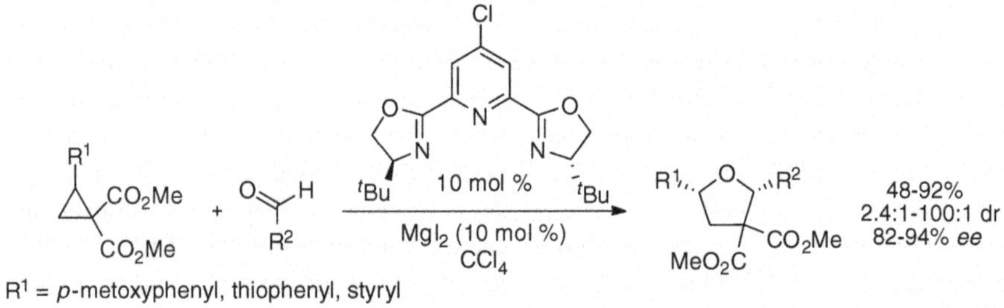

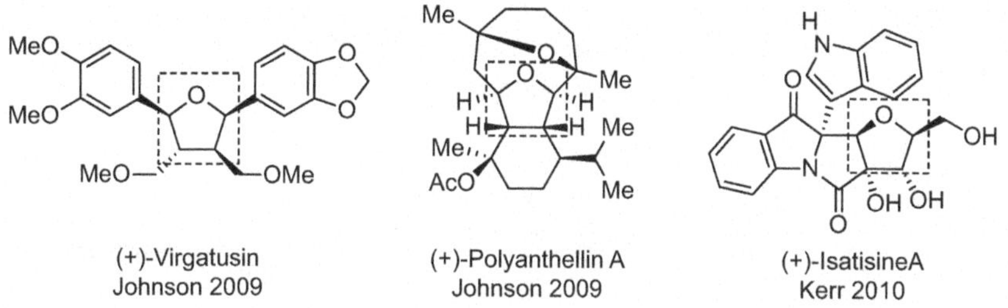

Fig. 1 Natural products synthesized via the intermolecular formal [3+2] cycloaddition of diester cyclopropanes and carbonyl compounds

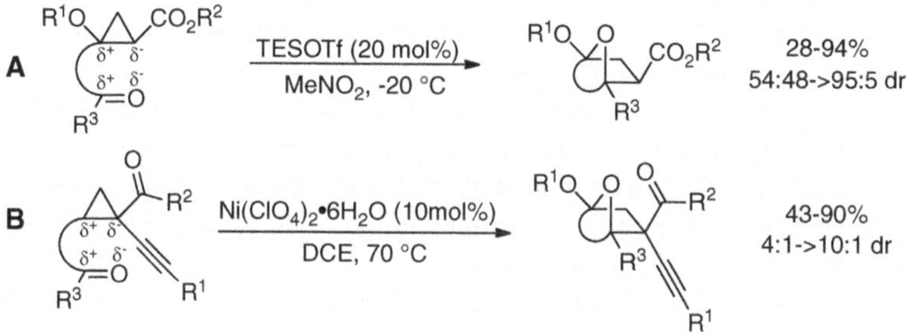

Scheme 5 Extending the scope of formal [3+2] cycloaddition reaction to oxy- and alkynyl-substituted cyclopropanes

Further recent extensions of this reaction include highly diastereoselective formal cycloadditions catalyzed by $AlCl_3$ [27] and the use of cyclopropanes bearing a quaternary donor site [28]. In the case of vinyl-substituted cyclopropanes, activation with a palladium catalyst became possible due to the formation of a stable π-allyl intermediate [29]. The broad applicability of the method was further demonstrated in the total synthesis of natural products, including (+)-virgatusin [30], (+)-polyanthellin A [31, 32], and (+)-isatisine A [33, 34] (Fig. 1). Finally, Wang and co-workers developed intramolecular variations of this reaction to give both fused and bridged polycyclic systems and applied the method to a formal synthesis of platensimycin [35].

The seminal work of Johnson and co-workers had enhanced tremendously the range of applications of formal [3+2] cycloadditions to access tetrahydrofurans. Nevertheless, it remains limited to the use of donor–acceptor cyclopropanes bearing an alkyl (aryl/alkenyl) group and diester substituents. Recently, Wang and co-workers reported two intramolecular approaches with other types of cyclopropanes: the first one involves oxycyclopropanes used in intramolecular cycloadditions to access bridged cyclopropanes (Scheme 5**A**) [36], whereas the other made use of ketone-substituted alkynyl cyclopropanes (Scheme 5**B**) [37]. In the latter case, more saturated furan derivatives could also be accessed via an alternative [4+2] annulation process if a gold catalyst was used.

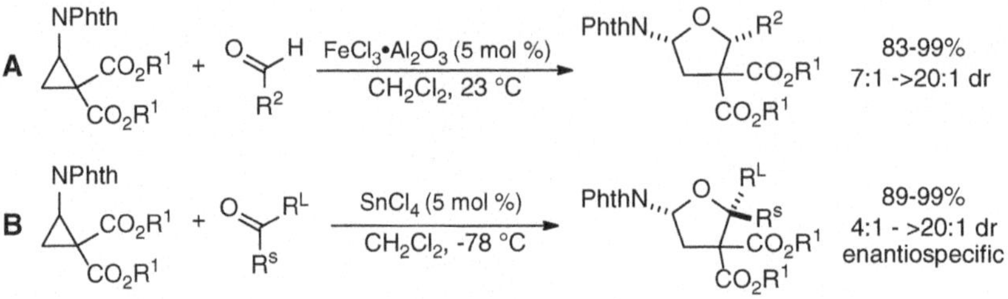

Scheme 6 Formal [3+2] cycloaddition of aminocyclopropanes

In 2012, Waser and co-workers reported the first use of amino-substituted cyclopropanes in the formal [3+2] cycloaddition with both aldehydes and ketones (Scheme 6). The reaction with aldehydes proceeded at room temperature with an iron catalyst and gave racemic products when starting from enantiopure cyclopropanes (**A**) [38]. In contrast, the tin-catalyzed annulation with ketones was enantiospecific (**B**) [39]. The obtained amino-substituted tetrahydrofurans are important heterocycles, as they constitute the core of natural DNA and RNA, as well as numerous synthetic drugs.

Apart from the introduction of polarizing group, the introduction of an exo double bond is another important approach to increase the reactivity of cyclopropanes [9]. In fact Nakamura and Yamago already demonstrated in 1990 that the formal cycloaddition of methylene cyclopropane acetal with aldehydes and ketones occurred spontaneously upon heating to 80°C (Eq. 2) [40]. A trimethylenemethane intermediate can be proposed for this reaction, leading to a true cycloaddition after ring opening has occurred.

Equation 2. Thermal formal [3+2] cycloaddition of methylenecyclopropane

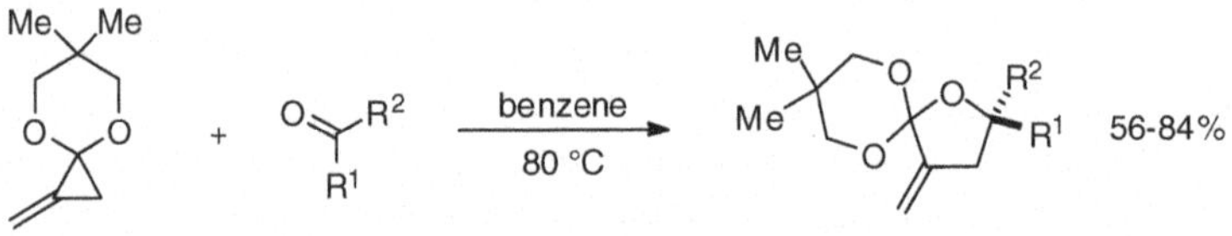

The use of a palladium catalyst allowed Yamamoto and co-workers to extend the scope of cycloaddition reactions between alkylidene cyclopropanes and aldehydes (Scheme 7) [41]. The reaction has been proposed to proceed via oxidative addition of Pd(0) onto the C–C bond of the cyclopropane to form a palladium-stabilized trimethylenemethane intermediate **II**. Nucleophilic addition onto the aldehyde to give a π-allyl intermediate **III** followed by reductive elimination then regenerates the catalyst. When compared to other precursors of trimethylenemethane in catalysis [42], alkylidene cyclopropanes are perfectly atom economical, but still require relatively high temperature to react.

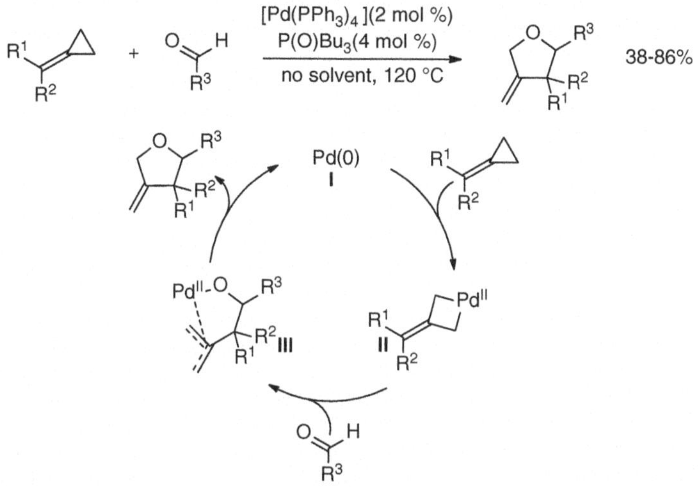

Scheme 7 Palladium-catalyzed formal [3+2] cycloaddition of alkylidenecyclopropanes

Finally, the use of Lewis acid to promote the cycloaddition between alkylidenecyclopropanes and carbonyl compounds has also been reported, but most reactions remain limited in scope or lead to mixture of products [43–46].

C–N Bond Formation

In a similar way as tetrahydrofurans are obtained via the [3+2] cycloaddition of cyclopropanes and carbonyls, pyrrolidines are generated from cyclopropanes and imines. It is consequently not surprising that many methods established in the case of carbonyls were later extended to imines.

Nevertheless, one of the first examples of the synthesis of pyrrolidines derived from oxindoles was developed by Carreira and co-workers based on a unique stepwise mechanism (Scheme 8) [47]. In this reaction catalyzed by MgI_2, nucleophilic attack by iodide was proposed as the first step. The generated enolate **I** would then add onto the imine, followed by cyclization via an SN^2 process. The broad potential of the method was further demonstrated in the total synthesis of spiroxindole alkaloids, including horsfiline [48], strychnofoline [49, 50], and spyrotryprostatin B [51, 52], as well as in the production of small molecule libraries with a pyrrolidine core [53–60].

Kerr and co-workers were the first to apply the principle of diester activation for the intramolecular annulation between imines and cyclopropanes (Scheme 9) [61]. The reaction proceeded in one-pot from the hydroxylamine derivatives and was catalyzed by $Yb(OTf)_3$. Interestingly, the formation of the *cis* or *trans* diastereoisomer depended on the order of addition of catalyst or aldehyde. Kerr and co-workers proposed that in the absence of aldehyde, nucleophilic attack of the nitrogen on the cyclopropane was the first step, followed by condensation with the

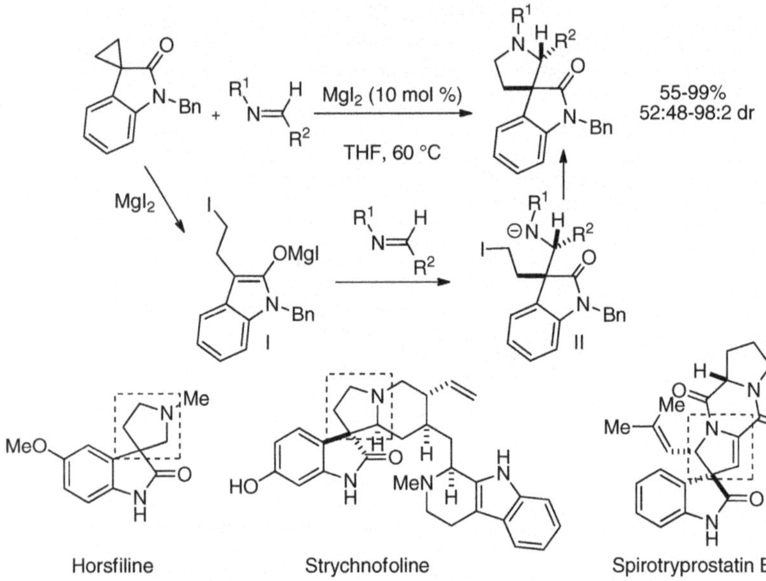

Scheme 8 MgI_2-catalyzed [3+2] annulation of cyclopropanes with imines and application in the total synthesis of spiroxindole alkaloids

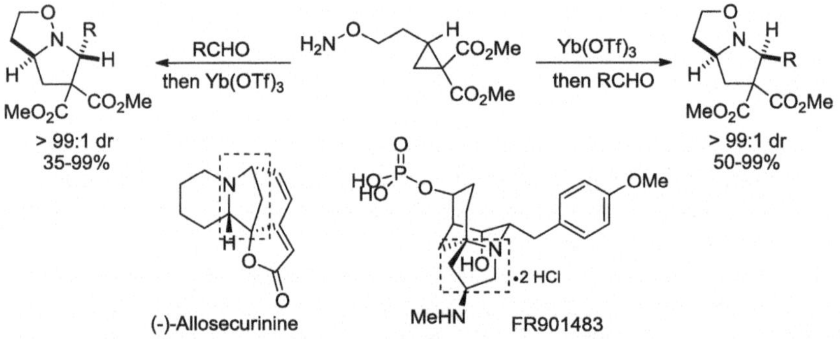

Scheme 9 Intramolecular [3+2] annulation of cyclopropanes with oximes and applications in the total synthesis of alkaloids

aldehyde and ring closing. When the aldehyde was added first, the formation of the oxime would occur initially, followed by attack on the cyclopropane, resulting in an inversion of the diastereoselectivity.

More recently, the methodology was also extended to the synthesis of bridged systems [62] and of bicyclopyrazolidines starting from hydrazines [63]. As the N–O or N–N bond is easily cleaved in the obtained products, they are easily further functionalized, as has been demonstrated by Kerr and co-workers in the total synthesis of (+)-allosecurinine [64] and FR901483 [65]. In 2010, Tomilov and co-workers have also reported a first example of intermolecular reaction between aryl-diester-substituted cyclopropanes and pyrazolines [66].

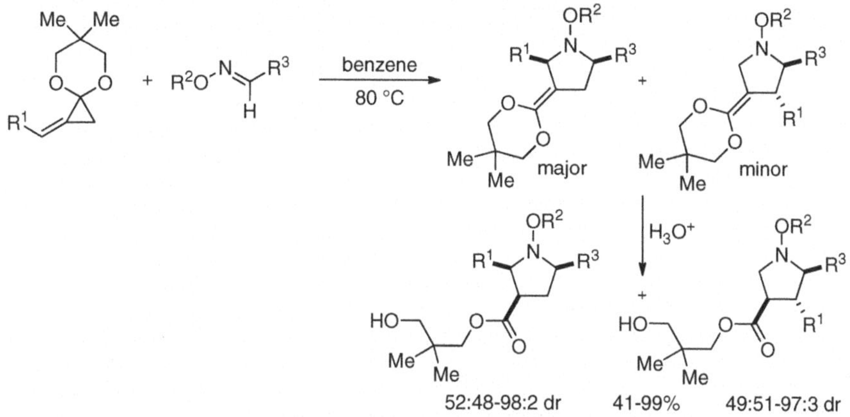

Scheme 10 Cycloaddition of alkylidenecyclopropanes and oximes under thermal conditions

In 2010, Johnson and co-workers reported that the dynamic kinetic asymmetric formal cycloaddition they have developed for aldehydes could also be used in the case of imines (Eq. 3) [67]. In this case, the choice of protecting group was key to obtain good asymmetric induction, diastereoselectivity, and yield in the reaction.

Equation 3. Dynamic kinetic asymmetric [3+2] formal cycloaddition

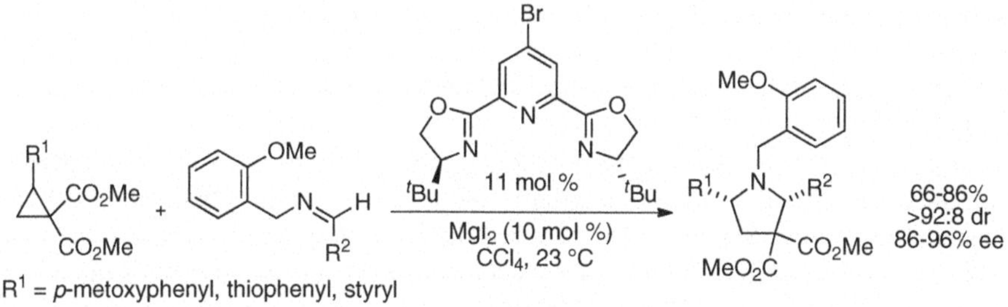

Like in the case of cycloadditions with carbonyls, alkylidenecyclopropanes have also been used for the reaction with imines. Nakamura and co-workers were again able to use alkylidenecyclopropane acetals for a thermal cycloaddition with oximes (Scheme 10) [68]. The reaction was proposed to proceed via a concerted cycloaddition of a trimethylenemethane singlet intermediate after cyclopropane opening. Interestingly, cycloaddition occurred on the two less substituted carbon atoms of the trimethylenemethane in contrast to the result with carbonyl compounds. The obtained keteneacetal can be easily hydrolyzed to the corresponding ester to give trisubstituted pyrrolidines. Later, the method could also be extended to sulfonyl and acyl imines as substrates [69]. As in the case of furans, the use of a palladium catalyst allowed Yamamoto and co-workers to significantly expand the scope of alkylidenecyclopropanes used in cycloaddition reactions with imines [70, 71]. More recently,

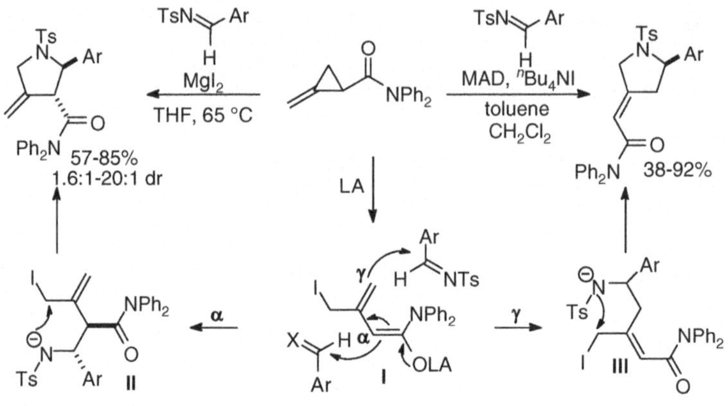

Scheme 11 MgI_2-catalyzed formal cycloaddition and other annulation reaction of methylidenecyclopropanes

Shi and co-workers have reported that non-activated alkylidenecyclopropanes could react thermally with imines in an intramolecular reaction [72].

An important progress in the use of alkylidenecyclopropanes activated by an electron-withdrawing group was realized by Lautens and co-workers using cooperative iodide–Lewis acid catalysis (Scheme 11) [73]. The use of MgI_2 led to the formation of the formal [3+2] cycloaddition products. The reaction probably proceeds via ring opening of the Lewis acid-activated cyclopropane by the iodide, followed by addition of the formed enolate **I** to the imine to give **II** and finally intramolecular SN^2 reaction leading to the pyrrolidine. Interestingly, the use of the bulky MAD Lewis acid led to the attack of the γ position instead and the formation of a different product via **III** [74].

In order to access enantiopure products, Lautens and co-workers subsequently introduced a chiral sulfoxide auxiliary on the imine and obtained excellent diastereoselectivity (Scheme 12**A**) [75, 76]. In 2007, they finally reported the first example of catalytic asymmetric formal cycloaddition using a chiral BOX ligand on the magnesium catalyst (Scheme 12**B**) [77].

Formal [3+2] Cycloadditions with Cumulenes

Formal cycloadditions with cumulenes, especially CO_2, are very important reactions with small heterocyclic substrates like epoxides or aziridines (vide infra). In contrast, these reactions have been only rarely studied with cyclopropanes, although the palladium-catalyzed reaction of alkylidenecyclopropanes with CO_2 was reported initially in 1979 by Inoue and co-workers (Eq. 4) [78]. In 2011, Shi and co-workers studied this transformation in greater detail and were able to significantly increase its scope [79]. Nevertheless, controlling the regiochemistry of the addition still remains a major challenge for this transformation.

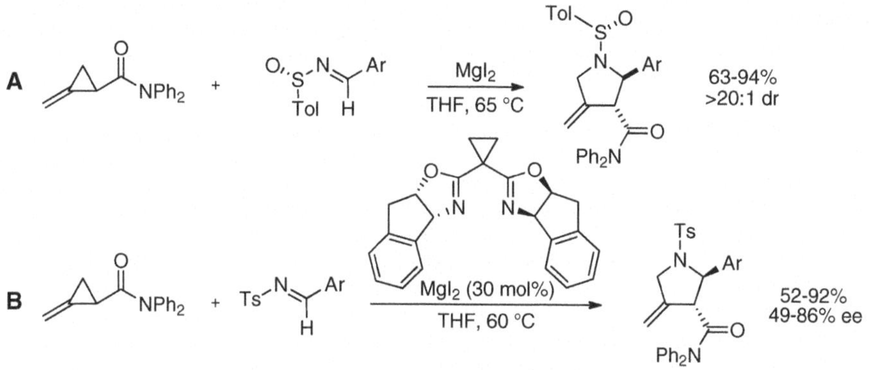

Scheme 12 Asymmetric approaches for the formal cycloaddition between methylidenecyclopropanes and imines

Equation 4. Formal [3+2] cycloaddition of alkylidenecyclopropanes with CO_2

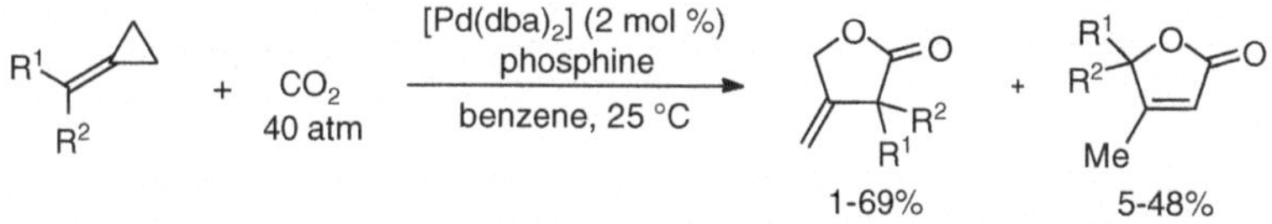

Until 2012, examples of cycloadditions of cyclopropanes with other heterocumulenes were rare, with single examples reported with carbon disulfide [80], phenylisocyanate [81], phenylisothiocyanate [82], diazenes [83, 84], and a special [3 +1+1] process involving isonitriles [85]. In 2012, Li and co-workers reported first the iron-mediated formal cycloaddition of aryl- and vinyl-cyclopropane diesters with isothiocyanates (Scheme 13**A**) [86]. In this work, the products were suggested to be thiolactams. However, Stoltz and co-workers reported shortly afterwards that the obtained product were more probably thioimidates, which were in their case obtained via the same transformation, but using a tin(II) catalyst (Scheme 13**B**) [87]. Stoltz and co-workers also reported the first cycloaddition reactions of carbodiimides and isocyanates to give amidines and imidates respectively.

2.1.2 Formal [3+*n*] Cycloadditions

Formal cycloaddition of cyclopropanes with larger partners have been much less investigated. Most research has focused on the [3+3] cycloaddition of donor–acceptor-substituted cyclopropanes with nitrones. The seminal studies on this reaction were reported by Kerr and co-workers in 2003, and the reaction was first called a homo [3+2] cycloaddition reaction (Scheme 14) [88–94]. The reaction proceeded with good yield and stereoselectivity to give 1,2-tetrahydrooxazines. The obtained heterocycles are interesting, as they are found at the core of natural products,

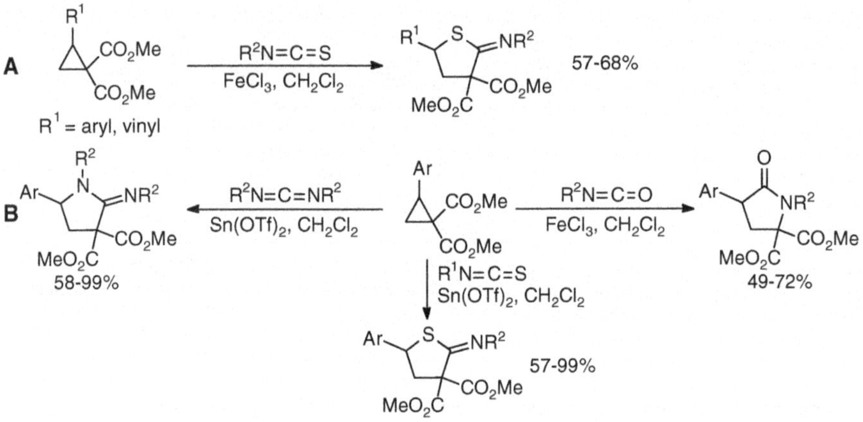

Scheme 13 Formal [3+2] cycloadditions with isothiocyanates, carbodiimides, and isocyanates

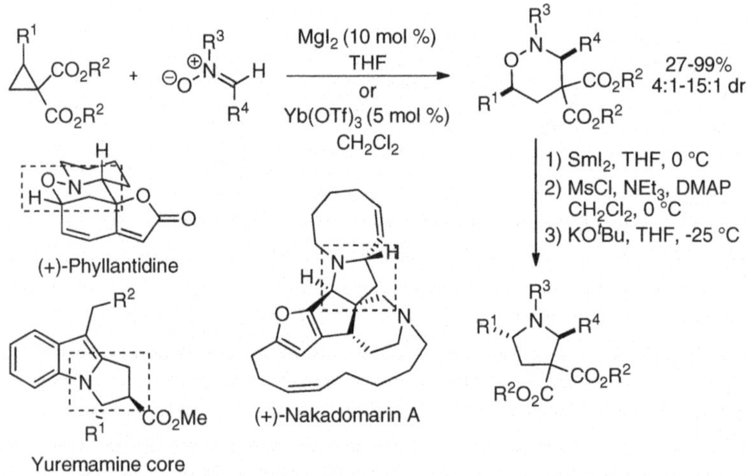

Scheme 14 Formal [3+3] cycloadditions of cyclopropanes with nitrones and applications in the total synthesis of natural alkaloids

such as phyllantidine, which was synthesized by Kerr using this methodology in 2006 [95]. Furthermore, the N–O bond can be easily reduced with samarium iodide. After activation of the alcohol and intramolecular nucleophilic substitution, ring-contracted pyrrolidines are obtained, which led to an alternative strategy to the direct [3+2] formal cycloaddition between cyclopropanes and imines discussed previously. This approach was successfully applied in an impressive synthesis of the alkaloid nakadomarin A [96, 97] and the core of the natural product yuremamine [98]. Interesting further extensions of the methodology include the use of cobalt complexes of alkynyl cyclopropanes diesters as a new approach for donor–acceptor activation of the three-membered ring [99], the use of nitrones derived from isatin to obtain important spiroxindole products [100], and the use of cyclic nitrones as substrates [101].

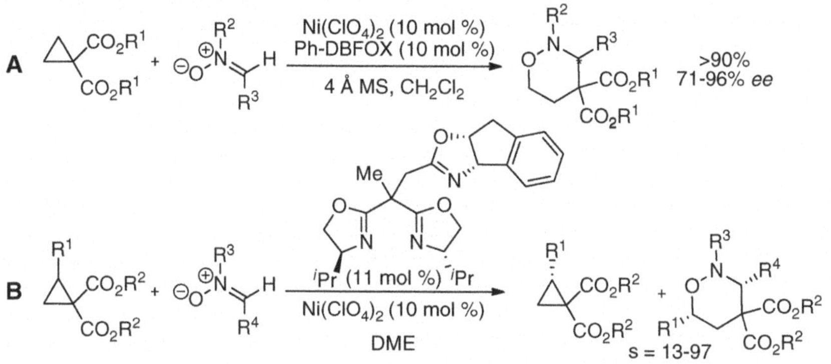

Scheme 15 Asymmetric formal [3+3] cycloadditions of cyclopropanes with nitrones

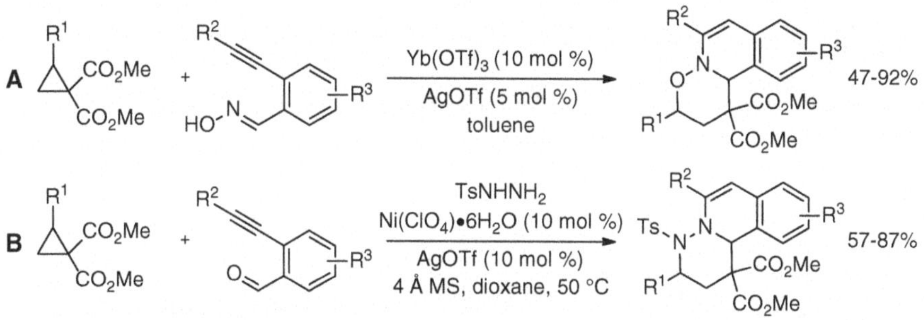

Scheme 16 Domino-cyclization cycloaddition from alkynes

An important breakthrough for the further development of formal [3+3] cycloaddition of cyclopropanes and nitrones was the discovery of a catalytic asymmetric variation of the reaction (Scheme 15). Using a Nickel-DBFOX catalyst, Sibi and co-workers were able to develop in 2005 an enantioselective cycloaddition using unsubstituted cyclopropanes (**A**) [102]. In 2007, Tang and co-workers reported a kinetic resolution of substituted cyclopropanes using a C1-symmetric modified BOX ligand on the nickel catalyst (**B**) [103].

In 2008, Charette and co-workers further demonstrated that the [3+3] cycloaddition between azomethine imines and donor–acceptor cyclopropanes was also possible (Eq. 5) [104]. In 2013, Tang and co-workers developed a highly enantioselective variation of this reaction using a C1-symmetric modified BOX ligand on the nickel [105]. Wu and co-workers developed domino reactions in which the nitrone [106] or the azomethine imine [107] is generated in situ by addition of a nucleophile on a triple bond (Scheme 16**A, B**). In the case of the azomethine imine, a three component reaction starting directly from an alkynyl aldehyde, a hydrazine, and the cyclopropane was possible.

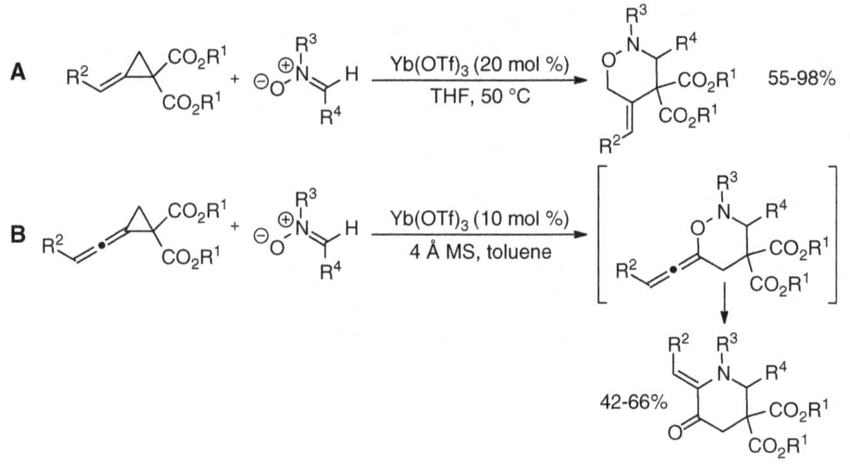

Scheme 17 [3+3] Cycloaddition with alkylidene- and vinylidenecyclopropanes

Equation 5. Formal [3+3] cycloadditions of cyclopropanes and azomethine imines

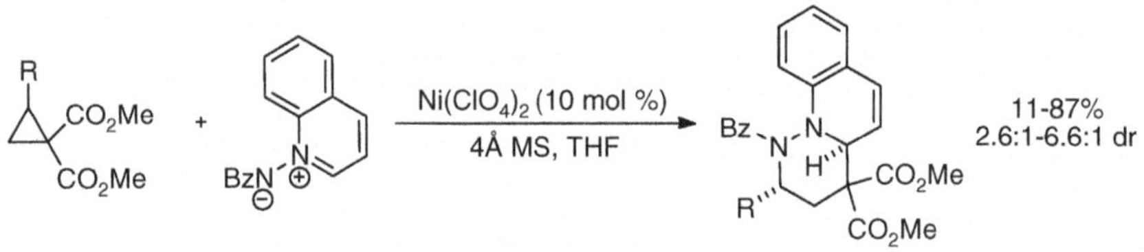

Finally, two recent examples make use of cyclopropanes bearing a further exo double bond (Scheme 17): Wang and co-workers reported the first use of alkylidenecyclopropane diesters in the formal cycloaddition with nitrones in 2009 (**A**) [108]. In 2010, Wu and Shi reported that the reaction with vinylidenecyclopropane diesters proceeded with different regiochemistry (**B**) [109]. The obtained allenes were unstable and rearranged to form the ketones.

In addition to [3+3] formal cycloadditions, there are few examples of reactions with larger partners, but they usually lead to more saturated heterocycles [110].

2.2 *Reactions with Epoxides*

In contrast to cyclopropanes, for which the most frequent reactions have been with isolated π systems such as carbonyls and imines, the chemistry of epoxides and aziridines is dominated by formal cycloadditions with CO and CO_2. These reactions are very important for the synthesis of heterocycles, and they would require a dedicated chapter to be described in detail. As this chemistry has already been

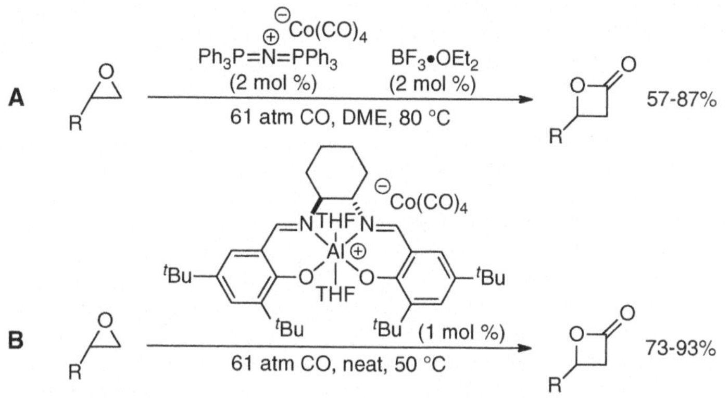

Scheme 18 Formal [3+1] cycloaddition of epoxides and carbon monoxide

described in several reviews [111–115], we will present only a few seminal studies and concentrate more on other transformations, which have been less in the focus of attention.

2.2.1 Formal [3+1] Cycloadditions

The formal cycloaddition of epoxides with carbon monoxide is an important reaction for the synthesis of β-lactones. One of the major challenges associated with this process is to prevent subsequent polymerization of the formed lactones. Except for scattered publications and patents describing this transformation in low yield, the first truly efficient protocol was reported by Alper and co-workers in 2001 (Scheme 18**A**) [116]. Key for success was the use of a zwitterionic cobalt catalyst and a Lewis acid as a co-catalyst. Coates and co-workers later developed a more efficient catalyst, in which the cation of the zwitterionic cobalt catalyst is itself a Lewis acid (Scheme 18**B**) [117, 118]. Best results were initially obtained with an aluminum salen complex, but later other Lewis acids were found to be even more efficient [119–121].

Following the discovery of the carbonylation reaction, intensive mechanistic studies have given a deeper insight in the catalytic cycle (Scheme 19) [122, 123]. The reaction is initiated by dissociation of a weakly bound ligand from aluminum to generate Lewis acidic complex **I**. Activation of the epoxide (**II**) is followed by nucleophilic attack of cobalt to give five-coordinated aluminum alkoxide complex **III**. Insertion of CO into the C–Co bond then gives intermediate **IV**, which reacts with CO to give complex **V**. The subsequent four-membered ring formation from **V** to give **VI** has been proposed to be rate limiting, and the intermediacy of **V** was supported by IR spectroscopy and kinetic studies. Finally, release of the product regenerates the active catalyst **I**. In accordance with the proposed mechanism, the reaction proceeded with high stereocontrol, and *cis* lactones were obtained starting from *trans* epoxides.

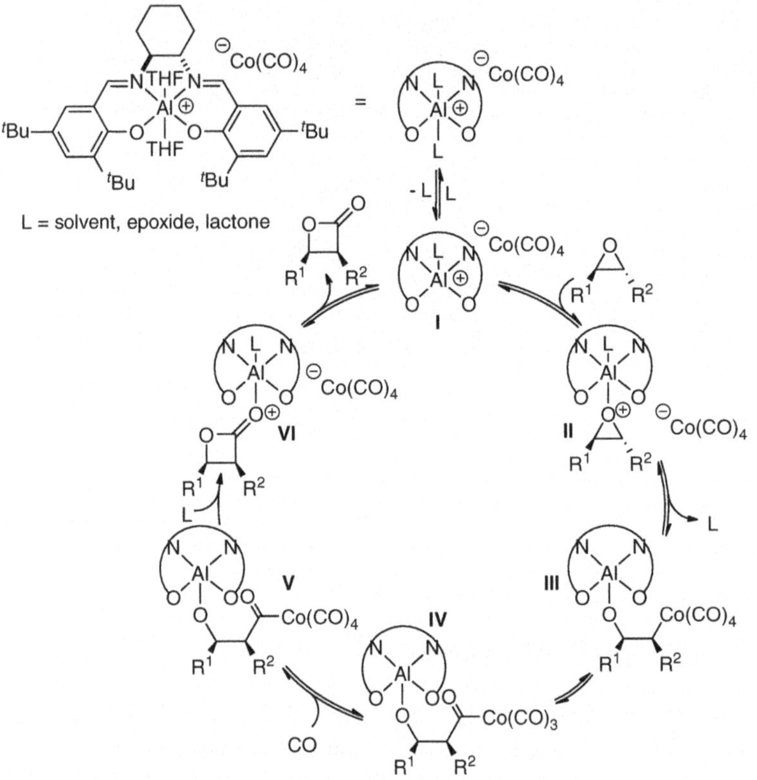

Scheme 19 Proposed catalytic cycle for the carbonylation of epoxides

Interesting recent extensions of the reaction are the use of alkylidenecyclopropanes as substrates [124], the synthesis of anhydrides via a double carbonyl insertion process [125], and the first example of carbonylative desymmetrization of meso-epoxides using a chiral chromium Lewis acid [126].

2.2.2 Formal [3+2] Cycloadditions with Isolated π Systems

In principle, the reaction of epoxides with two-carbon π systems can occur either via C–C bond cleavage or via C–O bond cleavage. In contrast to cyclopropanes, a lone pair is available on the oxygen of the epoxide and allows a concerted ring opening to give a carbonyl ylide intermediate, which can then undergo a concerted [3+2] cycloaddition with olefins or carbonyl compounds. In fact, the thermal or photochemical ring opening of epoxides was one of the first methods used to generate carbonyl ylides for cycloaddition reactions [127]. Nevertheless, ring opening occurs under relatively mild conditions only with specific substituents, especially cyano and aryl groups. Probably

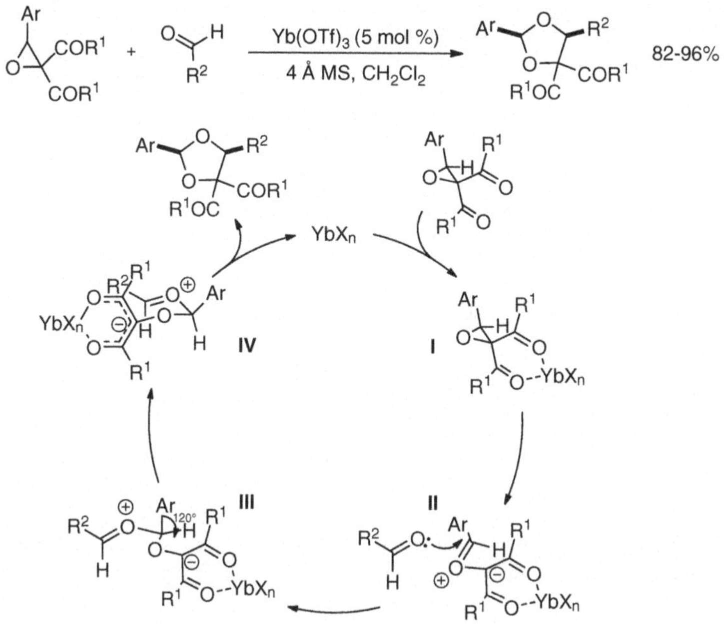

Scheme 20 Formal [3+2] cycloaddition of epoxides and aldehydes and proposed reaction mechanism

for this reason, other methods to generate carbonyl ylides are nowadays favored. As these reactions can be considered as "true cycloadditions," they will not be discussed here.

Surprisingly, Lewis acid activation of epoxides for (formal) cycloaddition reactions has not been investigated in detail until the work of Zhang and co-workers in 2011 (Scheme 20) [128]. Inspired by the successful design in the field of donor–acceptor cyclopropanes, they discovered that Lewis acid activation of diester-substituted epoxides was possible to give dioxolanes with excellent diastereoselectivity after cycloaddition with aldehydes. In contrast to what has been observed with cyclopropanes, racemization of the starting material was observed, indicating a probable carbonyl ylide intermediate **II**. As the reaction was accelerated with electron-rich aldehydes, Zhang and co-workers then proposed a stepwise process via intermediates **III** and **IV** to finally give the dioxolane. Using the same activation principles, Zhang and co-workers also developed a [3+2] formal cycloaddition with alkynes [129] and a [4+3] annulation between nitrones and alkynyl-substituted epoxides [130], but these reactions gave access to more saturated heterocycles. Finally, they reported in 2012 the formal [3+2] cycloaddition of cyclopropanes with indoles (Eq. 6) [131].

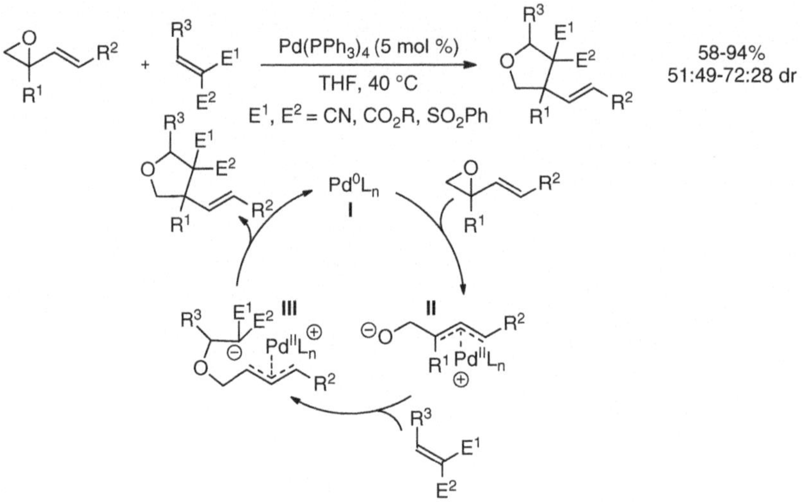

Scheme 21 Palladium-catalyzed formal [3+2] cycloaddition of vinyl epoxides and electron-poor olefins and proposed reaction mechanism

Equation 6. Formal [3+2] cycloaddition of epoxides and indoles

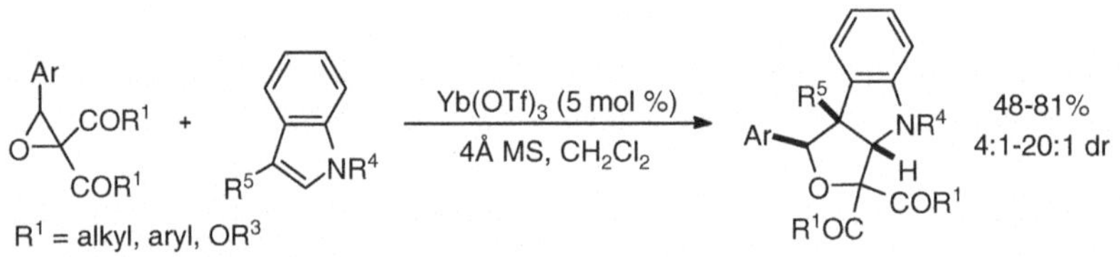

In principle, reactions proceeding via C–C cleavage and carbonyl ylides can be concerted cycloadditions. On the other hand, reactions involving C–O cleavage does not allow a continuous overlap of orbitals and are thus clearly formal cycloadditions. A first approach was developed in the special case of vinyl epoxides: based on the well-established access to palladium-π-allyl complex from vinyl epoxides (vide infra), Shim and Yamamoto reported in 1998 the formal [3+2] cycloaddition of this class of substrates with electron-poor olefins (Scheme 21) [132]. The reaction proceeded in good yield, but with low diastereoselectivity. The first step in the catalytic cycle was proposed to be the formation of the palladium π-allyl intermediate **II**. Michael addition of the alkoxide to give **III** followed by reductive elimination will then give the observed product and regenerate the Pd(0) catalyst **I**. In 1999, they then extended the methodology to the synthesis of oxazolidine by formal cycloaddition of vinyl epoxides and tosyl imines (Eq. 7) [133, 134].

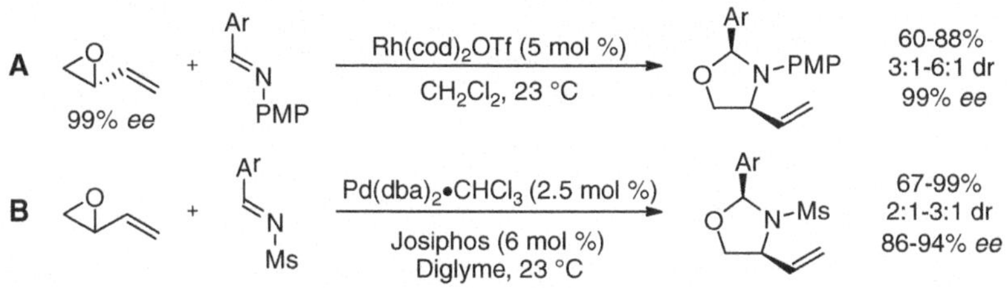

Scheme 22 Enantiospecific and enantioselective formal [3+2] cycloaddition of vinyl epoxides and imines

Equation 7. Palladium-catalyzed formal [3+2] cycloaddition of vinyl epoxides and imines

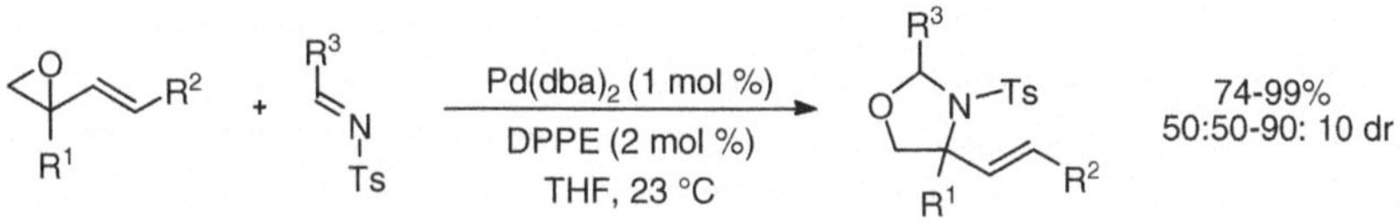

In 2009, Jarvo and co-workers developed the first asymmetric version of the reaction between vinyl epoxides and imines (Scheme 22) [135]. When using a rhodium catalyst, the reaction was enantiospecific. The retention of the stereochemistry indicated a mechanism involving double inversion. In contrast, when using a palladium catalyst, a dynamic kinetic asymmetric transformation (DYKAT) was possible, and enantioenriched products could be obtained from a racemic mixture. Both methods gave the oxazolidine in good enantiopurity, but only moderate diastereoselectivity. In 2011, Matsubara and co-workers reported that a nickel catalyst could also be used for the formal cycloaddition of vinyl epoxides and unsaturated ketones [136]. Finally, Hou and co-workers reported the first examples of palladium-catalyzed cycloaddition of nitro olefins and vinyl epoxides, which proceeded with up to 72% ee [137].

The main limitation of the palladium-based methods is the requirement for a π-allyl intermediate. In principle, a simple Lewis acid activation would have less limitation. Nevertheless, there are only two reports of Lewis acid-catalyzed formal [3+2] cycloaddition of epoxides with two-carbon π systems: Su and co-workers first reported the ytterbium-catalyzed cycloaddition of imines and epoxides to give oxazolidines in 2007 (Scheme 23**A**) [138]. In 2012, Zhang and co-workers studies the ring opening of diester-substituted epoxides more in detail and found out that the reaction could proceed either via C–C or C–O cleavage depending on the catalyst (Scheme 23**B**) [139]. With a nickel catalyst, C–C cleavage was observed, and the products were obtained in good yield and *cis* stereoselectivity. In contrast, C–O cleavage was favored in the presence of a tin(II) catalyst and the reaction proceeded with lower diastereoselectivity. The origin of the regioselectivity was rationalized based on calculation: the nickel catalyst favored chelation of the two ester carbonyl group, leading to C–C bond activation. In contrast, the tin catalyst is

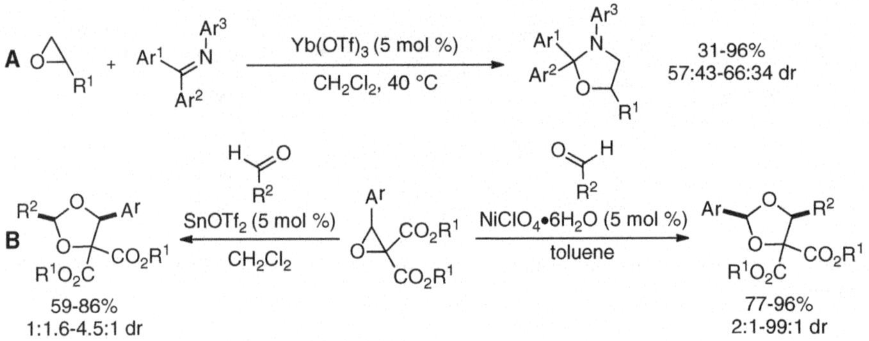

Scheme 23 Lewis acid-catalyzed formal [3+2] cycloaddition of vinyl epoxides with imines and aldehydes

bound preferentially to the oxygens of one carbonyl group and the epoxide, leading to C–O bond cleavage. Finally, a last approach was reported by Liu and co-workers in 2004 based on the oxidation of chalcone epoxides with aminium cations [140]. The obtained radical cation intermediate is very reactive and can be used in cycloaddition reactions with non-activated or electron-rich olefins.

2.2.3 Formal [3+2] Cycloadditions with Cumulenes

The most important cycloaddition of epoxide with cumulenes is by far the reaction with carbon dioxide (Eq. 8). The obtained carbonates can be easily polymerized to give polycarbonates, which are an important class of polymer. With the right catalyst, the polymer can also be obtained directly. More than 100 publications have been focused on this reaction, and a full description of this work goes far beyond the scope of this section. Fortunately, several recent reviews have been dedicated to this transformation [141–144]. One of the most successful classes of catalysts are cobalt, chromium, and aluminum salen complexes, which have also allowed the development of asymmetric variations of the reaction [145].

Equation 8. Formal [3+2] cycloaddition of epoxides and carbon dioxide

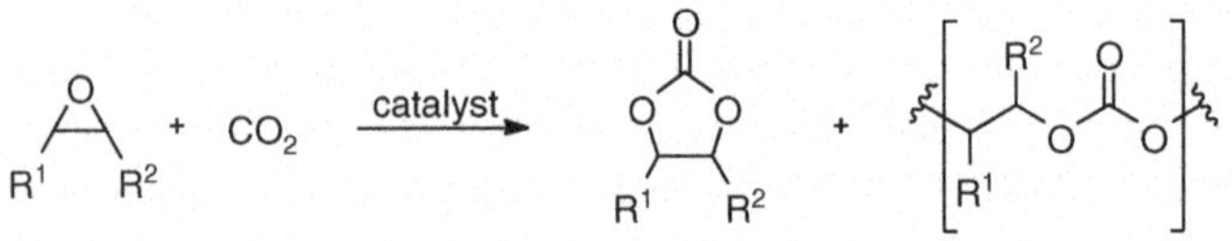

The reaction with cumulenes is not limited to CO_2. In particular, isocyanates, isothiocyanates, and carbodiimides react with epoxides to give the corresponding five-membered heterocycles. One of the main challenges in this transformation is to control the regioselectivity of the formal cycloaddition. Earlier work in this field

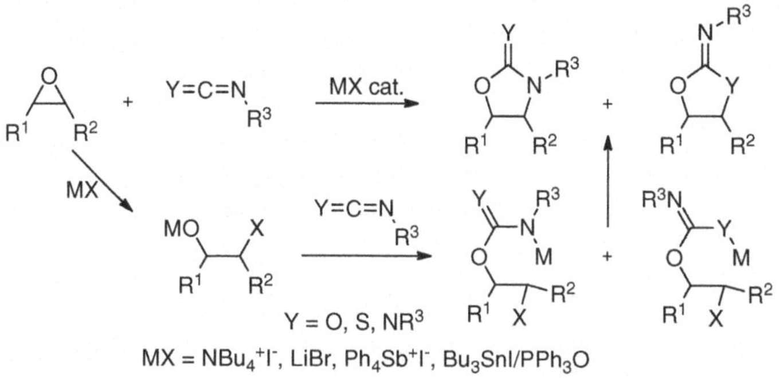

Scheme 24 Halide salt-catalyzed formal cycloaddition of epoxides and cumulenes

focused on the use of halide salts, such as tetrabutylammonium iodide [146], lithium bromide [147], tetraphenylstibonium iodide [148–150], or tributyltin iodide–Lewis base complexes [151, 152]. The halide has been proposed to play a key role for nucleophilic ring opening of the epoxide (Scheme 24). The formed alkoxide can then add on the cumulene and a SN^2 ring closure finally gives the heterocycles. More recently, the methodology has been used in the synthesis of libraries of bioactive compounds [153, 154], and the first example involving isoselenocyanates has been reported [155].

The only attempt of asymmetric induction using a ytterbium-Pybox catalyst was reported by Barros and Phillips in 2010 (Eq. 9) [156]. However, only moderate enantioselectivity was obtained and the yield was low due to the formation of regioisomers and chlorohydrin side products.

Equation 9. Enantioselective [3+2] cycloaddition of epoxides and isocyanates

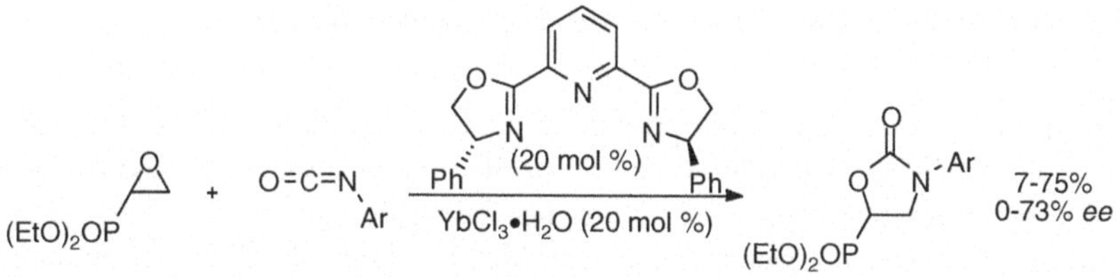

Like in the case of formal cycloaddition with two-carbon π systems, a successful solution to the challenge of regio- and stereoselectivity was found in the use of palladium catalysts with vinyl epoxides. In fact, the first reaction of this type was reported by Trost and Sudhakar with isocyanates in 1987 (Scheme 25**A**) [157]. Interestingly, the reaction was stereospecific when tosyl isocyanate was used, but became stereoconvergent with the use of isocyanates bearing a bulky aryl group [158, 159]. In this case, high *cis* stereoselectivity was observed regardless of the configuration of the epoxide. Isomerization of the π-allyl intermediate was proposed to rationalize this result. In 1997, Larksarp and Alper reported the first enantioselective variation of the

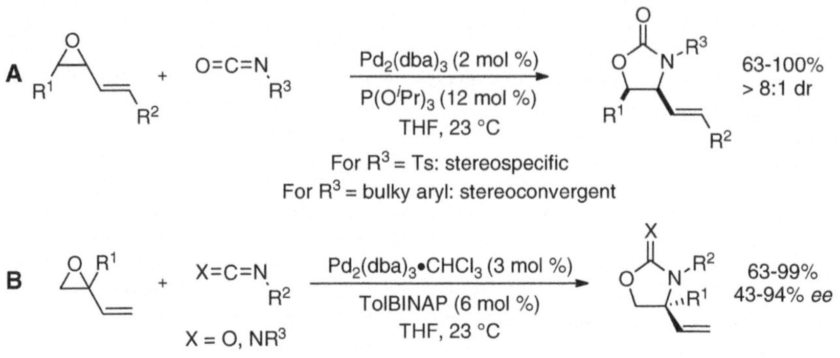

Scheme 25 Pd-catalyzed formal [3+2] cycloaddition of vinyl epoxides and cumulenes

method using TolBINAP as a ligand (Scheme 25**B**) [160, 161]. This reaction gave high enantioselectivity for both isocyanates and carbodiimides as substrates.

In addition to formal cycloadditions involving cumulenes with two heteroatoms, Baba and co-workers have reported a single example of cycloaddition of ketenes with epoxides (Eq. 10) [162]. Depending on substrate structure and solvent, the tetraphenylstibonium iodide-catalyzed reaction proceeded in high yield for the formation of either the γ-lactone or the ketene acetal product.

Equation 10. Formal [3+2] cycloaddition of epoxides and ketenes

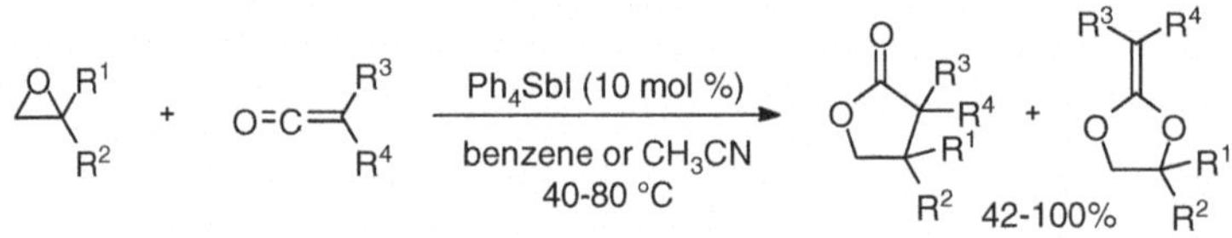

2.3 Reactions with Other Three-Membered Rings

Apart from epoxides and cyclopropanes, most investigations have focused on the use of aziridines, oxaziridines, and diaziridines substrates. As the use of the last two for the functionalization of olefins has been already discussed in Chap. 2 (Synthesis of Saturated Heterocycles via Metal-Catalyzed Alkene Diamination, Aminoalkoxylation, or Dialkoxylation Reactions) of this volume, the discussion will be here limited to aziridines. Not surprisingly, many parallels can be drawn with the reactions involving epoxides, and depending on the transformation, reports involving aziridines either inspired or take inspiration from similar work with epoxides.

2.3.1 Formal [3+1] Cycloadditions

The carbonylation of aziridines is an important method for the synthesis of β-lactams. The main research in this field was conducted by Alper and co-workers (Scheme 26). They first reported the rhodium-catalyzed carbonylation

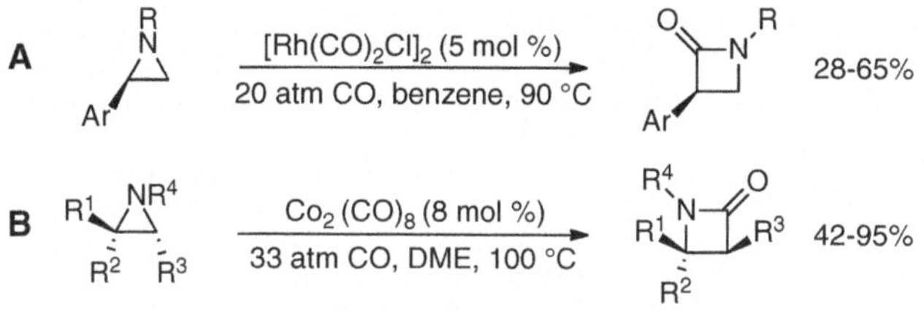

Scheme 26 Formal [3+1] cycloaddition of aziridines and carbon monoxide

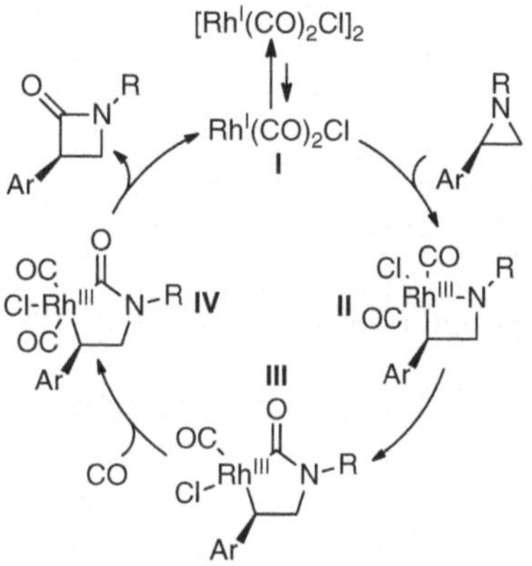

Scheme 27 Mechanism of the rhodium-catalyzed carbonylation of aziridines

of aziridines (Scheme 26**A**) [163–165]. The reaction was limited to aryl-substituted aziridines. High regioselectivity was observed for insertion in the benzylic C–N bond. Furthermore, the reaction proceeded with retention of the stereochemistry at the benzylic center. In 1996, they reported a cobalt-catalyzed carbonylation (Scheme 26**B**) [166]. The reaction was more general and proceeded this time with inversion of the stereochemistry and insertion in the less substituted double bond. This striking result can be explained by the different mechanism of the two reactions. Like for the carbonylation of epoxides (vide supra, Scheme 19), the reaction with cobalt most probably proceeds via nucleophilic attack of a cobaltate intermediate [167]. In the case of rhodium, oxidative insertion of **I** into the C–N bond occurs first to give **II** (Scheme 27). Hyperconjugation with an aromatic ring is essential for this step [168]. Carbonyl insertion followed by addition of carbon monoxide and reductive elimination then gives the lactam. Subsequently, the scope of the cobalt-catalyzed reaction was studied more in detail [169, 170]. Coates and co-workers also demonstrated that the Lewis acid cobaltate complex developed for epoxide carbonylation is also more efficient for aziridine carbonylation [118].

In addition to the most successful rhodium and cobalt catalysts, examples of carbonylation with stoichiometric nickel complexes were also reported [171, 172]. Finally, the use of palladium catalysis remains limited to methylene- [173] and vinyl- [174, 175, 223] substituted aziridines.

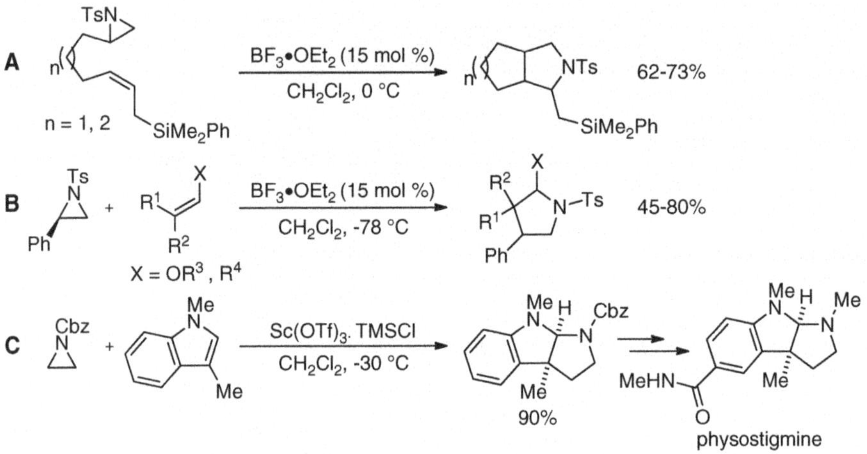

Scheme 28 Formal [3+2] cycloaddition of aziridines with olefins

2.3.2 Formal [3+2] Cycloadditions with Isolated π Systems

As for epoxides, the lone pair on nitrogen allows the thermal or photolytic opening of aziridines to generate an azomethine ylide. Again, this process is often limited to specific substituents on the aziridines and requires harsh reaction conditions. It can also be considered as a "true" cycloaddition and will therefore not be discussed in this section.

As in the case of cyclopropanes, the use of aziridines in formal [3+2] cycloadditions has increased tremendously during the last 15 years [176, 177]. The first breakthroughs were reported in 1999. Bergmeier and co-workers demonstrated that the intramolecular cycloaddition of tosyl aziridines and allyl silanes could be catalyzed by boron-trifluoride etherate (Scheme 28**A**) [178]. Also in 1999, Mann and co-workers used the same catalyst for the intermolecular cycloaddition of aryl-substituted aziridines with enol ethers [179], and later demonstrated that this system could also be applied to non-activated alkenes (Scheme 28**B**) [180]. In these early works, high diastereoselectivity could be achieved only in the case of the formation of bicyclic five–five ring systems. In 2000, Nakagawa and Kawahara then reported the scandium-catalyzed cycloaddition of unsubstituted Cbz protected aziridines with skatole and used the method in a formal synthesis of physostigmine (Scheme 28**C**) [181]. In 2001, Yadav and co-workers finally reported that scandium triflate was also an efficient catalyst for the reaction of aryl-substituted tosyl aziridines with enol ethers and allyl silanes [182].

In 2004, Johnson and co-workers then reported that diester-substituted *N*-aryl aziridines could be activated by Lewis acid for reaction with enol ethers (Scheme 29**A**) [183]. Due to the diester activation, the reaction now proceeds via C–C instead of C–N cleavage. For cyclic enol ethers, a [4+2] annulation process was observed, proceeding probably via a Friedel–Crafts reaction on the aryl ring.

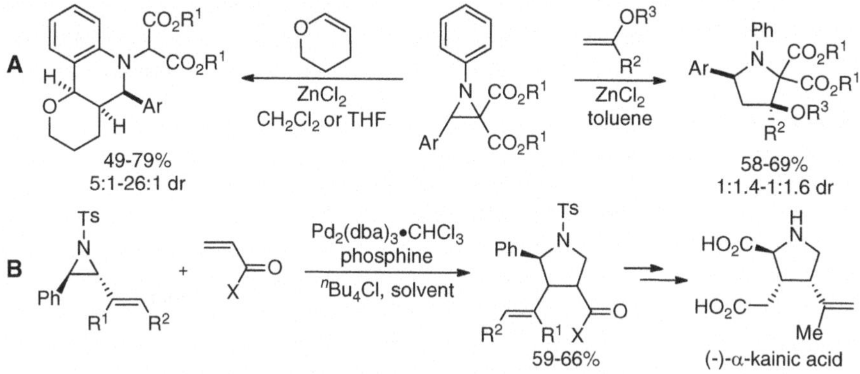

Scheme 29 The use of diester- and vinyl-substituted aziridines in formal cycloaddition reactions

The reaction could also be extended to norbornene as substrate. In the case of acyclic enol ethers, the formal [3+2] cycloaddition product was obtained with low diastereoselectivity. Although the reaction was usually performed with stoichiometric amount of zinc chloride as Lewis acid, two examples of reactions catalytic in zinc were also reported. In 2011, Zhang and co-workers demonstrated that the [3+2] cycloaddition product could be obtained for both cyclic and acyclic enol ethers when using tosyl aziridines and yttrium triflate as catalyst [184].

As has been seen for epoxide and cyclopropanes, the use of vinyl aziridines opened the way for π-allyl palladium chemistry. The first example of [3+2] cycloaddition with an isolated two-carbon π system was reported by Aggarwal and co-workers in 2011 (Scheme 29**B**) [185]. Depending on the reaction conditions and substrate structure, pyrrolidine products could be obtained with high diastereoselectivity. The synthetic utility of the method was further demonstrated in a formal total synthesis of the natural product (−)-α-kainic acid. Furthermore, Shipman and co-workers reported in 2012 the first example of intramolecular formal cycloaddition of methylene aziridines with alkenes [186].

The formal [3+2] cycloaddition of aziridines is not limited to olefins as partners. The reaction of aziridines with aldehydes and ketones was reported by Yadav and co-workers in 2004 using a silyl group to stabilize the carbocation obtained after C–N bond cleavage (Scheme 30**A**) [187]. Oxazolidine products were obtained in excellent yield, but moderate diastereoselectivity. In 2007, Singh and co-workers then studied the reaction of aryl-substituted tosyl aziridines with both carbonyls and imines (Scheme 30**B**) [188]. This reaction was possible using zinc triflate as catalyst and also proceeded via C–N bond cleavage. In 2011, Hanamoto and co-workers finally reported the formal [3+2] cycloaddition of trifluoromethyl-substituted tosyl aziridines with aldehydes (Scheme 30**C**) [189]. This reaction proceeded also with C–N cleavage, but with opposite regioselectivity and high diastereoselectivity. As the trifluoromethyl group is not able to stabilize a carbocation intermediate, the reaction starts most probably by a nucleophilic SN^2-like attack of the carbonyl on the less substituted carbon of the aziridine.

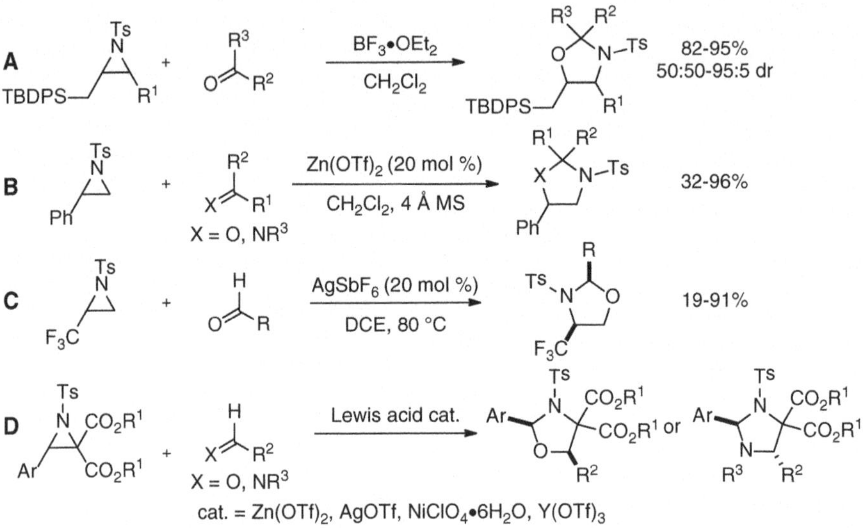

Scheme 30 Formal [3+2] cycloaddition of aziridines with carbonyls and imines

The use of diester-substituted aziridines allowed again cycloadditions involving C–C instead of C–N cleavage. Using different Lewis acids as catalysts, the groups of Zhang [190, 191] and Wang [192] reported the cycloadditions with both carbonyls and imines (Scheme 30**D**). Interestingly, good diastereoselectivity was observed for the formation of *cis*-oxazolidines and *trans*-imidazolidines.

2.3.3 Formal [3+2] Cycloadditions with Cumulenes

As in the case of epoxides, carbon dioxide is again the most attractive cumulene for reaction with aziridines. This reaction gives important oxazolidinones as products and constitutes an alternative to the reaction of epoxides with isocyanates. Even if early work demonstrated already in the 1970s and the 1980s that the cycloaddition of aziridines and CO_2 could be accelerated with halide salts [193, 194], progress has been much slower than in the case of epoxides, focusing mostly on technical improvements. Interesting recent results include the use of a chromium salen catalyst by Miller and Nguyen [195] and a palladium catalyst together with vinyl aziridines by Aggarwal and co-workers [196] (Scheme 31**A, B**). The former reaction gave excellent regioselectivity, whereas the latter reaction already proceeded at atmospheric pressure of carbon dioxide.

The reaction of aziridines with other cumulenes can also be catalyzed by halide salts [194]. In 1992, Baeg and Alper reported the first palladium-catalyzed formal cycloaddition of aziridines with carbodiimides [197] and later extended the protocol

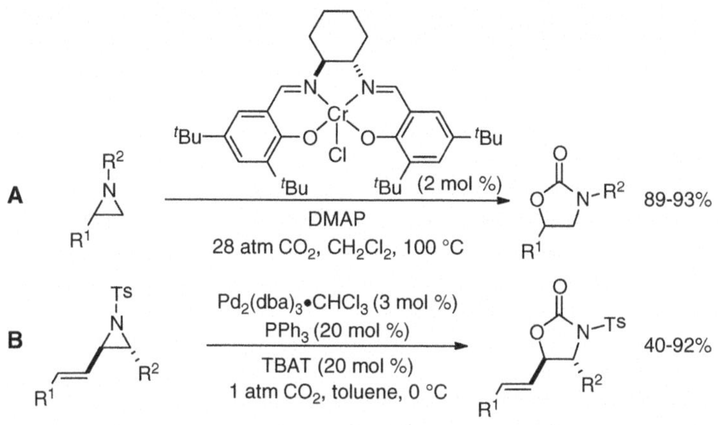

Scheme 31 Formal [3+2] cycloaddition of aziridines with carbon dioxide

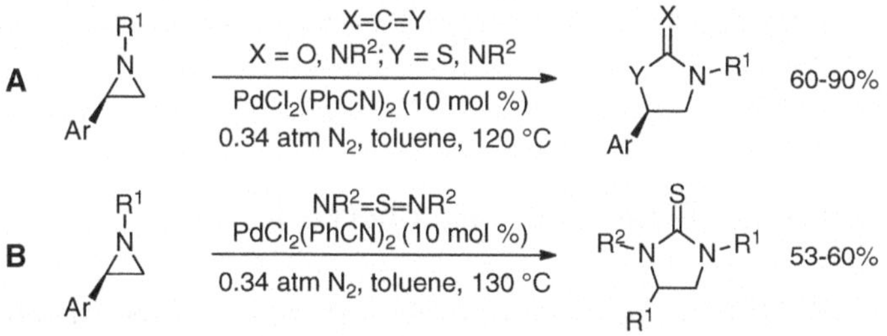

Scheme 32 Pd-catalyzed formal [3+2] cycloaddition of aziridines with cumulenes

to isocyanates and isothiocyanates (Scheme 32**A**) [198]. The regiochemistry of the cycloaddition was dependent of the cumulene structure. The reaction was enantiospecific. When sulfurdiimides were used, a thiourea was obtained instead of the expected product (Scheme 32**B**) [199]. Although the mechanism of this transformation was not yet fully elucidated, a labeling experiment showed that the extra carbon atom originated from the methylene group of the aziridine.

The use of vinyl aziridines together with a palladium catalyst allowed cycloaddition with cumulenes under milder conditions. Such a process was first reported by Alper and co-workers in 2000 [200]. In 2003, Trost and Fandrick reported an asymmetric variation of the cycloaddition of vinyl aziridines and isocyanates using the bis(phosphine) ligands developed in their laboratory (Eq. 11) [201]. As the reaction proceeded via a π-allyl palladium intermediate, a dynamic kinetic asymmetric cycloaddition became possible. In 2004, Dong and Alper reported a second asymmetric cycloaddition, but the enantioselectivity was moderate [202].

Equation 11. Pd-catalyzed dynamic kinetic asymmetric cycloaddition of vinyl aziridines and isocyanates

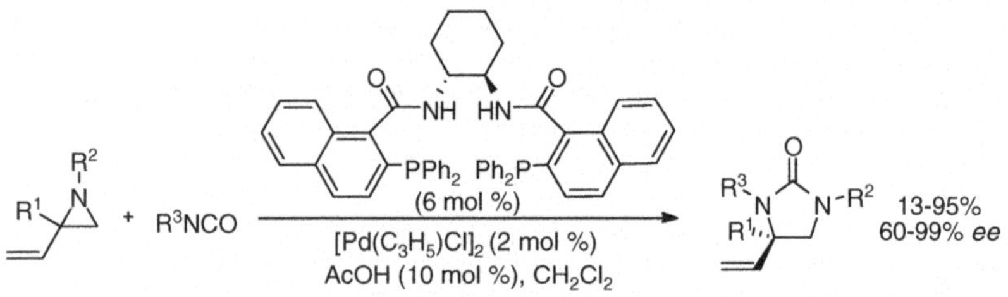

Recently, progress using other catalysts than palladium has also emerged. In 2008, Hou and co-workers reported the use of tributylphosphine as a catalyst for the cycloaddition of aziridines with carbon disulfide and isothiocyanates [203]. Finally, Sengoden and Punniyamurthy developed in 2013 the iron-catalyzed cycloaddition of aziridines with isoselenocyanates [204]. Interestingly, this reaction could be performed "on water" under air, without the care required for more sensitive catalysts.

3 Reactions Involving Four-Membered Rings

In comparison with the use of three-membered ring, the field of formal cycloaddition involving four-membered ring is still in its infancy. This is probably due to the smaller strain energy per bond, but also to the less developed synthetic methods used to access four-membered rings.

3.1 Reactions with Cyclobutanes

Neglected for a long time, the catalytic activation of cyclobutanes has come recently at the center of attention of the organic chemistry community [205]. Prior to 2008, only one example of cycloaddition involving a 1,2-donor–acceptor-substituted aminocyclobutane had been reported by Saigo and co-workers in 1991 (Eq. 12) [206]. A mixture of half aminal and acetal was obtained, which was subsequently completely hydrolyzed to the acetal. In this pioneering work, the diastereoselectivity was low and the scope of the reaction was limited. In 1993, Saigo and co-workers then reported a multi-step [4+2] annulation procedure for the synthesis of δ-lactones starting from acetal-ester-substituted cyclobutanes [207].

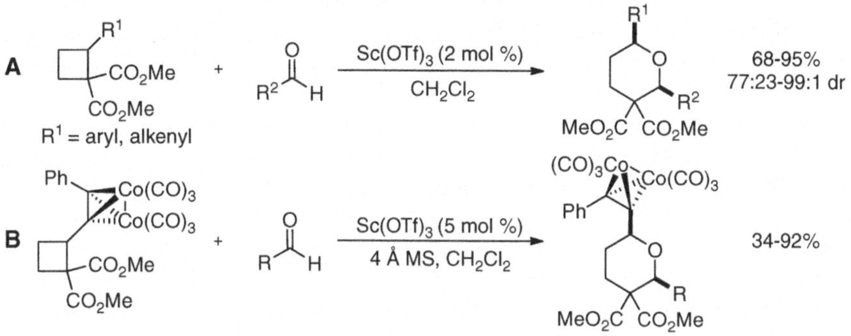

Scheme 33 Formal [4+2] cycloaddition of 1,2-donor–acceptor-substituted cyclobutanes with aldehydes

Equation 12. First example of formal [4+2] cycloaddition of cyclobutanes and carbonyls

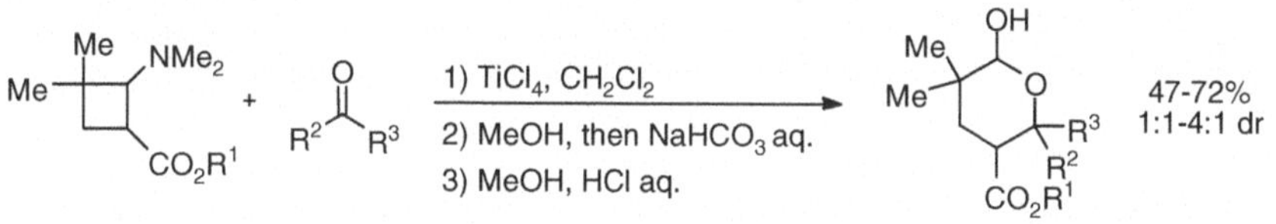

Surprisingly, it is only in 2009 that this type of transformation was studied more in detail. In this year, Parsons and Johnson reported an important breakthrough by using diester-aryl/alkenyl-substituted cyclobutanes in the reaction with aldehydes (Scheme 33**A**) [208]. The reaction was catalyzed by scandium triflate and gave tetrahydropyran products with good yield and excellent diastereoselectivity. In contrast to the similar reaction developed for cyclopropanes, racemization of the stereocenter was observed during the reaction. The required cyclobutanes were themselves synthesized by scandium-catalyzed [2+2] formal cycloaddition of olefins and methylidenemalonates, which allowed the development of a one-pot formal [2+2+2] process to access tetrahydropyrans. The same year, Pritschard, Christie, and co-workers used cobalt octacarbonyl complexes of acetylenes as cation-stabilizing groups on the cyclobutane (Scheme 33**B**) [209]. Using again scandium triflate as catalyst, *cis*-substituted tetrahydropyrans were obtained in good yield.

To further extend the scope of formal [4+2] cycloadditions, Pagenkopf and co-workers then studied oxygen-diester-substituted cyclopropanes as substrates (Scheme 34) [210–212]. The reaction was especially successful with bicyclic cyclobutanes. They first reported the cycloaddition with imines, which gave enamine products after elimination of the alcohol (**A**) [210]. In this case, the reaction of aryl-diester-substituted cyclobutanes gave stable piperidines as products. In a second work, they extended the reaction to aldehydes (**B**) [211]. In this case, stable acetal products were obtained with high

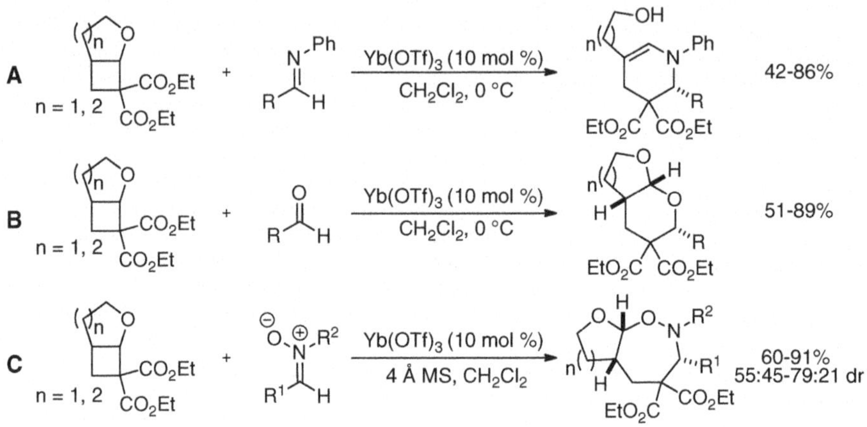

Scheme 34 Formal [4+2] cycloaddition of oxygen-diester-substituted cyclobutanes

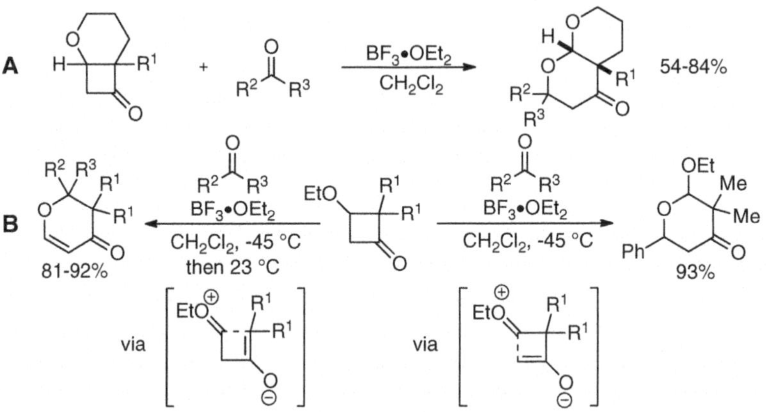

Scheme 35 Formal [4+2] cycloaddition of 3-oxo-cyclobutanones

diastereoselectivity. Finally, they developed in 2011 the formal [4+3] cycloaddition with nitrones (**C**) [212]. The reactions proceeded in good yield, but only with moderate diastereoselectivity. In 2012, Matsuo and co-workers further reported a one-pot reaction of 1,2-oxygen-diester-substituted cyclopropanes involving cycloaddition and intramolecular lactonization [213].

Three-donor-substituted cyclobutanones were introduced by Matsuo and co-workers for their use in formal [4+2] cycloaddition in 2008 (Scheme 35) [214]. In the case of bicyclic cyclobutanones, cleavage of the less substituted C–C bond was observed, leading to stable bicyclic acetals as products (**A**). For acyclic cyclobutanones, the regioselectivity of the reaction was dependent from the temperature (**B**). At −45°C, cleavage of the less substituted C–C bond was observed. In contrast, if the temperature was raised to room temperature, the

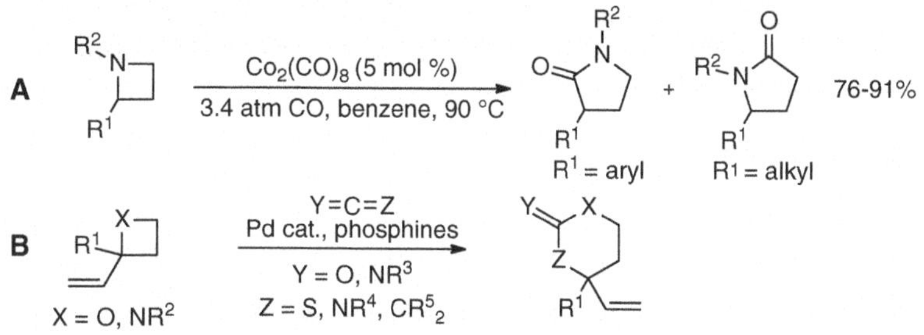

Scheme 36 Formal [4+2] cycloaddition of azetidines and oxetanes with carbon monoxides and cumulenes

reversed regiochemistry and elimination of ethanol was observed. This is probably due to the higher stability of the more substituted zwitterion intermediate. In 2012, Matsuo and co-workers also reported the use of cobalt octacarbonyl alkyne complexes as donor on cyclobutanones [215].

3.2 Reactions with Other Four-Membered Rings

Most of the cycloadditions involving other four-membered rings are based on reactions between oxetanes and azetidines and carbon monoxide or cumulenes. Baba and co-workers first reported that tetraphenylstibonium iodide was also a good catalyst for the cycloaddition of oxetanes and cumulenes [162, 216]. Alper and co-workers then studied the activation of both oxetanes and azetidines with cobalt and palladium catalysts (Scheme 36). Cobalt octacarbonyl was a good catalyst for the carbonylation of azetidines with carbon monoxide (**A**) [217]. Cleavage of the more substituted C–N bond was observed in the case of aryl substituents on the azetidine. For alkyl substituents, the other regioselectivity was observed. Reaction under milder conditions could be achieved using a palladium catalyst and vinyl azetidines or oxetanes [218–220]. This reaction proceeds via π-allyl palladium intermediates and was successful in the case of isocyanates, isothiocyanates, carbodiimides, ketenes, and ketimines as cumulenes. Finally, Mann and co-workers reported the reaction of tosyl azetidines and electron-rich olefins promoted by boron-trifluoride etherate [221]. In this case, a mixture of [4+2] cycloaddition and further elimination products was obtained.

4 Reactions with Larger Rings

Up to now, there are only very few studies on formal cycloaddition reactions involving C–C bond cleavage of larger rings which gives saturated heterocycles. This is probably due to the lack of ring strain, which makes these reactions less favorable. An interesting example has nevertheless been reported by Zhou and Alper, who developed the palladium-catalyzed formal [5+2] cycloaddition of vinyl pyrrolidines and isocyanates to give diazepin-2-ones (Eq. 13) [222].

Equation 13. Formal [5+2] cycloaddition of vinylpyrrolidines with isocyanates

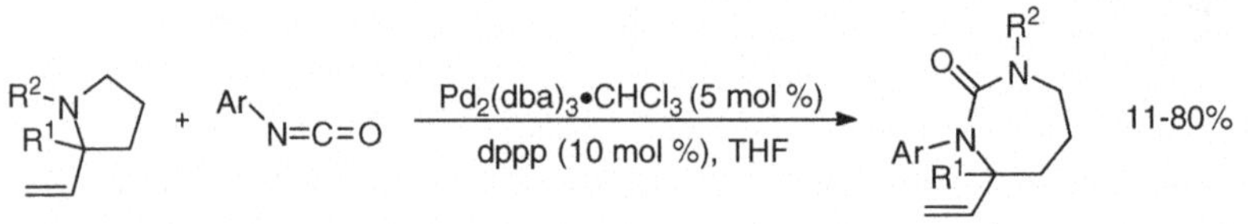

5 Conclusions

Formal cycloadditions proceeding by C–C bond cleavage are important synthetic tools, as they give access to more saturated heterocycles than "classical" cycloadditions involving π-systems. However, the activation of C–C bond is difficult, and the use of ring strain or strong polarized bonds has been necessary to develop efficient processes. During the last two decades, broadly applicable methods have appeared that build on the earlier pioneering work in this area. Cycloaddition of donor–acceptor cyclopropanes and two or three-atom π systems, as well as reactions of epoxides and aziridines with carbon monoxide or cumulenes are now burgeoning fields of research in organic chemistry. They have found important applications both in the synthesis of natural products and in the large-scale synthesis of commodity chemicals. Nevertheless, the field is still in its infancy when considering the nearly endless possible combinations of partners for formal cycloaddition reactions. Furthermore, only few successes have been reported for the simultaneous control of diastereo- and enantioselectivity. There is consequently a huge potential for both applications and further methodological developments in the field of formal cycloaddition reactions for saturated heterocycle synthesis.

References

1. Huisgen R (1968) Cycloadditions – definition, classification and characterization. Angew Chem Int Ed 7:321–328
2. Reissig HU (1988) Donor-acceptor substituted cyclopropanes – versatile building blocks in organic synthesis. Top Curr Chem 144:73–135
3. Reissig HU, Zimmer R (2003) Donor-acceptor-substituted cyclopropane derivatives and their application in organic synthesis. Chem Rev 103:1151–1196
4. Yu M, Pagenkopf BL (2005) Recent advances in donor-acceptor (DA) cyclopropanes. Tetrahedron 61:321–347
5. Carson CA, Kerr MA (2009) Heterocycles from cyclopropanes: applications in natural product synthesis. Chem Soc Rev 38:3051–3060
6. De Simone F, Waser J (2009) Cyclization and cycloaddition reactions of cyclopropyl carbonyls and imines. Synthesis 3353–3374

7. Mel'nikov MY, Budynina EM, Ivanova OA, Trushkov IV (2011) Recent advances in ring-forming reactions of donor-acceptor cyclopropanes. Mendeleev Commun 21:293–301
8. Tang P, Qin Y (2012) Recent applications of cyclopropane-based strategies to natural product synthesis. Synthesis 44:2969–2984
9. Nakamura I, Yamamoto Y (2002) Transition metal-catalyzed reactions of methylenecyclopropanes. Adv Synth Catal 344:111–129
10. Reissig HU (1981) Lewis-acid-promoted additions of carbonyl-compounds to donor-acceptor substituted cyclopropanes - a new synthesis of 2,3-dihydrofurane derivatives. Tetrahedron Lett 22:2981–2984
11. Shimada S, Hashimoto Y, Sudo A, Hasegawa M, Saigo K (1992) Diastereoselective ring-opening aldol-type reaction of 2,2-dialkoxycyclopropanecarboxylic esters with carbonyl-compounds. 1. Synthesis of cis 3,4-substituted gamma-lactones. J Org Chem 57:7126–7133
12. Shimada S, Hashimoto Y, Nagashima T, Hasegawa M, Saigo K (1993) Diastereoselective ring-opening aldol-type reaction of 2,2-dialkoxycyclopropanecarboxylic esters with carbonyl-compounds. 2. Synthesis of *cis*-2,3-substituted-gamma-lactones. Tetrahedron 49:1589–1604
13. Shimada S, Hashimoto Y, Saigo K (1993) Ring-opening aldol-type reaction of 2,2-dialkoxycyclopropanecarboxylic esters with carbonyl-compounds. 3. The diastereoselective synthesis of 2,3,4-trisubstituted gamma-lactones. J Org Chem 58:5226–5234
14. Sugita Y, Kawai K, Yokoe I (2000) Synthesis of tetrahydrofuro[2,3-b][1]benzopyranones by the ring-expansion reaction of methanochromanone with symmetric ketones. Heterocycles 53:657–664
15. Sugita Y, Kawai K, Yokoe I (2001) Diastereoselective ring-expansion reaction of methanochromanone with aldehydes: formation of trans-fused tetrahydrofuro-[2,3-b][1] benzo-pyranones and their isomerization. Heterocycles 55:135–144
16. Han Z, Uehira S, Tsuritani T, Shinokubo H, Oshima K (2001) Enolate formation from cyclopropyl ketones via iodide-induced ring opening and its use for stereoselective aldol reaction. Tetrahedron 57:987–995
17. Gupta A, Yadav VK (2006) A highly diastereoselective approach to tetrahydrofurans via [3+2] cycloadditions of silylmethyl-substituted cyclopropanes with aldehydes and ketones. Tetrahedron Lett 47:8043–8047
18. Dunn J, Motevalli M, Dobbs AP (2011) Donor cyclopropanes in synthesis: utilising silylmethylcyclopropanes to prepare 2,5-disubstituted tetrahydrofurans. Tetrahedron Lett 52:6974–6977
19. Pohlhaus PD, Johnson JS (2005) Enantiospecific Sn(II)- and Sn(IV)-catalyzed cycloadditions of aldehydes and donor-acceptor cyclopropanes. J Am Chem Soc 127:16014–16015
20. Pohlhaus PD, Johnson JS (2005) Highly diastereoselective synthesis of tetrahydrofurans via lewis acid-catalyzed cyclopropane/aldehyde cycloadditions. J Org Chem 70:1057–1059
21. Pohlhaus PD, Sanders SD, Parsons AT, Li W, Johnson JS (2008) Scope and mechanism for Lewis acid-catalyzed cycloadditions of aldehydes and donor-acceptor cyclopropanes: evidence for a stereospecific intimate ion pair pathway. J Am Chem Soc 130:8642–8650
22. Campbell MJ, Johnson JS, Parsons AT, Pohlhaus PD, Sanders SD (2010) Complexity-building annulations of strained cycloalkanes and C=O pi bonds. J Org Chem 75:6317–6325
23. Yankee EW, Cram DJ (1970) Carbanion-carbonium ion intermediates in racemizations and solvolyses of cyclopropanes. J Am Chem Soc 92:6328–6329
24. Sliwinska A, Czardybon W, Warkentin J (2007) Zwitterion from a cyclopropane with geminal donor and acceptor groups. Org Lett 9:695–698
25. Zhang JS, Shen W, Li M (2007) DFT study on the Sn^{II}-catalyzed diastereoselective synthesis of tetrahydrofuran from D-A cyclopropane and benzaldehyde. Eur J Org Chem 4855–4866
26. Parsons AT, Johnson JS (2009) Catalytic enantioselective synthesis of tetrahydrofurans: a dynamic kinetic asymmetric 3+2 cycloaddition of racemic cyclopropanes and aldehydes. J Am Chem Soc 131:3122–3123

27. Yang GS, Shen Y, Li K, Sun YX, Hua YY (2011) $AlCl_3$-promoted highly regio- and diastereoseletive 3+2 cycloadditions of activated cyclopropanes and aromatic aldehydes: construction of 2,5-diaryl-3,3,4-trisubstituted tetrahydrofurans. J Org Chem 76:229–233
28. Smith AG, Slade MC, Johnson JS (2011) Cyclopropane-aldehyde annulations at quaternary donor sites: stereoselective access to highly substituted tetrahydrofurans. Org Lett 13:1996–1999
29. Parsons AT, Campbell MJ, Johnson JS (2008) Diastereoselective synthesis of tetrahydrofurans via palladium(0)-catalyzed [3+2] cycloaddition of vinylcyclopropanes and aldehydes. Org Lett 10:2541–2544
30. Sanders SD, Ruiz-Olalla A, Johnson JS (2009) Total synthesis of (+)-virgatusin via $AlCl_3$-catalyzed 3+2 cycloaddition. Chem Commun 5135–5137
31. Campbell MJ, Johnson JS (2009) Asymmetric synthesis of (+)-polyanthellin A. J Am Chem Soc 131:10370–10371
32. Campbell MJ, Johnson JS (2010) Enantioselective synthesis of (+)-polyanthellin A via cyclopropane-aldehyde (3+2)-annulation. Synthesis 2841–2852
33. Karadeolian A, Kerr MA (2010) Total synthesis of (+)-isatisine A. Angew Chem Int Ed 49:1133–1135
34. Karadeolian A, Kerr MA (2010) Total synthesis of (+)-isatisine A. J Org Chem 75:6830–6841
35. Xing SY, Pan WY, Liu C, Ren J, Wang ZW (2010) Efficient construction of Oxa- and Aza-n.2.1 skeletons: Lewis acid catalyzed intramolecular 3+2 cycloaddition of cyclopropane 1,1-diesters with carbonyls and imines. Angew Chem Int Ed 49:3215–3218
36. Xing SY, Li Y, Li Z, Liu C, Ren J, Wang ZW (2011) Lewis acid catalyzed intramolecular 3+2 cross-cycloaddition of donor-acceptor cyclopropanes with carbonyls: a general strategy for the construction of acetal n.2.1 skeletons. Angew Chem Int Ed 50:12605–12609
37. Bai Y, Tao WJ, Ren J, Wang ZW (2012) Lewis acid catalyzed intramolecular 4+2 and 3+2 cross-cycloaddition of alkynylcyclopropane ketones with carbonyl compounds and imines. Angew Chem Int Ed 51:4112–4116
38. Benfatti F, de Nanteuil F, Waser J (2012) Iron-catalyzed 3+2 annulation of aminocyclopropanes with aldehydes: stereoselective synthesis of aminotetrahydrofurans. Org Lett 14:386–389
39. Benfatti F, de Nanteuil F, Waser J (2012) Catalytic enantiospecific 3+2 annulation of aminocyclopropanes with ketones. Chem Eur J 18:4844–4849
40. Yamago S, Nakamura E (1990) Thermal hetero 3+2 cycloaddition approach to functionalized tetrahydrofurans. J Org Chem 55:5553–5555
41. Nakamura I, Oh BH, Saito S, Yamamoto Y (2001) Novel [3+2] cycloaddition of alkylidenecyclopropanes with aldehydes catalyzed by palladium. Angew Chem Int Ed 40:1298–1300
42. Trost BM (1986) 3+2 Cycloaddition approaches to 5-membered rings via trimethylenemethane and its equivalents. Angew Chem Int Ed 25:1–20
43. Patient L, Berry MB, Kilburn JD (2003) Lewis acid-mediated addition of silylated methylenecyclopropane to aldehydes—synthesis of tetrahydrofuran derivatives. Tetrahedron Lett 44:1015–1017
44. Shao LX, Xu B, Huang JW, Shi M (2006) Synthesis of the indene, THF, and pyrrolidine skeletons by Lewis acid mediated cycloaddition of methylenecyclopropanes with aldehydes, N-tosyl aldimines, and acetals. Chem Eur J 12:510–517
45. Shi M, Xu B (2003) Lewis acid-catalyzed novel [3+2] cycloaddition of methylenecyclopropanes with activated aldehydes or ketones. Tetrahedron Lett 44:3839–3842
46. Shi M, Xu B, Huang JW (2004) Lewis acid-mediated cycloaddition of methylenecyclopropanes with aldehydes and imines: a facile access to indene, THF, and pyrrolidine skeletons via homoallylic rearrangement protocol. Org Lett 6:1175–1178
47. Alper PB, Meyers C, Lerchner A, Siegel DR, Carreira EM (1999) Facile, novel methodology for the synthesis of spiro[pyrrolidin-3,3′-oxindoles]: catalyzed ring expansion reactions of cyclopropanes by aldimines. Angew Chem Int Ed 38:3186–3189

48. Fischer C, Meyers C, Carreira EM (2000) Efficient synthesis of (+/−)-horsfiline through the MgI_2-catalyzed ring-expansion reaction of a spiro[cyclopropane-1,3′-indol]-2′-one. Helv Chim Acta 83:1175–1181
49. Lerchner A, Carreira EM (2002) First total synthesis of (+/−)-strychnofoline via a highly selective ring-expansion reaction. J Am Chem Soc 124:14826–14827
50. Lerchner A, Carreira EM (2006) Synthesis of (+/−)-strychnofoline via a highly convergent selective annulation reaction. Chem Eur J 12:8209–8219
51. Marti C, Carreira EM (2005) Total synthesis of (−)-spirotryprostatin B: synthesis and related studies. J Am Chem Soc 127:11505–11515
52. Meyers C, Carreira EM (2003) Total synthesis of (−)-spirotryprostatin B. Angew Chem Int Ed 42:694–696
53. Bertozzi F, Gustafsson M, Olsson R (2002) A novel metal iodide promoted three-component synthesis of substituted pyrrolidines. Org Lett 4:3147–3150
54. Huang WW, O'Donnell MM, Bi G, Liu JF, Yu LB, Baldino CM, Bell AS, Underwood TJ (2004) Synthesis of 1,2-disubstituted-3-alkylidenylpyrrolidines via a one-pot three-component reaction. Tetrahedron Lett 45:8511–8514
55. Carson CA, Kerr MA (2005) Diastereoselective synthesis of pyrrolidines via the $Yb(OTf)_3$ catalyzed three-component reaction of aldehydes, amines, and 1,1-cyclopropanediesters. J Org Chem 70:8242–8244
56. Christie SDR, Davoile RJ, Jones RCF (2006) Preparation of highly substituted pyrrolidines via an organometallic dipole. Org Biomol Chem 4:2683–2684
57. Huang WW, Chin J, Karpinski L, Gustafson G, Baldino CM, Yu LB (2006) Metal iodide mediated ring expansion of cyclopropanecarboxylic thioesters with imines. Tetrahedron Lett 47:4911–4915
58. Kang YB, Tang Y, Sun XL (2006) Scandium triflate catalyzed cycloaddition of imines with 1,1-cyclopropanediesters: efficient and diastereoselective synthesis of multisubstituted pyrrolidines. Org Biomol Chem 4:299–301
59. Noda H, Wiedemann SH, Matsunaga S, Shibasaki M (2008) A DyI_3-catalyzed Mannich-type reaction of 1-methylcyclopropanecarboxylate-type donors for the stereoselective synthesis of pyrrolidines with quaternary stereocenters. Chem Lett 37:1180–1181
60. Helan V, Mills A, Drewry D, Grant D (2010) A rapid three-component MgI_2-mediated synthesis of 3,3-pyrollidinyl spirooxindoles. J Org Chem 75:6693–6695
61. Jackson SK, Karadeolian A, Driega AB, Kerr MA (2008) Stereodivergent methodology for the synthesis of complex pyrrolidines. J Am Chem Soc 130:4196–4201
62. Dias DA, Kerr MA (2009) Domino synthesis of bridged bicyclic tetrahydro-1,2-oxazines: access to stereodefined 4-aminocyclohexanols. Org Lett 11:3694–3697
63. Lebold TP, Kerr MA (2009) Stereodivergent synthesis of fused bicyclopyrazolidines: access to pyrazolines and pyrrolidines. Org Lett 11:4354–4357
64. Leduc AB, Kerr MA (2008) Total synthesis of (−)-allosecurinine. Angew Chem Int Ed 47:7945–7948
65. Carson CA, Kerr MA (2009) Total synthesis of FR901483. Org Lett 11:777–779
66. Tomilov YV, Novikov RA, Nefedov OM (2010) Lewis acid catalyzed reactions of donor-acceptor cyclopropanes with 1-and 2-pyrazolines formation of substituted 2-pyrazolines and 1,2-diazabicyclo 330 octanes. Tetrahedron 66:9151–9158
67. Parsons AT, Smith AG, Neel AJ, Johnson JS (2010) Dynamic kinetic asymmetric synthesis of substituted pyrrolidines from racemic cyclopropanes and aldimines: reaction development and mechanistic insights. J Am Chem Soc 132:9688–9692
68. Yamago S, Nakamura M, Wang XQ, Yanagawa M, Tokumitsu S, Nakamura E (1998) Thermal hetero [3 + 2] cycloaddition of dipolar trimethylenemethane to O-alkyloximes. Straightforward synthetic routes to substituted pyrrolidines and prolines. J Org Chem 63:1694–1703

69. Yamago S, Yanagawa M, Nakamura E (1999) Thermal hetero 3+2 cycloaddition of dipolar trimethylenemethane to N-sulfonyl and N-acyl imines. Synthesis 'of gamma-amino acid derivatives. Chem Lett 879–880
70. Oh BH, Nakamura I, Saito S, Yamamoto Y (2001) Palladium-catalyzed [3+2] cycloaddition of alkylidenecyclopropanes with imines. Tetrahedron Lett 42:6203–6205
71. Oh BH, Nakamura I, Saito S, Yamamoto Y (2003) Synthesis of 3-methylenepyrrolidines by palladium-catalyzed 3+2 cycloaddition of alkylidenecyclopropanes with imines. Heterocycles 61:247–257
72. Chen K, Zhang Z, Wei Y, Shi M (2012) Thermally induced 3+2 cyclization of aniline-tethered alkylidenecyclopropanes: a facile synthetic protocol of pyrrolo 1,2-a indoles. Chem Commun 48:7696–7698
73. Lautens M, Han WS (2002) Divergent selectivity in MgI_2-mediated ring expansions of methylenecyclopropyl amides and imides. J Am Chem Soc 124:6312–6316
74. Taillier C, Bethuel Y, Lautens M (2007) Use of a sterically demanding Lewis acid to direct ring expansion of monoactivated methylenecyclopropanes. Tetrahedron 63:8469–8477
75. Scott ME, Han W, Lautens M (2004) A highly diastereoselective MgI_2-mediated ring expansion of methylenecyclopropanes. Org Lett 6:3309–3312
76. Scott ME, Lautens M (2008) Synthesis of highly functionalized pyrrolidines via a selective iodide-mediated ring expansion of methylenecyclopropyl amides. J Org Chem 73:8154–8162
77. Taillier C, Lautens M (2007) Enantioselective catalytic ring expansion of methylenecyclopropane carboxamides promoted by a chiral magnesium Lewis acid. Org Lett 9:591–593
78. Inoue Y, Hibi T, Satake M, Hashimoto H (1979) Reaction of methylenecyclopropanes with carbon-dioxide catalyzed by palladium(0) complexes - synthesis of 5-membered lactones. J Chem Soc Chem Commun 982
79. Chen K, Jiang M, Zhang Z, Wei Y, Shi M (2011) Palladium(0)-catalyzed reaction of cyclopropylidenecycloalkanes with carbon dioxide. Eur J Org Chem 7189–7193
80. Bruckner C, Reissig HU (1985) Thiophene derivatives by novel rearrangements of siloxy-substituted cyclopropanecarboxylates. Angew Chem Int Ed 24:588–589
81. Graziano ML, Iesce MR (1987) Ring-opening reactions of cyclopropanes. 1. Formal [3+2] cyclo-addition of trans-ethyl 2,2-dimethoxy-3-methylcyclopropane-1-carboxylate to phenyl isocyanate. J Chem Res S:362–363
82. Graziano ML, Cimminiello G (1989) Ring-opening reactions of cyclopropanes. 2. investigation on the reactivity of ethyl 2,2-dimethoxy-cyclopropane-1-carboxylates towards phenyl isothiocyanate. J Chem Res S:42–43
83. Graziano ML, Iesce MR, Cermola F (1996) Ring-opening reactions of cyclopropanes. 5. Reactivity of ethyl trans-2,2-dimethoxy-3-methylcyclopropane-1-carboxylate towards electrophilic diazenes. J Chem Res S:82–83
84. Korotkov VS, Larionov OV, Hoftneister A, Magull J, de Meijere A (2007) $GaCl_3$-Catalyzed insertion of diazene derivatives into the cyclopropane ring. J Org Chem 72:7504–7510
85. Korotkov VS, Larionov OV, de Meijere A (2006) $Ln(OTf)_3$-catalyzed insertion of aryl isocyanides into the cyclopropane ring. Synthesis 3542–3546
86. Wang HN, Yang W, Liu H, Wang W, Li H (2012) $FeCl_3$ promoted highly regioselective 3+2 cycloaddition of dimethyl 2-vinyl and aryl cyclopropane-1,1-dicarboxylates with aryl isothiocyanates. Org Biomol Chem 10:5032–5035
87. Goldberg AFG, O'Connor NR, Craig RA, Stoltz BM (2012) Lewis acid mediated (3+2) cycloadditions of donor-acceptor cyclopropanes with heterocumulenes. Org Lett 14:5314–5317
88. Young IS, Kerr MA (2003) A homo [3+2] dipolar cycloaddition: the reaction of nitrones with cyclopropanes. Angew Chem Int Ed 42:3023–3026
89. Young IS, Kerr MA (2004) Three-component homo 3+2 dipolar cycloaddition. A diversity-oriented synthesis of tetrahydro-1,2-oxazines and FR900482 skeletal congeners. Org Lett 6:139–141

90. Ganton MD, Kerr MA (2004) Magnesium iodide promoted reactions of nitrones with cyclopropanes: a synthesis of tetrahydro-1,2-oxazines. J Org Chem 69:8554–8557
91. Wanapun D, Van Gorp KA, Mosey NJ, Kerr MA, Woo TK (2005) The mechanism of 1,3-dipolar cycloaddition reactions of cyclopropanes and nitrones - a theoretical study. Can J Chem Rev Can Chim 83:1752–1767
92. Karadeolian A, Kerr MA (2007) Examination of homo-[3+2]-dipolar cycloaddition: mechanistic insight into regio- and diastereoselectivity. J Org Chem 72:10251–10253
93. Sapeta K, Kerr MA (2007) The cycloaddition of nitrones with homochiral cyclopropanes. J Org Chem 72:8597–8599
94. Carson CA, Young IS, Kerr MA (2008) The reaction of nitrones with cyclopropanes: a convenient preparation of tetrahydro-1,2-oxazines. Synthesis 485–489
95. Carson CA, Kerr MA (2006) Total synthesis of (+)-phyllantidine. Angew Chem Int Ed 45:6560–6563
96. Young IS, Williams JL, Kerr MA (2005) Diastereoselective synthesis of pyrrolidines using a nitrone/cyclopropane cycloaddition: synthesis of the tetracyclic core of nakadomarin A. Org Lett 7:953–955
97. Young IS, Kerr MA (2007) Total synthesis of (+)-nakadomarin A. J Am Chem Soc 129:1465–1469
98. Johansen MB, Kerr MA (2008) Expedient synthesis of pyrrolo[1,2-a]indoles: preparation of the core of yuremamine. Org Lett 10:3497–3500
99. Lebold TP, Carson CA, Kerr MA (2006) The Nicholas-type activation of cyclopropanes toward reactions with nitrones in the homo-[3+2]-dipolar cycloaddition. Synlett 364–368
100. Yang HB, Shi M (2012) $Yb(NTf_2)_3$-catalyzed 3+3 cycloaddition between isatin ketonitrones and cyclopropanes to construct novel spiro tetrahydro-1,2-oxazine oxindoles. Org Biomol Chem 10:8236–8243
101. Gorbacheva EO, Tabolin AA, Novikov RA, Khomutova YA, Nelyubina YV, Tomilov YV, Ioffe SL (2013) Six-membered cyclic nitronates as 1,3-dipoles in formal [3 + 3]-cycloaddition with donor–acceptor cyclopropanes. Synthesis of new type of bicyclic nitrosoacetals. Org Lett 15:350–353
102. Sibi MP, Ma ZH, Jasperse CP (2005) Enantioselective addition of nitrones to activated cyclopropanes. J Am Chem Soc 127:5764–5765
103. Kang YB, Sun XL, Tang Y (2007) Highly enantioselective and diastereoselective cycloaddition of cyclopropanes with nitrones and its application in the kinetic resolution of 2-substituted cyclopropane-1,1-dicarboxylates. Angew Chem Int Ed 46:3918–3921
104. Perreault C, Goudreau SR, Zimmer LE, Charette AB (2008) Cycloadditions of aromatic azomethine imines with 1,1-cyclopropane diesters. Org Lett 10:689–692
105. Zhou Y-Y, Li J, Ling L, Liao S-H, Sun X-L, Li Y-X, Wang L-J, Tang Y (2013) Highly enantioselective [3+3] cycloaddition of aromatic azomethine imines with cyclopropanes directed by π–π stacking interactions. Angew Chem Int Ed 52:1452–1456
106. Ding QP, Wang ZY, Wu J (2009) Tandem cyclization-[3+3] cycloaddition reactions of 2-alkynylbenzaldoxime: synthesis of fused 1,2-dihydroisoquinolines. Tetrahedron Lett 50:198–200
107. Yu XX, Qiu GS, Liu JP, Wu J (2011) Synthesis of 2,3,4,11b-tetrahydro-1H-pyridazino 6,1-a isoquinolines via the three-component reaction of 2-alkynylbenzaldehydes, a sulfonohydrazide and dimethyl cyclopropane-1,1-dicarboxylate. Synthesis 2268–2274
108. Hu B, Zhu JL, Xing SY, Fang J, Du D, Wang ZW (2009) A highly site-, regio-, and stereoselective Lewis acid catalyzed formal [3+3] cycloaddition of methylenecyclopropane-1,1-diesters with C,N-diarylnitrones. Chem Eur J 15:324–327
109. Wu L, Shi M (2010) $Yb(OTf)_3$-catalyzed construction of indole derivatives through formal 3+3 cycloaddition of 1,1-vinylidenecyclopropanediesters with nitrones. Chem Eur J 16:1149–1152
110. Ivanova OA, Budynina EM, Grishin YK, Trushkov IV, Verteletskii PV (2008) Donor-acceptor cyclopropanes as three-carbon components in a [4+3] cycloaddition reaction with 1,3-diphenylisobenzofuran. Angew Chem Int Ed 47:1107–1110

111. Khumtaveeporn K, Alper H (1995) Transition-metal mediated carbonylative ring expansion of heterocyclic-compounds. Acc Chem Res 28:414–422
112. Louie J (2005) Transition metal catalyzed reactions of carbon dioxide and other heterocumulenes. Curr Org Chem 9:605–623
113. Church TL, Getzler Y, Byrne CM, Coates GW (2007) Carbonylation of heterocycles by homogeneous catalysts. Chem Commun 657–674
114. Omae I (2011) Transition metal-catalyzed cyclocarbonylation in organic synthesis. Coord Chem Rev 255:139–160
115. Kas'yan LI, Pal'chikov VA, Bondarenko YS (2011) Five-membered oxaza heterocyclic compounds on the basis of epoxides and aziridines. Russ J Org Chem 47:797–841
116. Lee JT, Thomas PJ, Alper H (2001) Synthesis of beta-lactones by the regioselective, cobalt and Lewis acid catalyzed carbonylation of simple and functionalized epoxides. J Org Chem 66:5424–5426
117. Getzler Y, Mahadevan V, Lobkovsky EB, Coates GW (2002) Synthesis of beta-lactones: a highly active and selective catalyst for epoxide carbonylation. J Am Chem Soc 124:1174–1175
118. Mahadevan V, Getzler Y, Coates GW (2002) Lewis acid$^+$ $CO(CO)_4^-$ complexes: a versatile class of catalysts for carbonylative ring expansion of epoxides and aziridines. Angew Chem Int Ed 41:2781–2784
119. Schmidt JAR, Mahadevan V, Getzler Y, Coates GW (2004) A readily synthesized and highly active epoxide carbonylation catalyst based on a chromium porphyrin framework: expanding the range of available beta-lactones. Org Lett 6:373–376
120. Schmidt JAR, Lobkovsky EB, Coates GW (2005) Chromium(III) octaethylporphyrinato tetracarbonylcobaltate: a highly active, selective, and versatile catalyst for epoxide carbonylation. J Am Chem Soc 127:11426–11435
121. Kramer JW, Lobkovsky EB, Coates GW (2006) Practical beta-lactone synthesis: epoxide carbonylation at 1 atm. Org Lett 8:3709–3712
122. Stirling A, Iannuzzi M, Parrinello M, Molnar F, Bernhart V, Luinstra GA (2005) beta-Lactone synthesis from epoxide and CO: reaction mechanism revisited. Organometallics 24:2533–2537
123. Church TL, Getzler Y, Coates GW (2006) The mechanism of epoxide carbonylation by Lewis acid$^+$ $Co(CO)_4^-$ catalysts. J Am Chem Soc 128:10125–10133
124. Kurahashi T, de Meijere A (2005) Cyclopropyl building blocks for organic synthesis, part 121. C-C bond activation by octacarbonyldicobalt: 3+1 cocyclizations of methylenecyclopropanes with carbon monoxide. Angew Chem Int Ed 44:7881–7884
125. Rowley JM, Lobkovsky EB, Coates GW (2007) Catalytic double carbonylation of epoxides to succinic anhydrides: catalyst discovery, reaction scope, and mechanism. J Am Chem Soc 129:4948–4960
126. Ganji P, Ibrahim H (2012) The first asymmetric ring-expansion carbonylation of meso-epoxides. Chem Commun 48:10138–10140
127. Huisgen R (1977) Electrocyclic ring-opening reactions of ethylene oxides. Angew Chem Int Ed 16:572–585
128. Chen ZL, Wei L, Zhang JL (2011) Lewis acid catalyzed carbon-carbon bond cleavage of aryl oxiranyl diketones: synthesis of *cis*-2,5-disubstituted 1,3-dioxolanes. Org Lett 13:1170–1173
129. Liu R, Zhang M, Zhang J (2011) Highly regioselective Lewis acid-catalyzed [3+2] cycloaddition of alkynes with donor-acceptor oxiranes by selective carbon-carbon bond cleavage of epoxides. Chem Commun 47:12870–12872
130. Wang T, Zhang JL (2011) Chemoselective C-C bond cleavage of epoxide motifs: gold(I)-catalyzed diastereoselective 4+3 cycloadditions of 1-(1-alkynyl)oxiranyl ketones and nitrones. Chem Eur J 17:86–90
131. Zhang JM, Chen ZL, Wu HH, Zhang JL (2012) $Ni(ClO_4)_2$-catalysed regio- and diastereoselective 3+2 cycloaddition of indoles and aryl oxiranyl-dicarboxylates/diketones: a facile access to furo 3,4-b indoles. Chem Commun 48:1817–1819

132. Shim JG, Yamamoto Y (1998) Palladium-catalyzed regioselective 3+2 cycloaddition of vinylic oxiranes with activated olefins. A facile synthesis of tetrahydrofuran derivatives. J Org Chem 63:3067–3071
133. Shim JG, Yamamoto Y (1999) A novel and effective route to 1,3-oxazolidine derivatives. Palladium-catalyzed regioselective 3+2 cycloaddition of vinylic oxiranes with imines. Tetrahedron Lett 40:1053–1056
134. Shim JG, Yamamoto Y (2000) A new synthetic route to 1,3-oxazolidines via palladium-catalyzed regioselective 3+2 cycloaddition of vinylic oxiranes with imines. Heterocycles 52:885–895
135. Shaghafi MB, Grote RE, Jarvo ER (2011) Oxazolidine synthesis by complementary stereospecific and stereoconvergent methods. Org Lett 13:5188–5191
136. Sako S, Kurahashi T, Matsubara S (2011) Nickel-catalyzed 3+2 cycloaddition of alpha, beta-unsaturated ketones with vinyl oxiranes. Chem Lett 40:808–809
137. Wu WQ, Ding CH, Hou XL (2012) Pd-catalyzed diastereo- and enantioselective 3+2 -cycloaddition reaction of vinyl epoxide with nitroalkenes. Synlett 1035–1038
138. Yu CM, Dai XP, Su WK (2007) Ytterbium(III) triflate catalyzed 3+2 cycloaddition of N-arylimines and epoxides: a novel and solvent-free synthesis of substituted 1,3-oxazolidines. Synlett 646–648
139. Chen ZL, Tian ZQ, Zhang JM, Ma J, Zhang JL (2012) C-O versus C-C bond cleavage: selectivity control in Lewis acid catalyzed chemodivergent cycloadditions of aryl oxiranyl-dicarboxylates with aldehydes, and theoretical rationalizations of reaction pathways. Chem Eur J 18:8591–8595
140. Huo CD, Jia XD, Zhang W, Yang L, Lu JM, Liu ZL (2004) Cation radical 3+2 cycloaddition of chalcone epoxides: a facile synthesis of highly substituted tetrahydrofurans. Synlett 251–254
141. Darensbourg DJ (2007) Making plastics from carbon dioxide: Salen metal complexes as catalysts for the production of polycarbonates from epoxides and CO_2. Chem Rev 107:2388–2410
142. North M, Pasquale R, Young C (2010) Synthesis of cyclic carbonates from epoxides and CO_2. Green Chem 12:1514–1539
143. Lu XB, Darensbourg DJ (2012) Cobalt catalysts for the coupling of CO_2 and epoxides to provide polycarbonates and cyclic carbonates. Chem Soc Rev 41:1462–1484
144. Pescarmona PP, Taherimehr M (2012) Challenges in the catalytic synthesis of cyclic and polymeric carbonates from epoxides and CO_2. Catal Sci Technol 2:2169–2187
145. Decortes A, Castilla AM, Kleij AW (2010) Salen-complex-mediated formation of cyclic carbonates by cycloaddition of CO_2 to epoxides. Angew Chem Int Ed 49:9822–9837
146. Speranza GP, Peppel WJ (1958) Preparation of substituted 2-oxazolidones from 1,2-epoxides and isocyanates. J Org Chem 23:1922–1924
147. Herweh JE, Kauffman WJ (1971) 2-Oxazolidones via the lithium bromide catalyzed reaction of isocyanates with epoxides in hydrocarbon solvents. Tetrahedron Lett 12:809–812
148. Baba A, Fujiwara M, Matsuda H (1986) Unusual cycloaddition of oxiranes with isocyanates catalyzed by tetraphenylstibonium iodide; selective formation of 3,4-disubstituted oxazolidinones. Tetrahedron Lett 27:77–80
149. Fujiwara M, Baba A, Matsuda H (1988) Selective alpha-cleavage cyclo-addition of oxiranes with heterocumulenes catalyzed by tetraphenylstibonium iodide. J Heterocyc Chem 25:1351–1357
150. Fujiwara M, Baba A, Matsuda H (1990) Mechanistic studies of tetraphenylstibonium iodide-catalyzed cycloaddition of oxiranes with heterocumulenes. Bull Chem Soc Jpn 63:1069–1073
151. Shibata I, Baba A, Iwasaki H, Matsuda H (1986) Cycloaddition reaction of heterocumulenes with oxiranes catalyzed by organotin iodide-Lewis base complex. J Org Chem 51:2177–2184
152. Baba A, Seki K, Matsuda H (1990) Stereospecific cycloaddition of heterocumulenes to oxiranes catalyzed by organotin halide-complexes. J Heterocyc Chem 27:1925–1930

153. Demaray JA, Thuener JE, Dawson MN, Sucheck SJ (2008) Synthesis of triazole-oxazolidinones via a one-pot reaction and evaluation of their antimicrobial activity. Bioorg Med Chem Lett 18:4868–4871
154. Brnardic EJ, Fraley ME, Garbaccio RM, Layton ME, Sanders JM, Culberson C, Jacobson MA, Magliaro BC, Hutson PH, O'Brien JA, Huszar SL, Uslaner JM, Fillgrove KL, Tang C, Kuo Y, Sur SM, Hartman GD (2010) 3-Aryl-5-phenoxymethyl-1,3-oxazolidin-2-ones as positive allosteric modulators of mGluR2 for the treatment of schizophrenia: hit-to-lead efforts. Bioorg Med Chem Lett 20:3129–3133
155. Sengoden M, Punniyamurthy T (2012) Role of temperature in 3+2-cycloaddition of isoselenocyanates with oxiranes using BF_3•Et_2O. RSC Adv 2:2736–2738
156. Barros MT, Phillips AMF (2010) The first enantioselective 3+2 cycloaddition of epoxides to arylisocyanates: asymmetric synthesis of chiral oxazolidinone phosphonates. Tetrahedron Asymmetry 21:2746–2752
157. Trost BM, Sudhakar AR (1987) *Cis* hydroxyamination equivalent. Application to the synthesis of (−)-acosamine. J Am Chem Soc 109:3792–3794
158. Trost BM, Sudhakar AR (1988) A stereoselective contrasteric conversion of epoxides to cis-oxazolidin-2-ones. J Am Chem Soc 110:7933–7935
159. Trost BM, Hurnaus R (1989) On the mechanism of Pd(0) catalyzed formation of oxazolidin-2-ones from vinyl epoxides. Tetrahedron Lett 30:3893–3896
160. Larksarp C, Alper H (1997) Palladium(0)-catalyzed asymmetric cycloaddition of vinyloxiranes with heterocumulenes using chiral phosphine ligands: an effective route to highly enantioselective vinyloxazolidine derivatives. J Am Chem Soc 119:3709–3715
161. Larksarp C, Alper H (1998) Highly enantioselective synthesis of 1,3-oxazolidin-2-imine derivatives by asymmetric cycloaddition reactions of vinyloxiranes with unsymmetrical carbodiimides catalyzed by palladium(0) complexes. J Org Chem 63:6229–6233
162. Fujiwara M, Imada M, Baba A, Matsuda H (1988) Ph_4SbI-catalyzed selective formation of gamma-lactones and delta-lactones from oxiranes or oxetanes with ketenes. J Org Chem 53:5974–5977
163. Alper H, Urso F, Smith DJH (1983) Regiospecific metal-catalyzed ring expansion of aziridines to beta-lactams. J Am Chem Soc 105:6737–6738
164. Calet S, Urso F, Alper H (1989) Enantiospecific and stereospecific rhodium(I)-catalyzed carbonylation and ring expansion of aziridines - asymmetric-synthesis of beta-lactams and the kinetic resolution of aziridines. J Am Chem Soc 111:931–934
165. Lu SM, Alper H (2004) Carbonylative ring expansion of aziridines to beta-lactams with rhodium-complexed dendrimers on a resin. J Org Chem 69:3558–3561
166. Piotti ME, Alper H (1996) Inversion of stereochemistry in the $Co_2(CO)_8$-catalyzed carbonylation of aziridines to beta-lactams. The first synthesis of highly strained trans-bicyclic beta-lactams. J Am Chem Soc 118:111–116
167. Ardura D, Lopez R (2007) A theoretical investigation of the Co(CO)4–catalyzed carbonylative ring expansion of N-benzoyl-2-methylaziridine to beta-lactams: reaction mechanism and effect of substituent at the aziridine C-alpha atom. J Org Chem 72:3259–3267
168. Ardura D, Lopez R, Sordo TL (2006) A theoretical study of rhodium(I) catalyzed carbonylative ring expansion of aziridines to beta lactams: crucial activation of the breaking C-N bond by hyperconjugation. J Org Chem 71:7315–7321
169. Davoli P, Moretti I, Prati F, Alper H (1999) Carbonylation of silylated hydroxymethyl aziridines to beta-lactams. J Org Chem 64:518–521
170. Davoli P, Forni A, Moretti I, Prati F, Torre G (2001) On the effect of ring substituents in the carbonylation of aziridines. Tetrahedron 57:1801–1812
171. Chamchaang W, Pinhas AR (1988) A one-pot conversion of an aziridine to a beta-lactam using nickel tetracarbonyl. J Chem Soc Chem Commun 710–711
172. Chamchaang W, Pinhas AR (1990) The conversion of an aziridine to a beta-lactam. J Org Chem 55:2943–2950

173. Alper H, Hamel N (1987) Regiospecific synthesis of α-methylene-β-lactams by a homogeneous palladium catalyzed ring expansion-carbonylation reaction. Tetrahedron Lett 28:3237–3240
174. Tanner D, Somfai P (1993) Palladium-catalyzed transformation of a chiral vinylaziridine to a β-lactam. An enantioselective route to the carbapenem (+)-PS-5. Bioorg Med Chem Lett 3:2415–2418
175. Fontana F, Tron GC, Barbero N, Ferrini S, Thomas SP, Aggarwal VK (2010) Stereoselective synthesis of *trans*-beta-lactams by palladium-catalysed carbonylation of vinyl aziridines. Chem Commun 46:267–269
176. Dauban P, Malik G (2009) A masked 1,3-dipole revealed from aziridines. Angew Chem Int Ed 48:9026–9029
177. Cardoso AL, Melo T (2012) Aziridines in formal 3+2 cycloadditions: synthesis of five-membered heterocycles. Eur J Org Chem 6479–6501
178. Bergmeier SC, Fundy SL, Seth PP (1999) Synthesis of bicyclic proline analogs using a formal 3+2 intramolecular aziridine-allylsilane cycloaddition reaction. Tetrahedron 55:8025–8038
179. Ungureanu I, Bologa C, Chayer S, Mann A (1999) Phenylaziridine as a 1,3-dipole. Application to the synthesis of functionalized pyrrolidines. Tetrahedron Lett 40:5315–5318
180. Ungureanu I, Klotz P, Mann A (2000) Phenylaziridine as a masked 1,3 dipole in reactions with nonactivated alkenes. Angew Chem Int Ed 39:4615–4617
181. Nakagawa M, Kawahara M (2000) A concise synthesis of physostigmine from skatole and activated aziridine via alkylative cyclization. Org Lett 2:953–955
182. Yadav JS, Reddy BVS, Pandey SK, Srihari P, Prathap I (2001) Scandium triflate-catalyzed 1,3-dipolar cycloaddition of aziridines with alkenes. Tetrahedron Lett 42:9089–9092
183. Pohlhaus PD, Bowman RK, Johnson JS (2004) Lewis acid-promoted carbon−carbon bond cleavage of aziridines: divergent cycloaddition pathways of the derived ylides. J Am Chem Soc 126:2294–2295
184. Li L, Wu XX, Zhang JL (2011) Lewis acid-catalyzed formal 3+2 cycloadditions of N-tosyl aziridines with electron-rich alkenes via selective carbon-carbon bond cleavage. Chem Commun 47:5049–5051
185. Lowe MA, Ostovar M, Ferrini S, Chen CC, Lawrence PG, Fontana F, Calabrese AA, Aggarwal VK (2011) Palladium-mediated annulation of vinyl aziridines with Michael acceptors: stereocontrolled synthesis of substituted pyrrolidines and its application in a formal synthesis of (−)-alpha-kainic acid. Angew Chem Int Ed 50:6370–6374
186. Griffin K, Montagne C, Cam Thuy H, Clarkson GJ, Shipman M (2012) Lewis acid promoted intramolecular (3+2) 'cycloadditions' of methyleneaziridines with alkene and alkyne acceptors. Org Biomol Chem 10:1032–1039
187. Yadav VK, Sriramurthy V (2005) Silylmethyl-substituted aziridine and azetidine as masked 1,3-and 1,4-dipoles for formal 3+2 and 4+2 cycloaddition reactions. J Am Chem Soc 127:16366–16367
188. Gandhi S, Bisai A, Prasad BAB, Singh VK (2007) Studies on the reaction of aziridines with nitriles and carbonyls: synthesis of imidazolines and oxazolidines. J Org Chem 72:2133–2142
189. Maeda R, Ishibashi R, Kamaishi R, Hirotaki K, Furuno H, Hanamoto T (2011) $AgSbF_6$-promoted cycloaddition reaction of 2-trifluoromethyl-N-tosylaziridine with aldehydes. Org Lett 13:6240–6243
190. Wu XX, Lia L, Zhang JL (2011) Nickel(II)-catalyzed diastereoselective 3+2 cycloaddition of N-tosyl-aziridines and aldehydes via selective carbon-carbon bond cleavage. Chem Commun 47:7824–7826
191. Wu XX, Zhang JL (2012) $Y(OTf)_3$-catalyzed diastereoselective 3+2 cycloaddition of N-Tosyl-aziridines and imines; efficient synthesis of multisubstituted imidazolidines. Synthesis 44:2147–2154

192. Jiang Z, Wang J, Lu P, Wang YG (2011) Diasteroselective synthesis of oxazolidines and imidazolidines via the Lewis acid catalyzed C-C cleavage of aziridines. Tetrahedron 67:9609–9617
193. Soga K, Hosoda S, Nakamura H, Ikeda S (1976) New synthetic route to 2-oxazolidones. J Chem Soc Chem Commun 617
194. Nomura R, Nakano T, Nishio Y, Ogawa S, Ninagawa A, Matsuda H (1989) Regioselective cycloaddition of 1,2-disubstituted aziridines to heterocumulenes catalyzed by organoantimony halides. Chem Ber 122:2407–2409
195. Miller AW, Nguyen ST (2004) (Salen)chromium(III)/DMAP: an efficient catalyst system for the selective synthesis of 5-substituted oxazolidinones from carbon dioxide and aziridines. Org Lett 6:2301–2304
196. Fontana F, Chen CC, Aggarwal VK (2011) Palladium-catalyzed insertion of CO_2 into vinylaziridines: new route to 5-vinyloxazolidinones. Org Lett 13:3454–3457
197. Baeg JO, Alper H (1992) Regiospecific palladium-catalyzed cycloaddition of aziridines and carbodiimides. J Org Chem 57:157–162
198. Baeg JO, Bensimon C, Alper H (1995) The first enantiospecific palladium-catalyzed cycloaddition of aziridines and heterocumulenes - novel synthesis of chiral 5-membered ring heterocycles. J Am Chem Soc 117:4700–4701
199. Baeg JO, Alper H (1994) Novel palladium(II)-catalyzed cyclization of aziridines and sulfur diimides. J Am Chem Soc 116:1220–1224
200. Butler DCD, Inman GA, Alper H (2000) Room temperature ring-opening cyclization reactions of 2-vinylaziridines with isocyanates, carbodiimides, and isothiocyanates catalyzed by $Pd(OAc)_2/PPh_3$. J Org Chem 65:5887–5890
201. Trost BM, Fandrick DR (2003) Dynamic kinetic asymmetric cycloadditions of isocyanates to vinylaziridines. J Am Chem Soc 125:11836–11837
202. Dong C, Alper H (2004) $CeCl_3$ promoted asymmetric cycloaddition of isocyanates with 2-vinylaziridines. Tetrahedron Asymmetry 15:1537–1540
203. Wu J-Y, Luo Z-B, Dai L-X, Hou X-L (2008) Tributylphosphine-catalyzed cycloaddition of aziridines with carbon disulfide and isothiocyanate. J Org Chem 73:9137–9139
204. Sengoden M, Punniyamurthy T (2013) "On water": efficient iron-catalyzed cycloaddition of aziridines with heterocumulenes. Angew Chem Int Ed 52:572–575
205. Seiser T, Saget T, Tran DN, Cramer N (2011) Cyclobutanes in catalysis. Angew Chem Int Ed 50:7740–7752
206. Shimada S, Saigo K, Nakamura H, Hasegawa M (1991) Novel 4+2 -type reaction of 2-(dimethylamino)cyclobutanecarboxylic esters with carbonyl-compounds. Chem Lett 1149–1152
207. Shimada S, Tohno I, Hashimoto Y, Saigo K (1993) Diastereoselective synthesis of cis-4,5-substituted delta-lactones by the reaction of 2-methoxy-2-(trimethylsiloxy)cyclobutane-carboxylic esters with carbonyl-compounds. Chem Lett 1117–1120
208. Parsons AT, Johnson JS (2009) Formal 4+2 cycloaddition of donor-acceptor cyclobutanes and aldehydes: stereoselective access to substituted tetrahydropyrans. J Am Chem Soc 131:14202–14203
209. Allart EA, Christie SDR, Pritchard GJ, Elsegood MRJ (2009) Preparation of highly substituted tetrahydropyrans via a metal assisted dipolar cycloaddition reaction. Chem Commun 7339–7341
210. Moustafa MMA, Pagenkopf BL (2010) Ytterbium Trif late catalyzed synthesis of alkoxy-substituted donor-acceptor cyclobutanes and their formal 4+2 cycloaddition with imines: stereoselective synthesis of piperidines. Org Lett 12:4732–4735
211. Moustafa MMA, Stevens AC, Machin BP, Pagenkopf BL (2010) Formal 4+2 cycloaddition of alkoxy-substituted donor-acceptor cyclobutanes and aldehydes catalyzed by $Yb(OTf)_3$. Org Lett 12:4736–4738
212. Stevens AC, Palmer C, Pagenkopf BL (2011) The formal 4+3 cycloaddition between donor-acceptor cyclobutanes and nitrones. Org Lett 13:1528–1531

213. Okado R, Nowaki A, Matsuo J, Ishibashi H (2012) Formal 4+2 cycloaddition of di-*tert*-butyl 2-ethoxycyclobutane-1,1-dicarboxylate with ketones or aldehydes and tandem lactonization. Chem Pharm Bull 60:21–22
214. Matsuo J, Sasaki S, Tanaka H, Ishibashi H (2008) Lewis acid-catalyzed intermolecular 4+2 cycloaddition of 3-alkoxycyclobutanones to aldehydes and ketones. J Am Chem Soc 130:11600–11601
215. Kawano M, Kiuchi T, Matsuo J, Ishibashi H (2012) Formal 4+2 cycloaddition of cyclobutanones bearing alkyne-cobalt complex at their 3-positions. Tetrahedron Lett 53:432–434
216. Fujiwara M, Baba A, Matsuda H (1989) The cycloaddition of heterocumulenes to oxetanes in the presence of catalytic amounts of tetraphenylstibonium iodide. J Heterocyc Chem 26:1659–1663
217. Roberto D, Alper H (1989) Novel synthesis of pyrrolidinones by cobalt carbonyl catalyzed carbonylation of azetidines - a new ring-expansion carbonylation reaction of 2-vinylazetidines to tetrahydroazepinones. J Am Chem Soc 111:7539–7543
218. Larksarp C, Alper H (1999) Synthesis of 1,3-oxazine derivatives by palladium-catalyzed cycloaddition of vinyloxetanes with heterocumulenes. Completely stereoselective synthesis of bicyclic 1,3-oxazines. J Org Chem 64:4152–4158
219. Inman GA, Butler DCD, Alper H (2001) Mild $Pd(OAc)_2/PPh_3$ catalyzed cyclization reactions of 2-vinylazetidines with heterocumulenes: an atom-economy synthesis of tetrahydropyrimidinone, tetrahydropyrimidinimine, and thiazinanimine analogs. Synlett 914–919
220. Martorell A, Inman GA, Alper H (2003) Regioselective palladium-catalysed cycloaddition reactions of 1-alkyl-2-vinylazetidines with ketenimines and ketenes. J Mol Catal A Chem 204:91–96
221. Ungureanu I, Klotz P, Schoenfelder A, Mann A (2001) The reactivity of N-tosylphenylaziridine versus N-tosylphenylazetidine in heterocyclization reactions. Tetrahedron Lett 42:6087–6091
222. Zhou HB, Alper H (2003) Synthesis of seven-membered ring diazepin-2-ones via palladium-catalyzed highly regioselective cyclization of 2-vinylpyrrolidines with aryl isocyanates. J Org Chem 68:3439–3445
223. Spears GW, Nakanishi K, Ohfune Y (1991) Novel entry to a 3,4-disubstituted 2-azetidinone derivative via palladium-assisted carbonylation of a 2-substituted 3-vinylaziridine. Synlett 91–92

Index

The manufacturer's authorised representative in the EU is Springer Nature Customer Service Centre GmbH, Europaplatz 3, 69115 Heidelberg, Germany. If you have any concerns regarding our products, please contact ProductSafety@springernature.com

Printed and bound by CPI Group (UK) Ltd, Croydon, CR0 4YY
30/09/2026
02225644-0020